MODERN PRESTRESSED CONCRETE

MODERN PRESTRESSED CONCRETE

H. KENT PRESTON, P. E.
Chief Product Engineer, Construction Materials
CF&I Steel Corporation

NORMAN J. SOLLENBERGER, P. E.
Chairman, Department of Civil and Geological Engineering
Princeton University

McGRAW-HILL BOOK COMPANY
New York San Francisco Toronto London Sydney

MODERN PRESTRESSED CONCRETE

 Library of Congress Catalog Card Number 67-11877

50823

1234567890MP72106987

Preface

Since its introduction into the United States in 1950, the many advantages of prestressed concrete have caused its use to spread more rapidly than that of any other building material. This book is designed to furnish the structural engineer with all the information required for the design of safe, economical prestressed concrete structures.

Basic principles, design procedures, and numerical examples are presented in terms of simple arithmetic and the formulas for stress and moment which are familiar to all structural engineers. The design examples are based on the latest editions of the Codes, Specifications, and Recommended Practices available at the time the book was written.

Construction methods and equipment are discussed and illustrated in a manner that will enable the engineer to design members which can be fabricated economically and also help the fabricator to understand the factors which require his special attention. High-strength tendons are described and their properties analyzed to ensure proper use in design and during construction of these materials, which are relatively new to the construction industry. The practical suggestions based on experience in the United States will be helpful to both the designer and the fabricator. The typical structures illustrated include some of the prestressed concrete sections which have been standardized by the fabricators in the industry or by committees established for this purpose.

The authors wish to express appreciation to Robert E. West, a partner in the consulting engineering firm of West, Preston and Sollenberger, for his help in clarifying some of the design examples as well as for performing the actual calculations in the same examples.

Although every effort has been made to make the information presented herein as accurate and complete as possible, the authors neither expressly nor impliedly warrant that this book is devoid of errors or omissions, and do not accept responsibility for the consequences of such errors and omissions.

H. Kent Preston
Norman J. Sollenberger

Contents

Contents

Notations

Complete uniformity of notation throughout this book could not be achieved because several different codes and specifications are involved and their notations do not entirely coincide. The majority of notations listed here follow Proposed Definitions and Notations for Prestressed Concrete* by the Joint ACI-ASCE Committee 323 on Prestressed Concrete. A few of the authors' own have been added. The second group of notations listed here is in accordance with ACI 318-63. Where other notations are used, the definition of the symbols are given at the point where they are used.

Cross-sectional constants:

A_c = area of entire concrete section (steel area not deducted)
c.g.c. = center of gravity of entire concrete section
c.g.s. = center of gravity of steel area
y_b = distance from bottom fiber to c.g.c.
y_t = distance from top fiber to c.g.c.
e = eccentricity of c.g.s. with respect to c.g.c.
I_c = moment of inertia of entire concrete section about c.g.c.
Z_b = section modulus of bottom fiber referred to c.g.c.
Z_t = section modulus of top fiber referred to c.g.c.

Loads, moments, and forces:

w_G = dead weight per unit length of prestressed member itself
w_S = additional dead load per unit length
w_D = total dead load per unit length = $w_G + w_S$
w_T = uniform load per unit length due to parabolic path of tendons
w_L = live load per unit length
P_L = concentrated live load
M_G = bending moment due to w_G
M_S = bending moment due to w_S
M_D = bending moment due to w_D
M_L = bending moment due to w_L and/or P_L
M_e = bending moment due to eccentricity of prestress force

* *J. Am. Concrete Inst.*, October, 1952

M_u = bending moment under ultimate load condition
V_L = live-load shear
V_u = ultimate load shear
F = effective prestress force after deduction of all losses
F_I = initial prestress force
F_o = prestress force after release of tendons from external anchors (applicable to pretensioned members)
F_A = arbitrarily assumed final tension

Concrete stresses:

$f^b{}_F, f^t{}_F$ = stress in bottom (top) fiber due to F
$f^b{}_{F_I}, f^t{}_{F_I}$ = stress in bottom (top) fiber due to F_I
$f^b{}_{F_o}, f^t{}_{F_o}$ = stress in bottom (top) fiber due to F_o
$f^b{}_G, f^t{}_G$ = stress in bottom (top) fiber due to w_G
$f^b{}_S, f^t{}_S$ = stress in bottom (top) fiber due to w_S
$f^b{}_D, f^t{}_D$ = stress in bottom (top) fiber due to w_D
$f^b{}_L, f^t{}_L$ = stress in bottom (top) fiber due to w_L and/or P_L
$f^b{}_T, f^t{}_T$ = stress in bottom (top) fiber due to w_T
f_{tp} = permissible tensile stress
f^t, f^s, f^{ts} = see Fig. 5-3 and explanation in adjacent text. Applicable to composite sections with different strengths of concrete
$f^b{}_{FA}, f^t{}_{FA}$ = stress in bottom (top) fiber due to F_A

Deflection or camber:

$\triangle_{F_I}$ = deflection due to F_I
$\triangle_{F_o}$ = deflection due to F_o
$\triangle_F$ = deflection due to F
$\triangle_G$ = deflection due to w_G
$\triangle_S$ = deflection due to w_S
$\triangle_D$ = deflection due to w_D
$\triangle_L$ = deflection due to w_L and/or P_L

Summation of stress, camber, etc.:

The summation of two or more stresses can be written in two ways thus:

$f^b{}_F + f^b{}_G + f^b{}_L$ can also be written $f^b{}_{F+G+L}$.
$\triangle_F + \triangle_G + \triangle_L$ can also be written $\triangle_{F+G+L}$.

Steel stresses:

f_{F_I} = stress in tendons due to F_I
f_{F_o} = stress in tendons due to F_o
f_F = stress in tendons due to F

Notations from ACI 318-63:

2600—Notation

$a = A_s f_{su}/0.85 f'_c b$
A_b = bearing area of anchor plate of post-tensioning steel
A'_b = maximum area of the portion of the anchorage surface that is geometrically similar to and concentric with the area of the anchor plate of the post-tensioning steel
A_s = area of prestressed tendons
A_{sf} = area of reinforcement to develop compressive strength of overhanging flanges in flanged members
A_{sr} = area of tendon required to develop the web
A'_s = area of unprestressed reinforcement
A_v = area of web reinforcement placed perpendicular to the axis of the member
b = width of compression face of flexural member
b' = minimum width of web of a flanged member
d = distance from extreme compression fiber to centroid of the prestressing force
f'_c = compressive strength of concrete (see Section 301)
f'_{ci} = compressive strength of concrete at time of initial prestress
f_{cp} = permissible compressive concrete stress on bearing area under anchor plate of post-tensioning steel
f_d = stress due to dead load, at the extreme fiber of a section at which tension stresses are caused by applied loads
f_{pc} = compressive stress in the concrete, after all prestress losses have occurred, at the centroid of the cross section resisting the applied loads, or at the junction of the web and flange when the centroid lies in the flange. (In a composite member f_{pc} will be the resultant compressive stress at the centroid of the composite section, or at the junction of the web and flange when the centroid lies within the flange, due to both prestress and to the bending moments resisted by the precast member acting alone.)
f_{pe} = compressive stress in concrete due to prestress only, after all losses, at the extreme fiber of a section at which tension stresses are caused by applied loads
f'_s = ultimate strength of prestressing steel
f_{se} = effective steel prestress after losses
f_{su} = calculated stress in prestressing steel at ultimate load
f_{sy} = nominal yield strength of prestressing steel
f'_y = strength of unprestressed reinforcement (see Section 301)

F_{sp} = ratio of splitting tensile strength to the square root of compressive strength (see Section 505)
h = total depth of member
I = Moment of inertia of section resisting externally applied loads*
K = wobble friction coefficient per foot of prestressing steel
L = length of prestressing steel element from jacking end to any point x
M = bending moment due to externally applied loads*
M_{cr} = net flexural cracking moment
M_u = ultimate resisting moment
$p = A_s/bd$; ratio of prestressing steel
$p' = A'_s/bd$; ratio of unprestressed steel
$q = pf_{su}/f'_c$
s = longitudinal spacing of web reinforcement
T_o = steel force at jacking end
T_x = steel force at any point x
t = average thickness of the compression flange of a flanged member
V = shear due to externally applied loads*
V_c = shear carried by concrete
V_{ci} = shear at diagonal cracking due to all loads, when such cracking is the result of combined shear and moment
V_{cw} = shear force at diagonal cracking due to all loads, when such cracking is the result of excessive principal tension stresses in the web
V_d = shear due to dead load
V_p = vertical component of the effective prestress force at the section considered
V_u = shear due to specified ultimate load
y = distance from the centroidal axis of the section resisting the applied loads to the extreme fiber in tension
α = total angular change of prestressing steel profile in radians from jacking end to any point x
ε = base of Naperian logarithms
μ = curvature friction coefficient
ϕ = capacity reduction factor (see Section 1504)

* The term "externally applied loads" shall be taken to mean the external ultimate loads acting on the member, excepting those applied to the member by the prestressing tendons. (Authors Note: "Externally applied loads" can be defined as "all loads except the dead weight of the member.")

CHAPTER 1

Basic Principles

1-1. Characteristics of Concrete. In many localities concrete is the cheapest material available for carrying compressive loads. On the other hand its tensile strength is only 10 to 15 per cent of its compressive strength, and this value is often reduced to zero by shrinkage cracks which occur during curing.

About A.D. 1900 engineers began to use concrete for flexural members by putting steel bars on the tension side of the member to carry the entire tensile load. This gave a member with a low-cost material to carry compressive stresses plus steel with a minimum amount of fabrication to carry tensile stresses. One of the chief disadvantages of reinforced concrete is the fact that all the concrete on the tensile side of the neutral axis is a useless dead weight except that it serves as a connection between the compressive concrete and the steel.

Prestressed concrete goes two steps beyond reinforced concrete. First, all the concrete on the tensile side of the neutral axis is put under an initial compressive stress of such a magnitude that all design loads which are to be applied to the structure in the future can reduce this stress but will not put the concrete in tension. Second, the prestress is applied in such a manner that it creates a moment of opposite sign to those which will be produced by applied dead and live loads. In the ideal design this negative moment carries all the dead load and creates the maximum allowable compressive stress in the concrete on the tensile side of the member. As far as the stresses in the prestressed member itself are concerned, it is then a weightless structure with all its moment carrying capacity available for live load—an especially important feature on long spans.

1-2. Plain Concrete. Figure 1-1 shows a plain, unreinforced, unprestressed concrete beam. It is a solid rectangular beam 9 in. high by 6 in. wide on a 15-ft 0-in. span. Live load is 184 lb per ft. The following calculations cover section properties, bending moments, unit stresses, and summation of stresses.

Area $= A_c = 9 \times 6 = 54$ sq in.

Dead weight $= w_G = 54 \times {}^{150}\!/_{144} = 56$ lb per ft

Section modulus of top fiber = Z_t
Section modulus of bottom fiber = Z_b
$Z_t = Z_b = 6(9^2)/6 = 81$
Dead-weight bending moment = M_G
$M_G = 56(15^2) \times 12/8 = 18{,}900$ in.-lb
Stress in top fiber due to dead weight = $f^t{}_G$
Stress in bottom fiber due to dead weight = $f^b{}_G$
$f^t{}_G = 18{,}900 \div 81 = +233$
$f^b{}_G = 18{,}900 \div 81 = -233$
(In prestressed concrete calculations a compressive stress in the concrete is indicated by a plus sign and a tensile stress by a minus sign.)

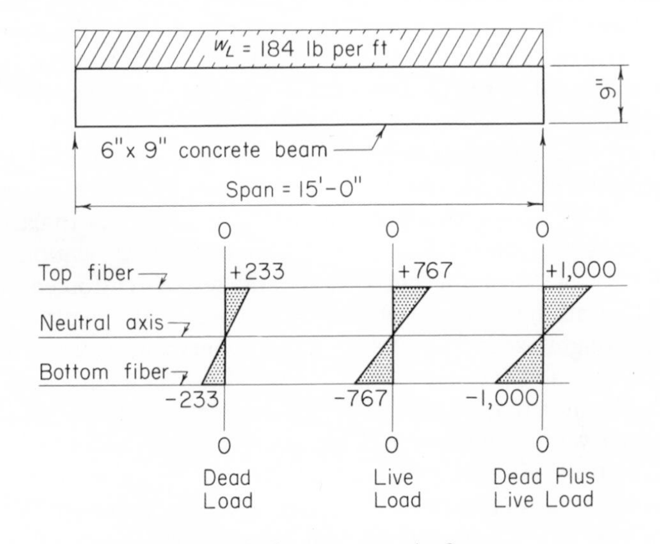

FIG. 1-1. Stresses in a plain concrete beam.

Live-load bending moment = M_L
$M_L = 184(15^2) \times 12/8 = 62{,}100$ in.-lb
Stress in top fiber due to live load = $f^t{}_L$
Stress in bottom fiber due to live load = $f^b{}_L$
$f^t{}_L = 62{,}100 \div 81 = +767$ psi
$f^b{}_L = 62{,}100 \div 81 = -767$ psi

Summation of stresses:
In top fiber: $f^t{}_G + f^t{}_L = +233 + 767 = +1{,}000$ psi
In bottom fiber: $f^b{}_G + f^b{}_L = -233 - 767 = -1{,}000$ psi

Obviously this beam would fail in tension before all the load was applied.

1-3. Centroidal Prestressing. Figure 1-2 shows the same beam as Fig.

1-1 except that it is prestressed by a force of 54,000 lb acting on the c.g.c. (center of gravity of the concrete section). This force creates a uniform compressive stress of +1,000 psi over the entire cross section of the beam. The stresses due to dead and live load are the same as in Fig. 1-1. The following calculations cover stress due to prestressing force and summation of stresses.

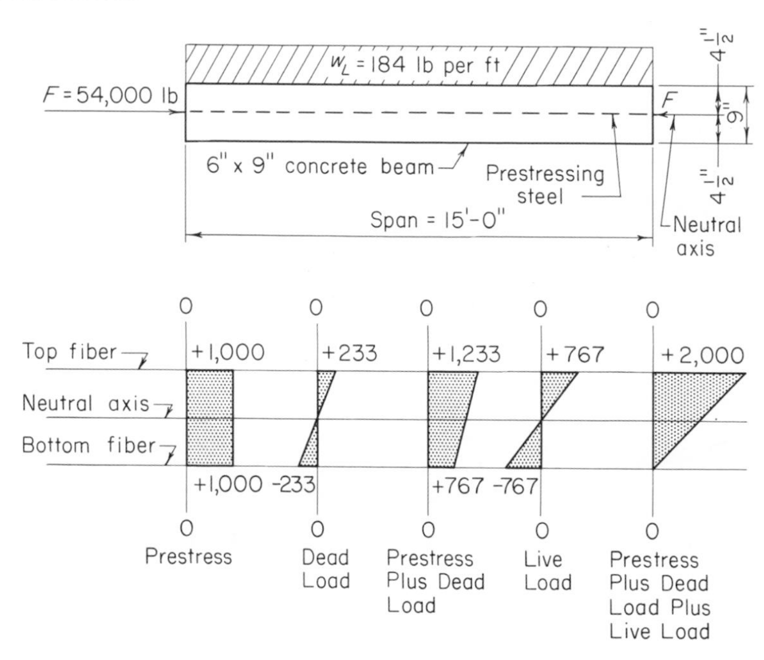

FIG. 1-2. Stresses in a beam with centroidal prestress.

f^t_F = stress in top fiber due to final prestressing force
f^b_F = stress in bottom fiber due to final prestressing force
$f^t_F = f^b_F = F/A_c = 54{,}000 \div 54 = +1{,}000$ psi

Summation of stresses:
In top fiber: $f^t_F + f^t_G + f^t_L = +1{,}000 + 233 + 767 = +2{,}000$ psi
In bottom fiber: $f^b_F + f^b_G + f^b_L = +1{,}000 - 233 - 767 = 0$

Under the same dead and live loads as the beam in Fig. 1-1, this beam has a net compressive stress of 2,000 psi in the top fiber and zero stress in the bottom fiber. A beam made of good concrete could safely carry these stresses.

1-4. Eccentric Prestressing. Figure 1-3 shows the same beam as Fig. 1-1 with the same prestressing force as the beam in Fig. 1-2 except that in

Fig. 1-3 the prestressing force is applied 1½ in. below the c.g.c. The distance from the c.g.s. (center of gravity of the steel creating the prestressing force) to the c.g.c. is called *e*, or eccentricity. Thus in Fig. 1-3, $e = 1½$ in. The stresses due to dead load are the same as in Fig. 1-1. The following calculations cover stress due to prestressing force and summation of stresses.

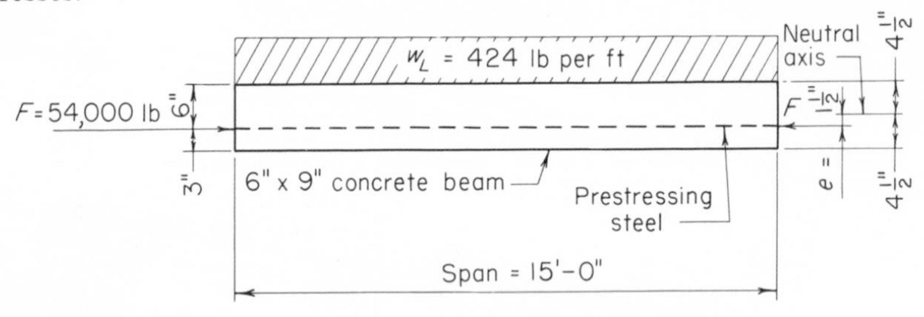

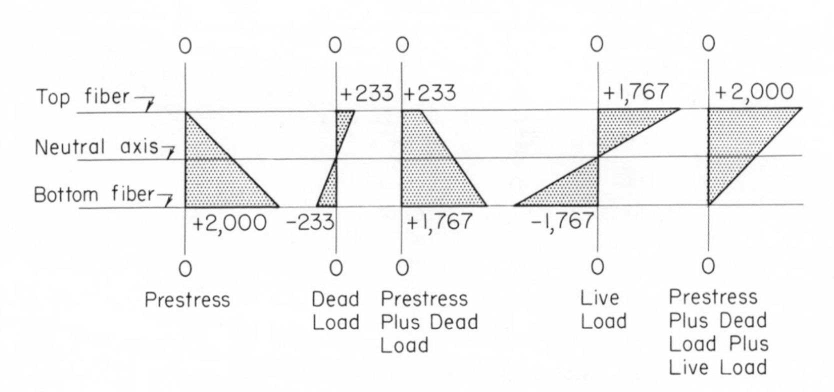

FIG. 1-3. Stresses in a beam with eccentric prestress.

In this example the stress in the concrete due to the prestressing force *F* is not uniform because the prestressing force is not applied on the c.g.c. Stresses in the concrete due to *F* can be found by the familiar method of replacing the eccentric force by an equal force on the c.g.c. and a couple. The force on the c.g.c. creates a uniform compressive stress over the entire section equal to the force divided by the area of the section, or F/A_c. The couple creates a bending moment equal to the force times the eccentricity, or *Fe*. Stress in the concrete due to the couple is equal to the moment divided by the section modulus of the concrete member, or Fe/Z. Thus the stress in the top or bottom fiber of the concrete member due to the eccentric prestressing force can be expressed by the equation

$$f_F = \frac{F}{A_c} \pm \frac{Fe}{Z} \quad (1\text{-}1)$$

The sign of Fe/Z is plus (indicating compressive stress) when the prestressing steel is on the same side of the c.g.c. as the fiber for which the stress is being computed and minus when on the opposite side.

Equation (1-1) can be used to find the stresses due to the prestressing force at any point in a *simple span* member. This is true no matter what path the prestressing steel follows and regardless of its elevation at the ends of the member. The value of e used in the equation is that measured at the point on the beam where the stresses are being computed. Equation (1-1) *does not apply* to beams continuous over one or more supports.

Use Eq. (1-1) to compute stresses in top and bottom fibers due to prestressing force.

$$f^t_F = \frac{F}{A_c} - \frac{Fe}{Z_t} \quad (1\text{-}1a)$$

$$f^t_F = \frac{54{,}000}{54} - \frac{54{,}000 \times 1.5}{81}$$

$$f^t_F = 1{,}000 - 1{,}000 = 0$$

$$f^b_F = \frac{F}{A_c} + \frac{Fe}{Z_b} \quad (1\text{-}1b)$$

$$f^b_F = \frac{54{,}000}{54} + \frac{54{,}000 \times 1.5}{81}$$

$$f^b_F = 1{,}000 + 1{,}000 = +2{,}000 \text{ psi}$$

In Fig. 1-2 the centroidal prestressing force of 54,000 lb created a compressive stress of 1,000 psi in the bottom fiber and the beam was able to support a total load of 56 + 184 = 240 lb per ft. Having a compressive stress of 2,000 psi, the beam in Fig. 1-3 should be able to support a total load of 2 × 240 = 480 lb per ft or a live load of 480 − 56 = 424 lb per ft. Check stresses using w_L = 424 lb per ft.

$$M_L = 424\ (15^2) \times {}^{12}\!/_{8} = 143{,}100 \text{ in.-lb}$$
$$f^t_L = 143{,}000 \div 81 = +1{,}767 \text{ psi}$$
$$f^b_L = 143{,}000 \div 81 = -1{,}767 \text{ psi}$$

Summation of stresses:

In top fiber: $f^t_F + f^t_G + f^t_L = 0 + 233 + 1{,}767 = +2{,}000$ psi

In bottom fiber: $f^b_F + f^b_G + f^b_L = +2{,}000 - 233 - 1{,}767 = 0$

The beam in Fig. 1-3 has the same cross section and total prestressing force as the beam in Fig. 1-2, yet it is carrying more than twice as much live load with the same net unit stress. This greater efficiency was achieved by locating the prestressing force in the area of the beam which is subject to tension under applied loads.

1-5. Buckling of Prestressed Concrete Members. In Figs. 1-2 and 1-3 the 54,000-lb prestressing force is shown as a load applied at each end of the concrete member. When a compression load is applied to a structural member, the effect of buckling must be considered in addition to the direct stresses. Buckling due to the prestressing load is not a consideration in prestressed concrete members where the tensioning element is closely encased by the concrete. The tendency to buckle under a compression load does exist, but if the member tries to buckle, it must deflect the tensioning elements as it moves. Since the tension in the tendons acts to keep them in a straight line, they resist the buckling of the concrete and a stable structure results. The same principle applies to members with curved tendons like those shown in Fig. 1-4.

Compression loads other than the prestressing force cause buckling which must be resisted by the stiffness of the structural member just as in any other building material. The tension in the tendons is sufficient only to balance the buckling due to the prestressing force.

1-6. Deflected Prestressing Steel. Since the prestressing steel in Fig. 1-3 is at the same elevation for the full length of the beam, the stresses due to F are the same at all points along the beam. Since there is no bending moment at the supports due to applied loads, the net stresses at the supports are those due to F. The e used in Fig. 1-3 was chosen to give maximum prestress in the bottom fiber without creating tension in the top fiber. If the e were increased, $f^t{}_F$ would become a tensile stress. At the center of the span a certain amount of tensile stress in the top fiber from F is permissible because it is offset by the dead-load moment which is always acting in a prestressed beam. Tensile stress in the top fiber at the end of the span is not desirable because there is no dead-load moment to relieve it. The benefits of greater eccentricity in the region where dead-load moments exist can be obtained by deflecting the prestressing steel as shown in Fig. 1-4.

In the calculations for Fig. 1-3 we found that $f^t{}_F = 0$ when $e = 1\frac{1}{2}$ in. At the center of the span the eccentricity can be increased until the prestressing force causes a tensile stress in the top fiber equal to the compressive stress from the dead-load bending moment. From previous calculations $f^t{}_G = +233$ psi. The additional e required to create an equivalent tensile stress can be found from the equation

$$f^t{}_G = \frac{Fe}{Z_t}$$

Substituting numerical values,

$$233 = \frac{54{,}000e}{81}$$

from which

$$e = 0.35 \text{ in.}$$

On this basis we can establish the eccentricity at the center of span as $e = 1.50 + 0.35 = 1.85$ in. With this eccentricity the beam should sup-

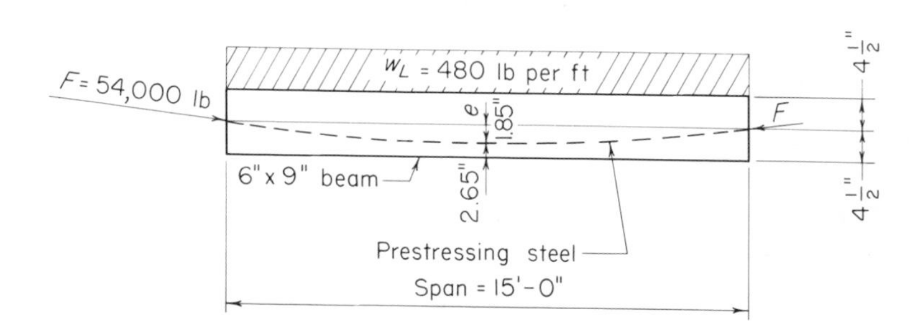

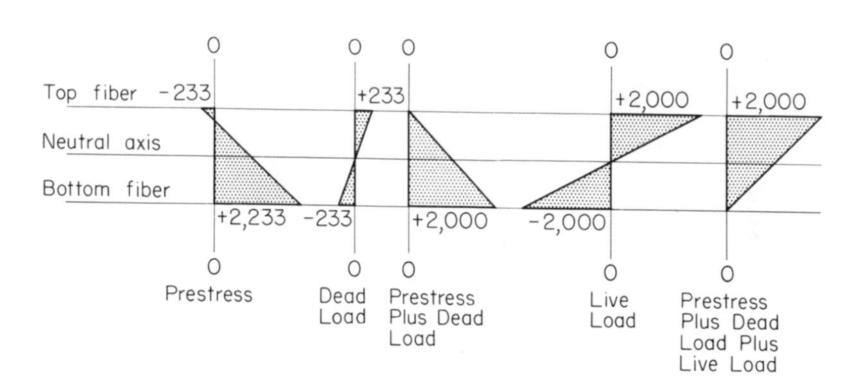

FIG. 1-4. Stresses in a beam with deflected tendons.

port a live load of 424 + 56 = 480 lb per ft without tensile stress. The following calculations are for the beam in Fig. 1-4.

Substituting in Eqs. (1-1*a*) and (1-1*b*)

$$f^t{}_F = \frac{54{,}000}{54} - \frac{54{,}000 \times 1.85}{81}$$

$$f^t{}_F = 1{,}000 - 1{,}233 = -233 \text{ psi}$$

$$f^b{}_F = \frac{54{,}000}{54} + \frac{54{,}000 \times 1.85}{81}$$

$$f^b_F = 1{,}000 + 1{,}233 = +2{,}233 \text{ psi}$$
$$M_L = 480\,(15^2) \times {}^{12}\!/\!_8 = 162{,}000 \text{ in.-lb}$$
$$f^t_L = 162{,}000 \div 81 = +2{,}000 \text{ psi}$$
$$f^b_L = 162{,}000 \div 81 = -2{,}000 \text{ psi}$$

Summation of stresses at center of span under dead load only:
In top fiber: $f^t_F + f^t_G = -233 + 233 = 0$
In bottom fiber: $f^b_F + f^b_G = +2{,}233 - 233 = +2{,}000$ psi

Summation of stresses at center of span under all loads:
In top fiber: $f^t_F + f^t_G + f^t_L = -233 + 233 + 2{,}000 = +2{,}000$ psi
In bottom fiber: $f^b_F + f^b_G + f^b_L = +2{,}233 - 233 - 2{,}000 = 0$

In Fig. 1-4 the steel tendons are shown on the c.g.c. at the ends of the beam, thus creating a uniform compressive stress in the concrete at the ends. They could just as readily be located above or below the c.g.c. to suit details of production or framing as long as they do not create stresses in excess of those permitted by the governing specification. For example, they could be located 1½ in. below the c.g.c. as in Fig. 1-3. The concrete stresses at the ends would then be zero in the top fiber and +2,000 psi in the bottom fiber. Although this lowering of the tendons at the ends of the beam does not cause excessive stresses, it will increase the camber because it increases the negative moment. Raising the tendons above the c.g.c. at the ends would decrease the camber.

CHAPTER 2

Properties Peculiar to Prestressed Concrete

2-1. Prestressing Methods. There are two kinds of tensioning elements for prestressed concrete: pretensioned and post-tensioned. The prestressing tendons in a given member will be all one kind or the other or a combination of the two, depending upon conditions.

The term pretensioned means that the tendons are tensioned to their full load before the concrete is placed. They are held under tension by anchors beyond the ends of the prestressed concrete member. After the concrete has been placed and allowed to cure to sufficient strength, the load in the tendons is transferred from the external anchors into the newly poured member, thus prestressing it. In the United States the standard tendons for pretensioned work are seven-wire uncoated stress-relieved prestressed concrete strands.

The term post-tensioned means that the tendons are tensioned after the concrete has been placed and allowed to cure. Frequently the tendon is placed inside a flexible metal hose, the entire assembly is placed in the form, and concrete is poured around it. After the concrete has cured, the tendon is tensioned and held under load by anchor fittings at its ends. Bond between the tendon and the concrete member is achieved by pumping the metal hose full of grout.

2-2. Prestressing Force. Examples in Figs. 1-2 to 1-4 used a prestressing force F equal to 54,000 lb. In the list of notations F is defined as "effective prestress force after deduction of all losses." In other words, F is the force that can be counted upon to remain active throughout the life of the structure. In the notations F_I is defined as "initial prestress force," which means the load to which the tendons are tensioned when the prestressed concrete member is being fabricated. F_I is always larger than F because several factors contribute to a reduction of tension in the tendons.

2-3. Shrinkage. Shrinkage of concrete is a process known to everyone familiar with reinforced concrete. As concrete cures, its total volume decreases slightly and it tends to shrink in all three dimensions. We are particularly concerned with the fact that it tends to shorten.

When the tension in pretensioned strands is transferred from external anchorages to the concrete members, the compressive force created in the concrete closes all the existing shrinkage cracks and both the concrete member and the strands shorten an amount equal to the sum of the widths of the shrinkage cracks. Although the rate of shrinkage is greatest when the member is first cast, it continues for some time after the prestressing operation. Since the continuing shrinkage cannot cause cracks in the concrete which is now in compression, it causes additional shortening. In the final analysis the pretensioned strands shorten an amount equal to the total shrinkage. This shortening causes a reduction of the tensile stress in the strands.

When a tendon is post-tensioned, it closes the existing shrinkage cracks during the tensioning operation. Since the initial tension F_I in this tendon is measured at the end of the tensioning operation after the existing cracks are closed, there will be no stress loss in the tendon due to the shrinkage which precedes the tensioning operation. In a post-tensioned tendon the only stress loss due to shrinkage is due to the shrinkage which takes place after the tensioning operation.

2-4. Elastic Compression. Elastic compression of the concrete member takes place as the prestressing force is applied. As in any elastic material its magnitude is a function of the modulus of elasticity of the concrete and the intensity of the compressive stress.

Since the pretensioned strand is under initial tension when the concrete member under zero stress is bonded to it, the strand shortens an amount equal to the elastic compression when the prestressing force is transferred from the anchorages to the concrete member.

If a post-tensioned member is stressed with only one tendon, all the elastic compression has taken place when the initial tension is measured at the end of the tensioning operation. In this case there will be no stress loss due to elastic shortening. When two or more tendons are post-tensioned in sequence, the first one tensioned will undergo the elastic shortening resulting from the tensioning of the remaining tendons. This is usually eliminated by computing the shortening which will result from tensioning the remaining tendons and overstretching the first tendons by this length.

2-5. Creep. Creep of concrete, often referred to as plastic flow, is inelastic shortening which takes place over a period of time. It is caused by a constant compressive stress, and its magnitude is a function of the intensity of that stress. Both pretensioned and post-tensioned tendons shorten an amount equal to the entire creep.

2-6. Relaxation. Relaxation of prestressing tendons is the fourth cause of stress loss. Since these tendons are used at an initial tension of 65 to 70 per cent of their ultimate strength, they, too, undergo a creep or plastic flow due to the constant high stress. Conditions, however, are somewhat different from those in the concrete. If a weight were hung on a tendon to create a stress equal to 70 per cent of its ultimate strength, the tendon would slowly elongate or creep just as the concrete slowly shortens. In an actual structure the tendon is not held at a constant load; it is not even held at a constant length. As the concrete member shortens from the factors described above, the tendon shortens too. There is a certain amount of plastic flow within the tendon which, since it cannot cause a gradual increase in length, shows up as a gradual drop in stress. This reduction in stress is called "relaxation."

In summary, F, the prestressing force remaining in the steel tendons after all losses, is less than F_I, the initial force in the tendons, by the following:

For pretensioned tendons:
100 per cent of shrinkage, 100 per cent of elastic shortening, 100 per cent of creep of concrete, and 100 per cent of relaxation of steel

For post-tensioned tendons:
Approximately 50 per cent of shrinkage, 100 per cent of creep of concrete, and 100 per cent of relaxation of steel

The stress losses just discussed make it necessary to use high-strength steel at high-unit stresses for the tendons. Reference to examples in later chapters shows that they allow for a stress loss of approximately 35,000 psi in a pretensioned tendon. An ordinary steel bar tensioned to its yield point, 33,000 psi, would lose its entire prestress by the time all stress losses had taken place. However, the initial tension in one grade of seven-wire strand is in the neighborhood of 189,000 psi. A stress loss of 35,000 psi reduces this to a final stress of 154,000 psi, which is sufficient to create the required prestressing force at reasonable cost.

CHAPTER 3

Design Procedure

In many ways it may seem that this chapter should follow the one on specifications. It was finally decided, however, that a basic knowledge of design procedure is needed to help the reader to understand the chapters on materials and methods of fabrication.

The object of this chapter is to present a step-by-step analysis of the design of a simple prestressed concrete beam. Definitions of the notations used are tabulated in Notations.

Design of a prestressed concrete beam for an actual structure involves two steps: choice of the shape and dimensions of the concrete member and analysis of the member under the specified loading conditions to check unit stresses and determine the amount and details of the prestressing steel. Facility in the choice of a proper section will come with experience. In addition to having sufficient section modulus to carry the applied loads, the concrete cross section must be properly proportioned to provide room for the tensioning elements. Once the section has been chosen, the following method of analysis can be applied.

During the calculations it may become apparent that the concrete section chosen is not adequate or that it is larger than necessary. If it is not adequate, a new section must be designed and the calculations repeated. In some cases it will prove economical to use an oversize member even at low stresses because it can be fabricated in existing forms. When this is not the case, an effort should be made to select a section in which the working stresses are reasonably near those permitted by the specification.

STEPS IN DESIGN PROCEDURE

Step 1. Compute the properties of the concrete cross section.

If the cross section is not uniform for the full length of the beam it is usually best to analyze the section at the point of maximum moment and then check at any other points which might be critical because of the change in section. Properties to be computed are:

A_c = area of entire concrete section (steel area not deducted)
y_t = distance from top fiber to c.g.c. (center of gravity of entire concrete section)
y_b = distance from bottom fiber to c.g.c.
I_c = moment of inertia of entire concrete section about c.g.c.
Z_t = section modulus of top fiber referred to c.g.c.
Z_b = section modulus of bottom fiber referred to c.g.c.
w_G = dead load of member per unit length

Step 2. Compute stresses in member due to its own dead weight.

M_G = bending moment due to w_G
$f^t{}_G$ = stress in top fiber due to M_G
$= M_G \div Z_t$
$f^b{}_G$ = stress in bottom fiber due to M_G
$= M_G \div Z_b$

Step 3. Compute stresses in member due to applied loads.

In most cases these will be made up of additional dead load such as roof deck or highway wearing surface and live load. (For the case where the prestressed member becomes a composite section with poured-in-place concrete see Chap. 5)

w_S = additional dead load
M_S = bending moment due to w_S
$f^t{}_S$ = stress in top fiber due to M_S
$= M_S \div Z_t$
$f^b{}_S$ = stress in bottom fiber due to M_S
$= M_S \div Z_b$
w_L = distributed live load per unit length
P_L = concentrated live load
M_L = bending moment due to w_L and/or P_L
$f^t{}_L$ = stress in top fiber due to M_L
$= M_L \div Z_t$
$f^b{}_L$ = stress in bottom fiber due to M_L
$= M_L \div Z_b$

Step 4. Determine the magnitude of the prestressing force and select the number of tendons.

The prestressing force must meet three general conditions.

1. It must provide sufficient compression stress to offset tension stress caused by the design loadings.

2. It must not create stresses, tension, or compression in excess of the allowable values when subjected to any loading condition equal to or less than the design loadings.

3. It must provide sufficient force to enable the beam to resist the flexure stresses when the beam is subjected to prescribed ultimate loadings.

Conditions 1 and 2 are somewhat related because both require that the stresses, under critical loading conditions, not exceed allowable working-stress design values. Condition 3 requires that a specified ultimate loading be resisted by stresses based on ultimate-strength design values. The fulfillment of conditions 1 and 2 does not imply that condition 3 is satisfied or vice versa. Therefore two separate and independent design requirements are present.

A first approximation of the number of strands is obtained from condition 1, which requires that the stresses due to prestressing be sufficient to offset the stresses caused by the design loadings. Conditions 2 and 3 also must be checked. Therefore, after a preliminary estimate of the magnitude of the prestressing force, the stresses due to the various loading combinations as well as the resistance to ultimate loadings must be examined.

As illustrated in Chap. 1, the stress in the concrete due to the prestressing force is a function of the magnitude of the force and also of its eccentricity e with regard to the c.g.c.

Since the beam under consideration is simply supported, the moments caused by the applied loadings create compression stresses in the top fibers and tension stresses in the bottom fibers.

The stresses due to design loadings in the bottom fibers at mid-span are assumed to be critical and in order to meet condition 1, the prestressing force must create a sufficient compression stress in the bottom of the beam to offset the tension stresses resulting from the bending moments. In other words, the stress $f^b{}_F$ in the bottom fiber at mid-span due to the prestressing force F must be at least equal in magnitude but opposite in direction to $f^b{}_G + f^b{}_S + f^b{}_L$.

From Eq. (1-1) the magnitude of the stress in the bottom fibers at mid-span due to the prestressing force is

$$f^b{}_F = \frac{F}{A_c} + \frac{Fe}{Z_b}$$

Setting the magnitude of the compression stress $f^b{}_F$ equal to the sum of the magnitudes of the tension stresses gives

$$\frac{F}{A} + \frac{Fe}{Z_b} = f^b{}_G + f^b{}_S + f^b{}_L$$

Use of this equation to determine appropriate values for F and e gives zero stress in the bottom fiber under full design load. For some structures the specifications permit a small tension stress f_{tp} under design-load conditions. In this case, the magnitude of the compression stress to be created by the prestressing force is the sum of the magnitudes of the tension stresses minus the allowable tension stress as shown in Eq. (3-1).

$$\frac{F}{A} + \frac{Fe}{Z_b} = f^b{}_G + f^b{}_S + f^b{}_L - f_{tp} \tag{3-1}$$

Equation (3-1) is a relationship between stress magnitudes in the bottom fibers at any point in a simple span member. The right-hand portion is the magnitude of the tension stresses which the prestressing force must neutralize by providing an equivalent amount of compression stresses. When substituting into the equation, use magnitudes only and disregard the usual convention of minus for tension and plus for compression.

Since there are two unknowns, F and e, in Eq. (3-1), it is necessary to estimate one and then solve for an approximate value for the other. The assumption that the actual location of the center of gravity of the prestressing steel, c.g.s., is a distance above the bottom of the beam equal to 15 per cent of the depth of the beam, or $0.15h$, is a good first guess. Therefore it is recommended that e be approximated by the formula

$$e_{\text{approx}} = y_b - 0.15h$$

With this value for e the first approximation for F is computed from Eq. (3-1), and the number of prestressing strands is determined.

Step 5. Locate the tendons in the cross section of the beam and compute the eccentricity provided.

In Step 4 the c.g.s. was assumed to be $0.85h$ from the top of the beam. In this step it is necessary to place each strand in a definite location that meets requirements for minimum spacing and concrete coverage. Once this is done the actual c.g.s. and eccentricity of the prestressing steel are computed. If the actual c.g.s. differs sufficiently from the first estimate of the location of the c.g.s. to change the number of tendons required, adjustment must be made.

Step 6. Check the top and bottom stresses at mid-span.

The allowable compression and tension stresses in the concrete are given in the appropriate specifications for two conditions (1) immediately after transfer of the prestressing force to the concrete, and (2) after all losses in the prestressing force with full design loading applied.

Establish f'_c and f'_{ci} and compute the allowable concrete stresses.

Check the design by computing the following actual stresses in the concrete:

$$f^t{}_{F_o+G}$$

$$f^b{}_{F_o+G}$$

$$f^t{}_{F+G+S+L}$$

$$f^b{}_{F+G+S+L}$$

The last value, $f^b{}_{F+G+S+L}$, should fall rather closely within the allowable value for the concrete, because this stress was chosen as critical when writing and solving Eq. (3-1). The other three should be compared with the allowable values, and adjustments must be made until all are acceptable.

Step 7. Check ultimate strength to make sure it meets the requirements in the specification. Check percentage of prestressing steel.

There is no constant ratio between the design strength and the ultimate strength of a prestressed concrete member as there is in structural steel. Two prestressed concrete members of different cross section can have exactly the same load-carrying capacity based on allowable design stresses yet have entirely different ultimate strengths.

If too much prestressing steel is used, failure of the member will occur by crushing of the concrete. This is not desirable because there is no warning of impending failure and the failure is of the explosive type. Check the amount of prestressing steel by the method outlined in the appropriate specification.

Step 8. Establish the path of the tendons and check critical points along the member under initial and final conditions.

All the preceding calculations have been concerned with the point of maximum moment. In some members other points may also be critical. These can usually be located by inspection of the moment diagrams, member properties, and location of prestressing steel.

There are two combinations of conditions which should be checked for maximum stress.

1. Final prestress plus full design load
2. Initial prestress plus dead load only (It is seldom necessary to check prestress alone without the benefit of dead load. As the prestressing force is applied, it creates a negative moment and the member develops a camber which raises the center portion off the form. Since the member is resting on each end, its dead-weight bending moment is effective.)

Step 9. Design of shear steel.

Present codes which specify the amount of web reinforcing steel are based partly on theory and partly on test data, and they differ in the methods of approach. The reader is referred to the various illustrative examples in Chap. 4, 5, and 11 for specific information concerning the application of the appropriate code.

Step 10. Compute camber and deflection.

Use the standard formulas for deflection of elastic members such as would be applied to a structural steel member. Compute camber at time of prestressing and for long-time loading.

The value for the modulus of elasticity for concrete E_c is selected for the short and long-time loadings by consulting the appropriate code. For normal designs, satisfactory accuracy is obtained by computing I as the moment of inertia of the cross section of the concrete. Effect of holes for tendons and the prestressing steel can be ignored.

Sometimes prestressed concrete members meet all the requirements for allowable stresses and ultimate strength but have excessive camber or deflection. When this occurs a new and stiffer section should be chosen.

CHAPTER 4

Numerical Example of Design Procedure for a Building Member

4-1. Conditions of Design. The purpose of this chapter is to explain the techniques involved in the design of a typical prestressed concrete member which will satisfy the requirements specified in ACI 318-63. (References such as "Sec. 909(b) ACI" are to ACI 318-63, and the sections referred to are reproduced in Appendix A.) Each step is numbered and the procedure is the same as that established in Chap. 3. Definitions of the symbols used are found at the beginning of the book listed under Notations.

A shape shown in Fig. 4-2 and a span of 75 ft. are chosen for this example because the condition is often encountered. Concrete having a final compression strength of 5,000 psi and prestressing tendons having a minimum ultimate tensile strength of approximately 270,000 psi are also chosen.

The cross section of the member is constant and the loading is symmetrical.

Span = 75 ft 0 in. center to center of supports
Live load = 40 psf, uniformly distributed
Roofing load = 8 psf, uniformly distributed

Members are to be 8-ft-wide single T's spaced 8 ft 0 in. center to center to form a roof, and it is estimated that a 32-in.-deep cross section will be required. The actual length of the member will be 75 ft 6 in. since it is assumed that the T's rest on 6-in. bearing pads.

The cross section assumed for first trial is shown in Fig. 4-2.

$$f'_c = 5{,}000 \text{ psi}$$
$$f'_{ci} = 4{,}000 \text{ psi}$$

Prestressing tendons chosen are ½-in.-diameter type 270K with properties shown in Table 7-1.

The depth-to-span ratio for this cross section is

$$\frac{L}{d} = \frac{75 \times 12}{32} = 28.2$$

which is larger than the $L/d = 20$ indicated in Sec. 909(b) ACI for simply supported reinforced concrete beams. Therefore, before this section can be considered adequate, the live-load deflection will have to be computed to assure that it does not exceed $L/180$ as specified in Sec. 909(e) ACI.

The values listed in Sec. 909(b) ACI pertain primarily to reinforced concrete members. For prestressed concrete members, these values usually can be increased by 50 per cent, and in some cases appreciably more, before the deflections given in Sec. 909(e) ACI are approached.

4-2. Design Procedure. In each of the ten steps recommended, a group of design values are computed. Since the total number of these values

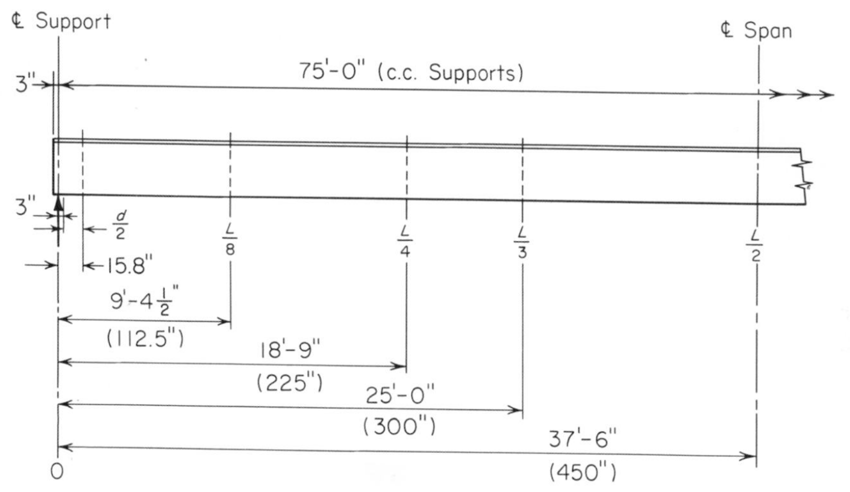

FIG. 4-1. Stations chosen for analysis of uniformly loaded beam.

is large, and since it is necessary to refer to them at various stages of the design, it is recommended that a tabular form be used to keep a systematic record.

The stations chosen for analysis are locations zero, $d/2$, $L/8$, $L/4$, $L/3$, and $L/2$, all measured from the center line of the support (except $d/2$, which is measured from the face of the support) along the length of the span as shown in Fig. 4-1. For uniformly distributed loadings, these stations will nearly always include all critical stations.

Station $d/2$ is chosen because it is one of the critical locations for shear computations. Section 2610(b) ACI says: "When applying Eq. (26-13) and (26-13A), the effective depth, d, shall be taken as the distance from the extreme compression fiber to the centroid of the prestressing tendons, or as 80 percent of the over-all depth of the member, whichever is the greater." In this example, later developments will show that the latter provision governs. Thus $d = 0.80 \times 32 = 25.6$ in. and $d/2 = 12.8$. Since Sec. 2610(b) says that the distance $d/2$ is to be measured from the face of the support, the distance from center of support is $3 + 12.8 = 15.8$ in.

Step 1. Compute properties of the concrete section* in Fig. 4-2. Take moments about the top (line TT).

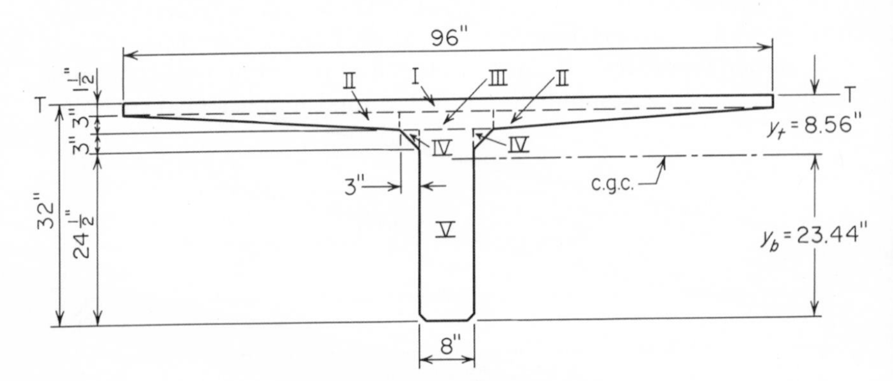

FIG. 4-2. Cross section of single T.

Section	A_c	y	Ay	Ay^2	I_o
I = 96 × 1.5	= 144	0.75	108	81	27
II = 2 × 41 × 3 × ½	= 123	2.50	308	770	62
III = 14 × 3	= 42	3.00	126	378	32
IV = 2 × 3 × 3 × ½	= 9	5.50	50	275	5
V = 8 × 27.5	= 220	18.25	4,015	73,280	13,860
Total	538		4,607	74,784	13,986
				13,986	
				I_T = 88,770	

$$y_t = \frac{Ay}{A} = \frac{4{,}607}{538} = 8.56 \text{ in.}$$

$$y_b = 32.00 - 8.56 = 23.44 \text{ in.}$$

$$I_c = 88{,}770 - (538)(8.56^2) = 49{,}350 \text{ in.}^4$$

$$Z_t = \frac{49{,}350}{8.56} = 5{,}765 \text{ in.}^3$$

$$Z_b = \frac{49{,}350}{23.44} = 2{,}105 \text{ in.}^3$$

$$w_G = 150/144 \times 538 = 561 \text{ lb per ft}$$

* The steel area is usually disregarded in computing section properties (see Art. 11-6).

Table 4-1. Summary of Conditions of Design

Span center-to-center supports	75 ft 0 in.
Length of bearing pads	6 in.
Additional dead load	8 psf
Live load	40 psf
Cross section	Fig. 4-2
Type of prestressing strands	Type 270K
Diameter of strands	½ in.
Quality of concrete	$f'_c = 5{,}000$ psi
Weight of concrete	150 lb per cu ft
Weight of T	561 lb per ft
Strength of concrete at transfer	$f'_{ci} = 4{,}000$ psi
Stations chosen for checking stresses	0, $d/2$, $L/8$, $L/4$, $L/3$, $L/2$ (see Fig. 4-1)

Table 4-2. Quantities Computed in Step 1

$A_c = 538$ in.^{2}	$I_c = 49{,}350$ in.^{4}
$y_t = 8.56$ in.	$Z_t = 5{,}765$ in.^{3}
$y_b = 23.44$ in.	$Z_b = 2{,}105$ in.^{3}
$w_G = 561$ lb per ft	

Step 2. Compute stresses in beam due to its own dead weight.

$$\text{Maximum } V_G = \frac{561(75)}{2} = 21{,}040 \text{ lb (at support)}$$

$$\text{Maximum } M_G = \frac{561(75^2)(12)}{8} = 4{,}733{,}000 \text{ in.-lb}$$

$$\text{Maximum } f^t{}_G = \frac{4{,}733{,}000}{5{,}765} = +822 \text{ psi (compression)}$$

$$\text{Maximum } f^b{}_G = \frac{4{,}733{,}000}{2{,}105} = -2{,}248 \text{ psi (tension)}$$

Step 3. Compute stresses in beam due to applied loads.

$$w_S = (8 \text{ psf})(8\text{-ft width}) = 64 \text{ lb per ft}$$

$$\text{Maximum } V_S = \frac{64(75)}{2} = 2{,}400 \text{ lb (at support)}$$

$$\text{Maximum } M_S = \frac{64(75^2)(12)}{8} = 540{,}000 \text{ in.-lb}$$

$$\text{Maximum } f^t{}_S = \frac{540{,}000}{5{,}765} = +93 \text{ psi}$$

$$\text{Maximum } f^b{}_S = \frac{540{,}000}{2{,}105} = -257 \text{ psi}$$

$$w_L = (40 \text{ psf})(8\text{-ft width}) = 320 \text{ lb per ft}$$

$$\text{Maximum } V_L = \frac{(320)(75)}{2} = 12{,}000 \text{ lb (at support)}$$

$$\text{Maximum } M_L = \frac{320(75^2)(12)}{8} = 2{,}700{,}000 \text{ in.-lb}$$

$$\text{Maximum } f^t{}_L = \frac{2{,}700{,}000}{5{,}765} = +467 \text{ psi}$$

$$\text{Maximum } f^b{}_L = \frac{2{,}700{,}000}{2{,}105} = -1{,}283 \text{ psi}$$

Table 4-3. Quantities Computed in Step 2 and Step 3

	M_G = 4,733,000 in.-lb	$f^t{}_G$ = +822 psi	$f^b{}_G$ = −2,248 psi
w_S = 64 lb per ft	M_S = 540,000 in.-lb	$f^t{}_S$ = +93 psi	$f^b{}_S$ = −257 psi
w_L = 320 lb per ft	M_L = 2,700,000 in.-lb	$f^t{}_L$ = +467 psi	$f^b{}_L$ = −1,283 psi
V_G (end) = 21,040 lb	V_S (end) = 2,400 lb	V_L (end) = 12,000 lb	

Step 4. Determine the magnitude of the prestressing force and select the number of tendons.

The number of strands required in the beam must provide resistance in flexure both to the design-load moment M_{G+S+L}, based on allowable working stresses as indicated in Sec. 2605 ACI, and to the ultimate moment M_u based on stresses indicated in Sec. 2608 ACI. The first approximation of the number of strands required will be estimated from the design-load conditions and then checked against the ultimate-flexure-moment requirements.

In a simply supported beam with uniform loading, the critical stresses caused by flexure are usually located in the bottom fiber at the mid-span. The sum of the stresses at this location produced by the design loadings and computed in Steps 2 and 3 is

$$f^b{}_{G+S+L} = f^b{}_G + f^b{}_S + f^b{}_L = -2{,}248 - 257 - 1{,}283 = -3{,}788 \text{ psi tension}$$

The prestressing force must neutralize this stress by providing in the concrete 3,788 psi compression. When tension stresses are allowed under design-loading conditions the prestressing force must neutralize 3,788 psi minus the allowable tension stresses.

Since the beam deflects upward when the prestressing force is transferred from the end anchors of the prestressing bed to the concrete, and since this results in the beam's being supported only at its ends, the dead load w_G is considered to be acting whenever the prestressing force is acting. Thus the stress due to dead load $f^b{}_G$ is always present and can be neutralized by the prestressing force. The stresses caused by the loads w_S and w_L may not be present at all times so the stresses

$$f^b{}_S + f^b{}_L = -257 - 1{,}283 = -1{,}540 \text{ psi}$$

may or may not be developed. Accordingly, in this section a variation of stress in the bottom fiber at mid-span of 1,540 psi is expected.

The magnitude of this variation compared with the allowable variation is an indication of the adequacy of the cross section of the beam. The maximum possible variation allowable in 5,000-psi concrete including allowable tension is the sum of the magnitudes of the allowable compression $0.45\,f'_c$ and the allowable tension $6\sqrt{f'_c}$ or

$$2{,}250 + 424 = 2{,}674 \text{ psi}$$

The expected variation of 1,540 psi is considerably less than 2,674 psi; therefore the cross section is more than adequate. For various reasons, such as possible excessive deflection or camber, the size of the cross section will not be reduced. However, if the expected stress variation approached or exceeded the 2,674 psi, inadequacy of the cross section would be indicated.

The stress in the bottom fibers caused by the prestressing force as indicated in Step 4, Chap. 3 must be sufficient to neutralize the tension stresses caused by the design loads minus the allowable tension stress. Therefore

$$\frac{F}{A_c} + \frac{Fe}{Z_b} \geq f^b{}_G + f^b{}_S + f^b{}_L - 6\sqrt{f'_c} \tag{4-1}$$

Equation (4-1) is a relationship between magnitudes only. The left-hand portion is the magnitude of the compression stresses caused by the prestressing force, and the right-hand portion is the magnitude of the tension stresses caused by the design loads minus the allowable tension stress.

In this example, the roof member is assumed to be protected from freezing; therefore, the allowable tension stress permitted by the code is included in the design.

In Eq. (4-1) all quantities are known except the prestressing force F and the eccentricity e. In solving this equation a value for e is estimated and then checked and revised.

An approximate value for e is obtained by assuming that the distance from the centroid of the prestressing steel to the bottom of the beam is $0.15h$. Then

$$e_{\text{approx}} = y_b - 0.15h = 23.44 - 4.80 = 18.64 \text{ in.}$$

With this value for e and with the left-hand portion of Eq. 4-1 set equal to the right-hand portion, a first estimate of the prestressing force F is obtained.

$$\frac{F}{538} + \frac{F \times 18.64}{2{,}105} = 2{,}248 + 257 + 1{,}283 - 424$$

$$F = 314{,}000 \text{ lb}$$

From Table 7-1 the final tension force for a ½-in.-diameter type 270K strand is 23,550 lb per strand. The number of strands is

$$N = \frac{F}{23{,}550} = \frac{314{,}000}{23{,}550} = 13.34 \text{ strands}$$

The 13.34 strands is the smallest number possible that will satisfy Eq. (4-1). Therefore it is necessary to choose the next larger whole number of strands.

Try 14 type 270K ½-in.-diameter strands.

For 14 strands,

$$F = 14 \times 23{,}550 = 329{,}600 \text{ lb}$$
$$A_s = 14 \times 0.1531 = 2.143 \text{ sq in.}$$

Table 7-1, for ½-in.-diameter type 270K strands specifies a minimum ultimate strength of 41,300 lb per strand or a minimum ultimate stress f'_s of

$$f'_s = 41{,}300 \div 0.1531 = 269{,}700 \text{ psi}$$

and an initial prestressing force of 70 per cent of f'_s or

$$269{,}700 \times 0.70 = 188{,}800 \text{ psi.}$$

In this example, it is assumed that the total loss of prestress in the strands due to the various causes indicated in Sec. 2607 ACI is 35,000 psi and that the partial loss resulting at transfer of the prestressing force from the external anchorages to the concrete is 15,000 psi. (See Art. 10-5 for computation of these quantities.) Therefore the prestressing force immediately after transfer is

$$F_o = A_s(188{,}800 - 15{,}000) = 2.143 \times 173{,}800 = 372{,}500 \text{ lb}$$

Table 4-4. Quantities Computed in Step 4

Number of strands = 14	$F_o = 372{,}500$ lb
$A_s = 2.143$ sq in.	$F = 329{,}600$ lb
$f'_s = 269{,}700$ psi	$F_o/F = 1.130$

Step 5. Locate the tendons in the cross section of the beam and compute the eccentricity provided.

Section 2617(b) ACI requires a clear spacing *between* strands of three times the strand diameter or 2 in. center to center for ½-in.-diameter strands. Although this provision applies only to the spacing at the ends of the member, it is frequently maintained at all locations along the beam. Since the width of the web is 8 in., a 2-in. spacing allows a maximum of three strands at each level if the minimum coverage of 1½ in. prescribed in Section 2616(a) ACI is provided.

The 2-in. center-to-center spacing is common in both horizontal and vertical directions even for strands of smaller diameter than ½ in. Most fixtures for deflecting strands, end templates, and anchor plates are set for this spacing.

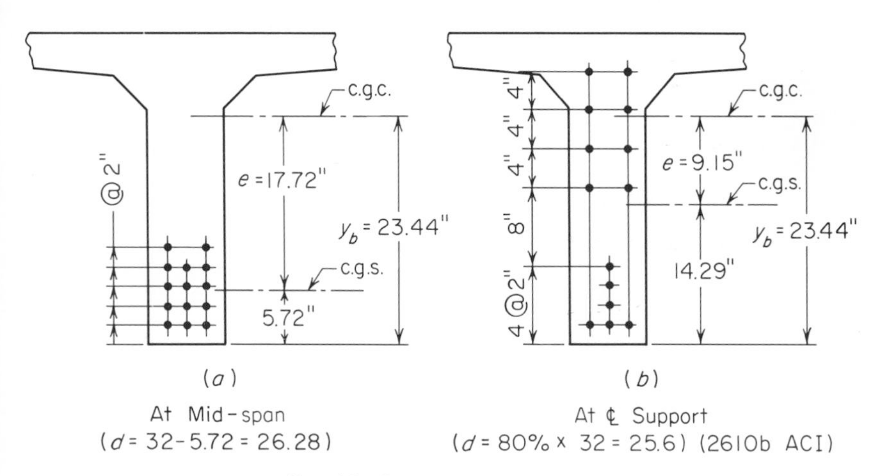

FIG. 4-3. Location of strands in T.

A strand pattern at mid-span is chosen as shown in Fig. 4-3*a*. For a 2-in. spacing this pattern provides the maximum eccentricity for the cross section.

Determine the center of gravity of the prestressing steel by taking moments about the bottom of the cross section.

$$
\begin{array}{rcl}
3 \times 2 &=& 6 \\
3 \times 4 &=& 12 \\
3 \times 6 &=& 18 \\
3 \times 8 &=& 24 \\
2 \times 10 &=& 20 \\
\hline
14 & & 80
\end{array}
$$

$$\text{c.g.s.} = \frac{80}{14} = 5.72 \text{ (from bottom)}$$

$$e = y_b - 5.72 = 23.44 - 5.72 = 17.72 \text{ in.}$$
$$d = y_t + e = 8.56 + 17.72 = 26.28 \text{ in.}$$

Table 4-5. Quantities Computed in Step 5

Strand locations = Fig. 4-3a
e = 17.72 in. at mid-span
d = 26.28 in. at mid-span

Step 6. Check the top and bottom stresses at mid-span.

Section 2605 ACI requires that the stresses in the concrete immediately after transfer of the tension force from the anchorage to the concrete, before losses due to creep and shrinkage, shall not exceed $0.60f'_{ci}$ in compression and $3\sqrt{f'_{ci}}$ in tension in the tension zone. For the member in this example the calculated stresses immediately after transfer include the weight of the beam.

Allowable compression stress immediately after transfer is

$$0.60f'_{ci} = 0.60 \times 4{,}000 = +2{,}400 \text{ psi}$$

Allowable tension stress immediately after transfer is

$$3\sqrt{f'_{ci}} = 3\sqrt{4{,}000} = -190 \text{ psi}$$

This limiting value of 190 psi tension will be assumed to apply not only in the tension zone but to all locations in the beam. If this value is exceeded, an appropriate amount of reinforcing steel is provided.

The stresses at the top and bottom of the beam at mid-span resulting from the prestressing force F_o are

$$f^t_{F_o} = \frac{F_o}{A_c} - \frac{F_o e}{Z_t} = \frac{372{,}500}{538} - \frac{372{,}500 \times 17.72}{5{,}765}$$
$$= +692 - 1{,}144 = -452 \text{ psi tension}$$

$$f^b_{F_o} = \frac{F_o}{A_c} + \frac{F_o e}{Z_b} = +692 + 3{,}133 = +3{,}825 \text{ psi compression}$$

The stresses at the top and bottom of the beam at mid-span resulting from the prestressing force F_o combined with the weight of the beam w_G are

$$f^t_{F_o+G} = f^t_{F_o} + f^t_G = -452 + 822 = +370 \text{ psi compression}$$
$$f^b_{F_o+G} = f^b_{F_o} + f^b_G = +3{,}825 - 2{,}248 = +1{,}577 \text{ psi compression}$$

Both of these stresses, +370 psi and +1,577 psi, are within the allowable −190 psi and +2,400 psi.

Section 2605 ACI also requires that the stresses in the concrete at design loads, after allowance for all prestress losses, shall not exceed $0.45f'_c$ in compression and, if the member is not exposed to freezing temperatures, $6\sqrt{f'_c}$ in tension. This tension is permitted only when bonded prestressed or unprestressed reinforcement is located so as to control cracking.

Allowable compression stress for design loads is

$$0.45f'_c = 0.45 \times 5{,}000 = +2{,}250 \text{ psi}$$

Allowable tension stress for design loads is

$$6\sqrt{f'_c} = 6\sqrt{5{,}000} = -424 \text{ psi}$$

The stresses at the top and bottom of the beam at mid-span caused by the prestressing force F are

$$f^t{}_F = \frac{F}{A_c} - \frac{Fe}{Z_t} = \frac{329{,}600}{538} - \frac{329{,}600 \times 17.72}{5{,}765}$$

$$= +612 - 1{,}012 = -400 \text{ psi tension}$$

$$f^b{}_F = \frac{F}{A_c} + \frac{Fe}{Z_b} = +612 + \frac{329{,}600 \times 17.72}{2{,}105} = +612 + 2{,}772$$

$$= +3{,}384 \text{ psi compression}$$

The stresses at the top and bottom of the beam at mid-span caused by the prestressing force F plus the design loads are

$$f^t{}_{F+G+S+L} = -400 + 822 + 93 + 467 = +982 \text{ psi compression}$$

$$f^b{}_{F+G+S+L} = +3{,}384 - 2{,}248 - 257 - 1{,}283 = -404 \text{ psi tension}$$

Both stresses, $+982$ psi and -404 psi, are within the allowable limits of $+2{,}250$ psi and -424 psi.

The tension stress ($f^b{}_{F+G+S+L} = -404$ psi) is quite close to the allowable (-424 psi). This means that the 14 prestressing strands are a minimum for the working-stress design.

The compression stress in the top flange ($f^t{}_{F+G+S+L} = 982$ psi) is considerably less than the allowable ($+2{,}250$ psi). One does not expect to develop high-compression stresses at mid-span in the top fiber of a tee-shaped member. The wide flanges are necessary to provide a roof, and therefore the area of the compression concrete exceeds the amount necessary to resist forces.

In addition to the stress situation at mid-span, there are other things which must be considered when evaluating a cross section. The flexibility of the member and the camber caused by the prestressing are two important items which will be checked later. From the foregoing calculations, it is observed that the section shown in Fig. 4-2 with 14 strands shows promise of being an excellent choice for the assumed design conditions.

The method used in Step 4 to estimate the number of strands required does not always give the desired prestressing force as closely as was the case in this example. If the stresses computed exceed the allowable stresses or are small in comparison with them, a variation in the number of prestressing strands must be made and the stresses must be checked.

Table 4-6. Quantities Computed in Step 6

Allowable stress in concrete after transfer	
Compression: $0.60 f'_{ci} = +2{,}400$ psi	
Tension: $3\sqrt{f'_{ci}} = -190$ psi	
Allowable stress in concrete with design loads	
Compression: $0.45 f'_c = +2{,}250$ psi	
Tension: $6\sqrt{f'_c} = -424$ psi	
$f^t{}_{F_o} = -452$ psi	$f^b{}_{F_o} = +3{,}825$ psi
$f^t{}_{F_o+G} = +370$ psi	$f^b{}_{F_o+G} = +1{,}577$ psi
$f^t{}_F = -400$ psi	$f^b{}_F = +3{,}384$ psi
$f^t{}_{F+G+S+L} = +982$ psi	$f^b{}_{F+G+S+L} = -404$ psi

Step 7. Check the ultimate flexural strength to assure compliance with Secs. 2608 and 2609 ACI. Check percentage of prestressing steel.

The determination of the ultimate-flexural-moment capacity of a beam as required by the code is calculated by means of one of two formulas. One formula is valid when the cross section is rectangular or the neutral axis lies within the top flange; the other applies when the neutral axis lies in the web of a flanged member. The code also states in a note in small print that if the average flange thickness t is greater than

$$1.4dp \frac{f_{su}}{f'_c} \tag{4-2}$$

the neutral axis usually lies within the flange of the cross section.

Values of p and f_{su} for substitution into Eq. (4-2) are computed as follows:

$$p = \frac{A_s}{bd} = \frac{14 \times 0.1531}{96 \times 26.28} = 0.00085$$

and $$f_{su} = f'_s \left(1 - \frac{0.5pf'_s}{f'_c}\right) \qquad \text{(26-6 ACI)}$$

$$= 269{,}700 \left(1 - \frac{0.5 \times 0.00085 \times 269{,}700}{5{,}000}\right) = 263{,}400 \text{ psi}$$

Evaluating Eq. (4-2),

$$1.4dp\frac{f_{su}}{f'_c} = 1.4 \times 26.28 \times 0.00085 \times \frac{263{,}400}{5{,}000} = 1.64 \text{ in.}$$

Since the average flange thickness of 3 in. is greater than 1.64 in., the neutral axis lies within the flange and the ultimate-flexural-moment capacity of the beam is

$$M_u = \phi[A_s f_{su} d(1 - 0.59q)] \qquad \text{(26-4 ACI)}$$

where $\phi = 0.90$, as specified in Sec. 1504 ACI, and d (at mid-span) = 26.28 in.

$$q = pf_{su}/f'_c \qquad \text{(ACI, Sec. 2600)}$$

$$= 0.00085 \times \frac{263{,}400}{5{,}000} = 0.0448$$

$$M_u = 0.90[2.143 \times 263{,}400 \times 26.28(1 - 0.59 \times 0.0448)]$$

$$= 13{,}050{,}000 \text{ in.-lb}$$

The ultimate moment U_M which the beam in this example is required to resist is computed from the ultimate loads specified in Sec. 1506 ACI.

$$U_M = 1.5M_D + 1.8M_L = 1.5(M_{G+S}) + 1.8M_L$$

$$= 1.5(4{,}733{,}000 + 540{,}000) + 1.8(2{,}700{,}000)$$

$$= 12{,}770{,}000 \text{ in.-lb}$$

The ultimate resisting moment capacity of the beam, $M_u = 13{,}050{,}000$ in.-lb, is greater than the required ultimate design moment, $U_M =$ 12,770,000 in.-lb.

According to Sec. 2609(a) ACI, $q = pf_{su}/f'_c$ must not exceed 0.30.

$$q = 0.00085 \times \frac{263{,}400}{5{,}000} = 0.0448 < 0.30$$

In order to prevent sudden failure, Sec. 2609(c) ACI provides that the ultimate-bending-moment capacity M_u shall exceed by 20 per cent the "cracking moment" calculated on the basis of a modulus of rupture of $7.5\sqrt{f'_c}$. The moment that causes flexural cracking must neutralize all of the compressive stress due to prestress and create a tensile stress equal to the modulus of rupture. Therefore

$$\text{Cracking moment} = (7.5\sqrt{f'_c} + f^b{}_F)Z_b$$

$$= (7.5\sqrt{5{,}000} + 3{,}384)2{,}105 = 8{,}240{,}000 \text{ in.-lb}$$

$$M_u \geq 1.2 \times \text{cracking moment}$$

$$13{,}050{,}000 > 1.2 \times 8{,}240{,}000 = 9{,}890{,}000$$

M_u is more than 20 per cent greater than the cracking moment.

If the ultimate moment is less than 1.2 times the cracking moment, it may be possible to correct the situation by adding untensioned reinforcing bars or strands in the tension zone.

Step 8. Establish the path of the tendons and check critical points along the member both for transfer and final prestressing forces.

Since the T beam is a simple span member, there is no bending moment M_G at the supports and the stresses in the beam are those due only to the prestressing force. If the prestressing force were held at the same eccentricity at the ends as at the center of the span, the stresses at the ends would be

$$f^t_{F_0} = -452 \text{ psi} \qquad f^t_F = -400 \text{ psi}$$
$$f^b_{F_0} = +3{,}825 \text{ psi} \qquad f^b_F = +3{,}384 \text{ psi}$$

Three of these are greater than the allowable; but by decreasing the eccentricity e at the ends, this situation can be rectified. It is accomplished by raising some of the strands at the ends according to a desired pattern and then deflecting them in the central portion of the beam to fit an already established mid-span pattern. Since considerable vertical force is required to effect this deflection, it is economical to hold the amount of strand deflection to a minimum—or, in other words, the eccentricity at the ends should be as large as possible.

The eccentricity chosen at the ends will be the maximum allowable for the permitted stresses in the concrete as indicated in Sec. 2605 ACI. A magnitude for e less than the smallest computed value for the following four conditions must be chosen in order to keep the stresses within the requirements.

The maximum allowable stress in the concrete is assumed to be present when computing the allowable magnitude of the eccentricity at the ends of the beam.

1. Top of beam at time of transfer

$$f^t_{F_0} = -190 \text{ psi} = \frac{F_0}{A_c} - \frac{F_0 e}{Z_t} = +692 - \frac{372{,}500e}{5{,}765}$$
$$e = 13.67$$

2. Bottom of beam at time of transfer

$$f^b_{F_0} = +2{,}400 \text{ psi} = \frac{F_0}{A_c} + \frac{F_0 e}{Z_b} = +692 + \frac{372{,}500e}{2{,}105}$$
$$e = 9.66 \text{ in.}$$

3. Top of beam subjected to design loads

$$f^t_F = -400 \text{ psi (does not govern)}$$

4. Bottom of beam subjected to design loads

$$f^b{}_F = +2{,}250 = \frac{F}{A_c} + \frac{Fe}{Z_b} = +612 + \frac{329{,}600e}{2{,}105}$$

$$e = 10.47 \text{ in.}$$

The maximum allowable value for eccentricity at the ends is 9.66 in.

Locate the strands in a pattern which is consistent with spacing requirements and which will provide an eccentricity at the center line of the supports of 9.66 in. or less. Figure 4-3*b* indicates a possible pattern which provides a c.g.s. at the supports of 14.29 in. above the bottom of the girder and an eccentricity of 9.15 in. This arrangement will be chosen for use here.

Note that the arrangement shown in Fig. 4-3*b* requires that only eight of the fourteen strands be subjected to hold-down forces in order to produce the required pattern at mid-span. The other six strands remain horizontal throughout the entire length of the member. Having three strands at one level in the upper portion of the stem of the beam was avoided so that concrete and vibrating equipment could be more easily inserted.

The choice of the path of the tendons between the ends and mid-span depends on several things. In order to keep the hold-down hardware to a minimum, only two hold-down points symmetrically placed are often used. This means that the path will be described by three straight lines, a horizontal line in the central portion and two sloping lines at the ends. For beams that are subjected to uniformly distributed loadings and where the design is controlled primarily by flexure stresses rather than shear, experience indicates that the hold-down points should be located near the one-third points.

The path of the tendons will have considerable effect on the camber of the beam. Under prestress plus dead load, the typical beam will have an upward camber because the bottom fibers are shortened by high compression stresses whereas the top fibers are subjected to small stresses. The magnitude of the camber is a function of the difference in stress between top and bottom fibers; therefore, if the tendons are raised at the ends the minimum amount, the camber will be a maximum. By varying the path of the tendons the designer can, within limits, control the magnitude of the camber.

The slope of the tendons in the end portions of the beam provides a vertical as well as a horizontal component. The horizontal component creates the prestress in the beam and the vertical component assists the concrete in resisting shearing forces; that is, the tendons carry an amount of shear equal to their vertical component. The more the path of the tendons is raised at the ends, the greater their ability to resist shear forces. The shear carried by the tendons is considered in the design of the web reinforcement, but changing the elevation of the tendons within the allowable limits seldom makes a significant change in the quantity of web reinforcement required.

The most important factor in choosing the path of the tendons is the bending-moment curve. The eccentricity of the tendons at any point along the beam must be adjusted to keep the sum of the bending moment and prestressing stresses within allowable values.

For this example, the path of the tendons is assumed to be a path made up of three straight lines with hold-down points at the one-third points along the span as shown in Fig. 4-4. The resulting stresses will be checked

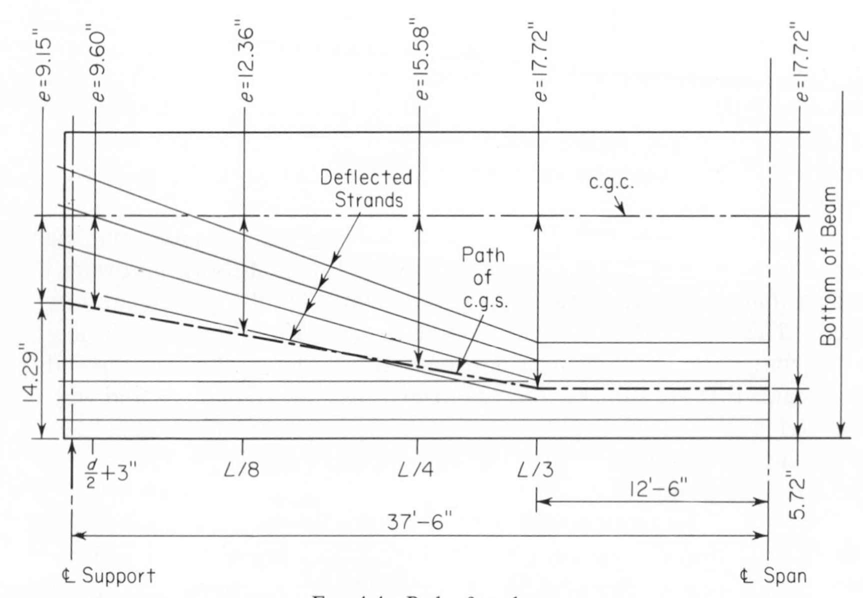

FIG. 4-4. Path of tendons.

at stations along the beam to insure that allowable stresses are not exceeded. Once this is done, possible changes in the chosen path can be considered.

Since the stresses at all points along the beam must be checked, it is necessary to compute systematically the stresses for arbitrarily chosen stations along the beam and to plot the values on graphs to insure allowable values for all points. The stations chosen are indicated at the beginning of Art. 4-2 and a tabular form of calculation is used (see Tables 4-7*b* and 4-8). Other stations should be checked if examination of the graphs indicates the need. Once he has some experience, a designer can locate the critical stations along the beam by inspection and compute stresses at these points without plotting a graph for each condition.

Table 4-7a. Summary of Essential Values for Use in Calculations for Tables 4-7b and 4-8

Design conditions	Section: Fig. 4-1. Span, L: 75 ft 0 in. Bearing: 6-in. pad $f'_c = 5{,}000$ psi	$w_G = 561$ plf $w_S = 64$ plf $w_L = 320$ plf $f'_{ci} = 4{,}000$ psi	Strand type: 270K Strand diameter: ½ in. area: 0.1531 Strand breaking load 41,300 lb $f'_s = 41{,}300 \div 0.1531$ $= 269{,}700$ psi
Stress limits	$0.45f'_c = 2{,}250$ psi $6\sqrt{f'_c} = -424$ psi	$0.6f'_{ci} = 2{,}400$ psi $3\sqrt{f'_{ci}} = -190$ psi	$f_{se} = 0.70f'_s - 35{,}000$ $= 153{,}800$ psi $f_{F_o} = 0.70f'_s - 15{,}000$ $= 173{,}800$ psi
Section properties	$A_c = 538$ sq in. $I_c = 49{,}350$ in.^{4}	$y_t = 8.56$ in. $y_b = 23.44$ in.	$Z_t = 5{,}765$ in.^{3} $Z_b = 2{,}105$ in.^{3}

The following four graphs, which are obtained by plotting the appropriate values from Tables 4-7*b* and 4-8, are required:

Fig. 4-5 which shows the difference between $f^b{}_F$ and $f^b{}_{G+S+L}$
Fig. 4-6 which shows the difference between $f^t{}_F$ and $f^t{}_{G+S+L}$
Fig. 4-7 which shows the difference between $f^b{}_{F_o}$ and $f^b{}_G$
($f^b{}_{F_o}$ = column 17 × F_o/F) (Table 4-8)
Fig. 4-8 which shows the difference between $f^t{}_{F_o}$ and $f^t{}_G$
($f^t{}_{F_o}$ = column 18 × F_o/F) (Table 4-8)

The net stresses in the concrete are indicated by the shaded portions of the graphs. The magnitude of these stresses is the net difference between the stresses caused by the prestressing forces and the loadings on the beam. As long as the values indicated by the shaded portions do not exceed the allowable stresses in the concrete the design is adequate.

In a pretensioned bonded member the load from the strand is transferred to the concrete by bond over a short distance called the "transfer length." Section 2610(b) ACI specifies a transfer length of 50 diameters for strand. Since this example is using ½-in. strand, the transfer length is 25 in. as shown in Figs. 4-5 through 4-8.

The graphs in Figs. 4-5 and 4-6 indicate that the stresses in the concrete do not exceed +2,250 psi compression nor −424 psi tension, which are the allowable concrete stresses when the beam is subjected to full design loadings. The graphs in Figs. 4-7 and 4-8 indicate that the stresses in the concrete do not exceed +2,400 psi compression nor −190 psi tension, which are the allowable concrete stresses when the beam is subjected to the loadings involved immediately after transfer of the prestressing force. Therefore it is concluded that this design is adequate as far as the stresses in the concrete are concerned.

Table 4-7b. Moments and Stresses for Design Loads

(1)	(2)	(3)	(4)	(5)	(6)	(7)	(8)	(9)	(10)
Station	Distance from support, in.	Moment from beam weight, in.-kips	Moment from additional dead load, in.-kips	Moment from live load, in.-kips	Total bending moment, in.-kips	Stress in bottom due to beam weight, psi	Stress in bottom due to total load, psi	Stress in top due to beam weight, psi	Stress in top due to total load, psi
	X	M_G	M_S	M_L	M_{G+S+L}	$f^b{}_G$	$f^b{}_{G+S+L}$	$f^t{}_G$	$f^t{}_{G+S+L}$
Support	0	0	0	0	0	0	0	0	0
$d/2$	15.8	326	37	186	549	−155	−261	+57	+95
$L/8$	112.5	2,070	236	1,181	3,487	−983	−1,657	+359	+604
$L/4$	225.0	3,550	405	2,026	5,981	−1,686	−2,842	+616	+1,037
$L/3$	300.0	4,210	480	2,400	7,090	−2,000	−3,369	+731	+1,230
$L/2$	450.0	4,733	540	2,700	7,973	−2,248	−3,787	+822	+1,382

Table 4-8. Moments and Stresses from Prestress Force

Station	(11)	(12)	(13)	(14)	(15)	(16)	(17)	(18)	(19)	(20)	(21)	(22)
	Effective final prestress force, kips	Compression at centroid due to final prestress, psi	Eccentricity, in.	Moment due to eccentricity of final prestress, in.-kips	Bending stress from eccentric prestress in bottom of beam, psi	Bending stress from eccentric prestress in top of beam, psi	Total stress in bottom fiber from prestress, psi	Total stress in top fiber from prestress, psi	Total stress in bottom fiber from prestress and all loads, psi	Total stress in top fiber from prestress and all loads, psi	Stress in bottom from prestress and beam weight at release, psi	Stress in top from prestress and beam weight at release, psi
	F	F/A	e	Fe	Fe/Z_b	Fe/Z_t	$f^b{}_F$	$f^t{}_F$	$f^b{}_{F+G+S+L}$	$f^t{}_{F+G+S+L}$	$f^b{}_{F_o+G}$	$f^t{}_{F_o+G}$
Support	39.6	74	9.15	362	172	−63	+246	+11	+246	+11	+278	+12
$d/2$	248.8	463	9.60	2,390	1,134	−414	+1,597	+49	+1,336	+144	+1,650	+112
$L/8$	329.6	612	12.36	4,070	1,935	−706	+2,547	−94	+890	+510	+1,895	+253
$L/4$	329.6	612	15.58	5,140	2,441	−891	+3,053	−279	+211	+758	+1,764	+301
$L/3$	329.6	612	17.72	5,840	2,772	−1,012	+3,384	−400	+15	+830	+1,824	+279

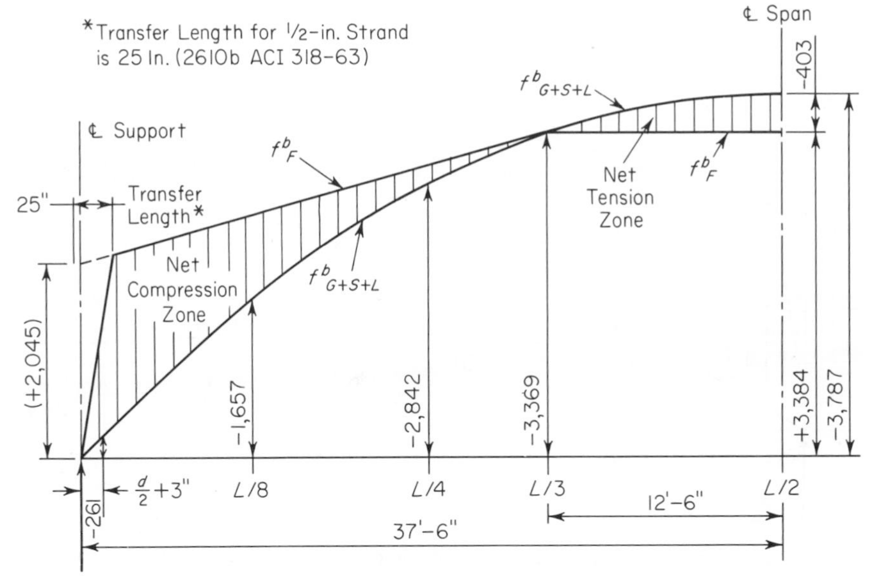

FIG. 4-5. Bottom fiber stress under prestress and all design loads.

What can be learned from Figs. 4-5 through 4-8 beside the fact that the section as prestressed is adequate?

Fig. 4-5. The tensile stress of 402 psi at center of span is close to the maximum allowable of 424 psi, which means that the prestressing force would have to be increased before the design load could be increased. At the support, the compressive stress of 2,045 psi is not much below the allowable 2,250 psi, which is usually a desirable condition from the fab-

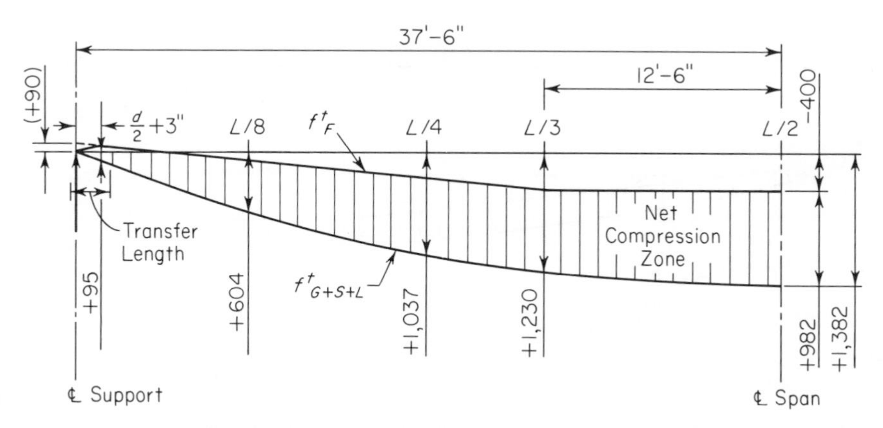

FIG. 4-6. Top fiber stress under prestress and all design tools.

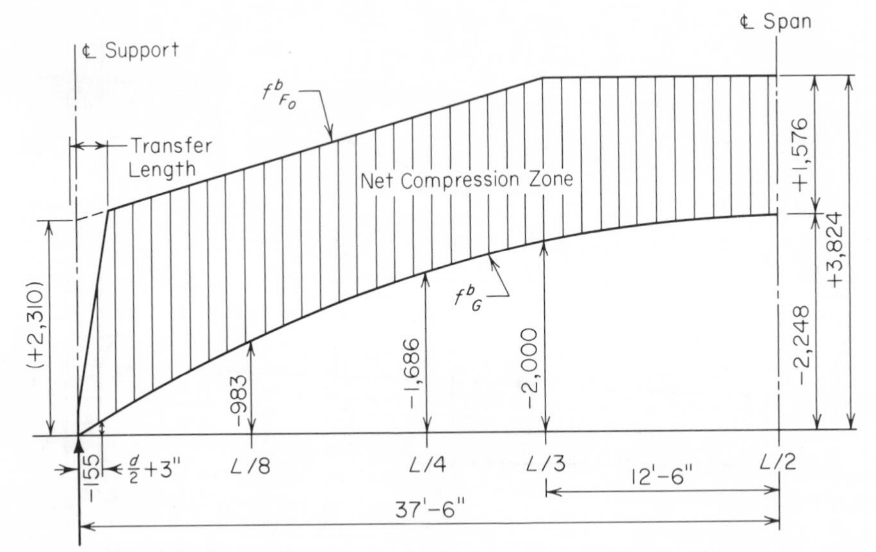

FIG. 4-7. Bottom fiber stress under initial prestress and beam weight only.

ricator's point of view. If the prestressing force were increased this stress could be held by raising the center of gravity of the strands at the ends.

Fig. 4-6. The net compressive stress of 982 psi is far below the allowable, but top fiber compressive stresses in T sections are generally low. The top flange is wide to provide a deck surface rather than for structural reasons.

Fig. 4-7. The compressive stress of 1,576 psi at center of span is well below the allowable of 2,400 psi. In Step 4 we deliberately chose a prestressing force just large enough to keep stresses within the allowable under design load. The live-load capacity of the T could be increased by

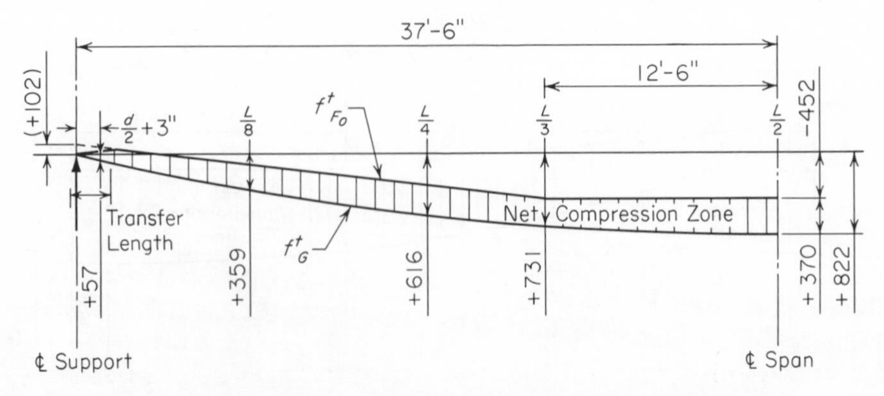

FIG. 4-8. Top fiber stress under initial prestress and beam weight only.

increasing the prestressing force and thus increasing the net compression at center of span to any value up to 2,400 psi. (Camber and/or deflection are frequently the limiting conditions rather than stress when stresses approach the allowable in comparatively shallow members.) Compressive stress at the end is near the allowable but could be kept under control by raising the strands if a larger prestressing force were desirable.

Fig. 4-8. This condition is seldom critical but should usually be checked at the end, the hold-down point, and the center of span.

As indicated in Figs. 4-5 to 4-8, all stresses caused by the uniformly distributed loadings, $f^b{}_{G+S+L}$, $f^t{}_{G+S+L}$, $f^b{}_G$, and $f^t{}_G$, plot as parabolic curves similar in shape to the moment diagram for the loads. Also, all stresses caused by the prestressing forces, $f^b{}_F$, $f^t{}_F$, $f^b{}_{F_0}$, and $f^t{}_{F_0}$, plot as straight lines with breaks or changes in slope at the hold-down points only (except at the ends of the strands within the anchorage transfer length).

In this example the hold-down points are located 12 ft 6 in. each way from the center of the span. If this location were moved, say, closer to the center of the span, the break in the straight lines would move the same amount whereas the ordinates at the end and at the center would remain the same. Straight lines drawn to the new hold-down point would define a new net compression or tension zone, and the effects of changing the hold-down position would be readily observed. A study of the graphs using this technique indicates that the hold-down location can vary from the chosen position by a considerable amount without causing excessive stresses.

Only one hold-down at the center of the span might be satisfactory for this example, but Fig. 4-5 indicates that excessive tension may appear a few feet from the center of the span. The single hold-down, however, requires twice the angle change in the prestressing strands as in the case of two hold-down locations, and it is desirable to minimize secondary stresses in the strands, such as those due to the pressure from hold-down rollers. The hold-down locations at the one-third points will be used in this example.

Step 9. Design of shear steel.

The requirements set forth in Sec. 2610 ACI are quite complex and cumbersome when compared with previous recommendations. Because of this, many engineers have attempted to develop methods less cumbersome but conservative that can be used when the cost of web reinforcement is not a significant factor. It is possible that the cost of making an analysis according to the present code will exceed by an appreciable margin the cost of extra web steel required by an approximate but conservative analysis.

The decision about whether or not an approximate analysis should be used can be made only after thorough experience with both approaches.

In many cases it is imperative that a careful design according to the code be made. In this step two designs are presented. Step 9A follows the code, and Step 9B presents a method suggested by Richard Elstner of Wiss, Janey, Elstner and Associates, Northbrook, Illinois, which is simpler and conservative.

Step 9A. Design of shear steel following the code.

According to Sec. 26-10 ACI, the area of web reinforcement A_v placed perpendicular to the axis of the member shall be not less than

$$A_v = \frac{(V_u - \phi V_c)s}{\phi d f_y} \qquad \text{(26-10 ACI)}$$

nor less than

$$A_v = \frac{A_s}{80}\frac{f'_s}{f_y}\frac{s}{d}\sqrt{\frac{d}{b'}} \qquad \text{(26-11 ACI)}$$

The minimum amount of shear steel A_v required in all members throughout their entire length is given by Eq. (26-11 ACI), unless it is shown by test, Sec. 26-10C ACI, that less is satisfactory. In this design it is assumed that testing the member is not feasible. Eq. (26-10 ACI) applies only when the shear forces are large enough to require more than the minimum web steel.

Eq. (26-10 ACI) states that the concrete carries an amount of shear V_c and that the difference between the shear force at ultimate loading V_u and ϕV_c must be resisted by web reinforcement stressed to the yield-point stress. The determination of a value for V_c requires an investigation of two types of shear failure that can occur in a prestressed beam.

The shear force resisted by the concrete V_c is taken as the lesser of V_{ci} or V_{cw} where

$$V_{ci} = 0.6b'd\sqrt{f'_c} + \frac{M_{cr}}{M/V - d/2} + V_d \qquad \text{(26-12 ACI)}$$

but not less than $1.7b'd\sqrt{f'_c}$ and where

$$V_{cw} = b'd(3.5\sqrt{f'_c} + 0.3f_{pc}) + V_p \qquad \text{(26-13 ACI)}$$

[Equations (26-12 ACI) and (26-13 ACI) apply to normal-weight concrete. If lightweight aggregate is being used, apply Eqs. (26-12A ACI) and (26-13A ACI).]

The expressions for V_{ci} and V_{cw} are completely independent of each other and are related to different regions of the member. V_{ci} governs in regions where both shear and flexure are acting together, and V_{cw} governs in regions of maximum shear and little flexure.

When solving for V_{ci} from Eq. (26-12 ACI), the net flexural cracking moment M_{cr} must be computed. The code defines M_{cr} as

$$M_{cr} = \frac{I}{y}(6\sqrt{f'_c} + f_{pe} - f_d) \tag{4-3}$$

At this point it is important to emphasize the concept that the shear force in the member is resisted by the concrete V_c and by the web steel V'_u. Since the code requires a minimum amount of web steel at all times, there is always present a resisting shear force $V'_{u,\min}$ in addition to the force resisted by the concrete V_c.

The design of the web steel is based on ultimate loading and the code requires that the ultimate shear force V_u be resisted by the various components after they are multiplied by a capacity reduction factor ϕ (see Sec. 1504).

By rewriting Eq. (26-10 ACI) the relation

$$\frac{V_u - \phi V_c}{\phi} = A_v f_y \frac{d}{s} \tag{4-4}$$

is obtained from which the value for the shearing force resisted by the web steel V'_u is defined as

$$V'_u = \frac{V_u - \phi V_c}{\phi} \tag{4-5}$$

Solving Eq. (4-5) for V_c gives

$$\phi V_c = V_u - \phi V'_u \tag{4-6}$$

The value for V'_u which corresponds to the minimum allowable web steel is noted as $V'_{u,\min}$ and is determined by rewriting Eq. (26-11 ACI).

$$\frac{A_s f'_s}{80}\sqrt{\frac{d}{b'}} = A_v f_y \frac{d}{s} \tag{4-7}$$

Since the right-hand portion of Eq. (4-4) is the same as that of Eq. (4-7), it can be stated that

$$V'_{u,\min} = \frac{A_s f'_s}{80}\sqrt{\frac{d}{b'}} \tag{4-8}$$

As previously stated, the code requires a minimum amount of web steel along the entire length of the beam which provides at all times a shear resisting ability of $V'_{u,\min}$. It is the designer's job to determine the locations in the member where extra web steel is needed in addition to that which provides $V'_{u,\min}$. This can be done by locating the regions where

$$\phi V_c > V_u - \phi V'_{u,\min} \tag{4-9}$$

Where Eq. (4-9) is true, $\phi V'_{u,\min}$ is sufficiently large—or, in other words, the minimum web steel A_v is adequate. Where Eq. (4-9) is not true, A_v must be greater than the minimum and Eq. (26-10 ACI) is used.

A comparison between the left-hand portion, ϕV_c, and the right-hand portion, $V_u - \phi V'_{u,\min}$, is obtained by plotting their values on a graph as ordinates against stations along the beam as abscissa (see Fig. 4-9). Whenever ϕV_c falls below $V_u - \phi V'_{u,\min}$ for a portion of the length of the beam, A_v is computed from Eq. (26-10 ACI) for that portion only. A_v is computed from Eq. (26-11 ACI) for the remaining portions of the beam.

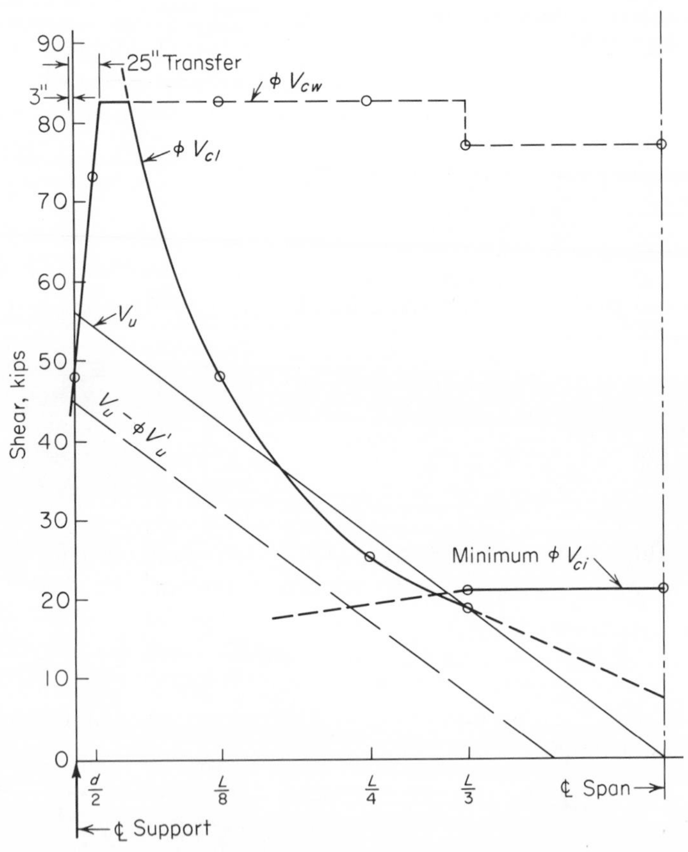

FIG. 4-9. Graphical representation of shear loads in beam and shear carrying capacity of concrete as computed by formulas from ACI 318-63.

The value for ϕV_c is always chosen as the smaller of ϕV_{ci} or ϕV_{cw}. Therefore, only ϕV_{ci} and ϕV_{cw} are plotted in Fig. 4-9, and their values are compared directly with $V_u - \phi V'_{u,\min}$.

In order to plot the curves shown in Fig. 4-9 it is necessary to compute, for each station along the beam, values for the following:

Calculation 1. $$V_{ci} = 0.6b'd\sqrt{f'_c} + \frac{M_{cr}}{M/V - d/2} + V_d \qquad \text{(26-12 ACI)}$$

where
$$M_{cr} = \frac{I}{y}(6\sqrt{f'_c} + f_{pe} - f_d) \tag{4-3}$$

and where V_{ci} is not less than $1.7b'd\sqrt{f'_c}$.

Calculation 2. $$V_{cw} = b'd(3.5\sqrt{f'_c} + 0.3f_{pc}) + V_p \tag{26-13 ACI}$$

Calculation 3. $$V_u = 1.5(V_G + V_S) + 1.8V_L \tag{15-1 ACI}$$

Calculation 4. $$V'_{u,\min} = \frac{A_s}{80}f'_s\sqrt{\frac{d}{b'}} \tag{4-8}$$

Calculation 5. $$V_u - \phi V'_{u,\min}$$

In some cases the foregoing symbols do not agree with the symbols used in previous steps. They are

$$M = 1.5M_S + 1.8M_L$$
$$V = 1.5V_S + 1.8V_L$$

[These are the values of M and V as defined by Sec. 2600 of ACI 318-63 and these values can be used in solving Eq. (26-12 ACI) for V_{ci}. However, in Eq. (26-12 ACI) these values appear as a ratio, M/V rather than as individual values. Since the ratio M/V is the same for all loading intensities to simplify the arithmetic the computations in Calculation 1 and Table 4-9 are based on the design load values of M and V rather than on ultimate load values. The derivation in Appendix E demonstrates that the ratio of M/V is the same for design loads as for ultimate loads.] The other symbols which are not consistent are

$$V_d = V_G$$
$$f_{pe} = f^b{}_F$$
$$f_{pc} = \frac{F}{A_c}$$

Calculations 1 through 5 for all locations along the beam are carried out in Tables 4-9 through 4-11. Most of the numerical values needed for these calculations will be found in Tables 4-7*a*, 4-7*b* and 4-8. Values for location $L/8$ are computed in the following sample calculations:

Calculation 1. Table 4-9. Computation of V_{ci} using Eq. (26-12 ACI).

Line 23. $$V_{S+L} = w_{S+L}(0.5L - X) = (64 + 320)(37.5 - 9.375) = 10{,}800 \text{ lb}$$

Line 24. $$M_{S+L} = 0.5w_{S+L}X(L - X)(12) = 0.5(64 + 320)9.375(75 - 9.375)12 = 1{,}417{,}000 \text{ in.-lb}$$

Line 25. M/V = line 24 ÷ line 23
$= 1{,}417 \div 10.8 = 131.2$ in.

Line 26. $d = y_t + e$
$= 8.56 + 12.36 = 20.92$ in.

Line 27. $M/V - d/2$ = line 25 − 0.5 line 26
$= 131.2 - 10.46 = 120.7$ in.

Line 28. $f_{pe} - f_d = f^b{}_F - f^b{}_G$
$= 2{,}547 - 983 = 1{,}564$ psi

Line 29. $6\sqrt{f'_c}$ + line 28 $= 424 + 1{,}564 = 1{,}988$ psi

Line 30. $\frac{M}{y}$ (line 29) $= Z_b$ (line 29)
$= 2{,}105 \times 1{,}988 = 4{,}185{,}000$ in.-lb $= M_{cr}$

Lines 28 to 30 represent the computation of M_{cr} from the equation

$$M_{cr} = \frac{I}{y}(6\sqrt{f'_c} + f_{pe} + f_d) \tag{4-3}$$

Line 31. $M_{cr} \div \left(\frac{M}{V} - \frac{d}{2}\right)$ = line 31 ÷ line 27
$= 4{,}185 \div 120.7 = 34.67$ kips

Line 31 is the numerical value for the middle term of Eq. (26-12 ACI)

Line 32. $V_d = V_G = w_G(0.5L - X)$
$= 561(37.5 - 9.375) = 15{,}780$ lb

Line 33. $0.6b'd\sqrt{f'_c}$ [First item of Eq. (26-12 ACI)]
$= 0.6 \times 8 \times 20.92 \times 70.7 = 7{,}100$ lb

Line 34. V_{ci} = lines 31 + 32 + 33 [Eq. (26-12 ACI)]
$= 34.67 + 15.78 + 7.1 = 57.55$ kips

Line 35. $V_{ci,\min} = 1.7b'd\sqrt{f'_c}$
$= 1.7 \times 8 \times 20.92 \times 70.7 = 20{,}120$ lb

Line 36. $\phi V_{ci} = 0.85 V_{ci}$
$= 0.85 \times 57.55 = 48.92$ kips

Line 37. $\phi V_{ci,\min} = 0.85 V_{ci,\min}$
$= 0.85 \times 20.12 = 17.10$ kips

Calculation 2. Table 4-10. Computation of V_{cw} using Eq. (26-13 ACI)

Line 38. d (From ACI318-63) "When applying Eq. (26-13) or (26-13A), the effective depth, d, shall be taken as the distance from the extreme compression fiber to the centroid of the prestressing tendons, or as 80% of the over-all depth of the member, whichever is greater."

$$d = 0.80 \times 32 = 25.60 > 20.92$$

Line 39. $0.3f_{pc} = 0.3F/A$
$= 0.3 \times 612 = 184$ psi

Line 40. Line 39 $+ 3.5\sqrt{f'_c} = 184 + (3.5 \times 70.7) = 431$ psi

Table 4-9. Computation of V_{ci} by Eq. (26-12 ACI)

1	Station	Support	$d/2$	$L/8$	$L/4$	$L/3$	$L/2$
2	Distance from support, in.	0.0	15.8	112.5	225.0	300.0	450.0
23	V_{S+L}, kips	14.40	13.89	10.80	7.20	4.80	0.00
24	M_{S+L}, in.-kips	0	223	1,417	2,431	2,880	3,240
25	M/V, in.	0.0	16.05	131.2	337.6	600.0	∞
26	$d = y_t + e$, in.	17.71	18.16	20.92	24.14	26.28	26.28
27	$M/V - d/2$, in.	......	6.97	120.7	325.5	586.9	∞
28	$f_{pc} - f_d = f^b{}_F - f^b{}_G$, psi	246	1,442	1,564	1,367	1,384	1,136
29	Line 28 $+ 6\sqrt{f'_c}$, psi	670	1,866	1,988	1,791	1,808	1,560
30	Z_b (line 29) $= M_{cr}$, in.-kips	1,410	3,928	4,185	3,770	3,806	3,284
31	Line 30 ÷ line 27, kips	∞	563.6	34.67	11.58	6.48	0.00
32	$V_d = V_G$, kips	21.04	20.30	15.78	10.52	7.01	0.00
33	$0.6b'd\sqrt{f'_c}$, kips	6.01	6.16	7.10	8.19	8.92	8.92
34	V_{ci} = line 31 + line 32 + line 33, kips	∞	590.06	57.55	30.29	22.41	8.92
35	V_{ci} (min) $= 1.7b'd\sqrt{f'_c}$, kips	17.03	17.45	20.12	23.20	25.27	25.27
36	$\phi V_{ci} = 0.85V_{ci}$, kips	∞	501.55	48.92	25.75	19.05	7.58
37	ϕV_{ci} (min) $= 0.85V_{ci}$ (min), kips	14.48	14.83	17.10	19.72	21.48	21.48

Line 41. Line 40 $(b'd) = 431 \times 8 \times 25.60 = 88{,}270$ lb

Line 42. V_p = vertical component of effective prestress in the tendons at the location being considered
$V_p = F$(tangent of angle between the center of gravity of tendons and the horizontal)

$$V_p = 329{,}600\left(\frac{17.72 - 9.15}{25 \times 12}\right) = 9{,}410 \text{ lb}$$

Line 43. V_{cw} = line 41 + line 42
$= 88.27 + 9.41 = 97.68$ kips

Line 44. $\phi V_{cw} = 0.85 V_{cw}$
$= 0.85 \times 97.68 = 83.03$ kips

Table 4-10. Computation of V_{cw} by Eq. (26-13 ACI)

1	Station	Support	$d/2$	$L/8$	$L/4$	$L/3$	$L/2$
2	Distance from support, in.	0.0	15.8	112.5	225.0	300.0	450.0
38	d (see text), in.	25.60	25.60	25.60	25.60	26.28	26.28
39	$0.3f_{pc} = 0.3F/A$, psi	22	139	184	184	184	184
40	$3.5\sqrt{f'_c}$ + line 39, psi	269	386	431	431	431	431
41	$b'd$ (line 40), kips	55.09	79.05	88.27	88.27	90.61	90.61
42	V_p (see text), kips	1.13	7.11	9.41	9.41	0.00	0.00
43	V_{cw} = line 41 + 42, kips	56.22	86.16	97.68	97.68	90.61	90.61
44	$\phi V_{cw} = 0.85 V_{cw}$, kips	47.79	73.24	83.03	83.03	77.02	77.02

Calculation 3. Table 4-11. Line 45. Computation of V_u. From ACI 318-63, Sec. 1506

$$U = 1.5D + 1.8L \text{ from which}$$
$$V_u = 1.5V_G + 1.5V_S + 1.8V_L$$
$$w_u = 1.5w_G + 1.5w_S + 1.8w_L$$
$$= 1.5(561) + 1.5(64) + 1.8(320)$$
$$= 1{,}513$$
$$V_u = w_u(0.5L - X)$$
$$= 1{,}513(37.5 - 9.375) = 42{,}570 \text{ lb}$$

Calculation 4. Table 4-11. Lines 46 and 47. Computation of $V'_{u,\min}$ = shear carried by minimum stirrups per Eq. (26-11 ACI)

From previous discussion in Step 9

$$V'_{u,\min} = \frac{A_s f'_s}{80}\sqrt{\frac{d}{b'}} \tag{4-8}$$

From ACI 318-63, Sec. 26-10, the d for use in Eqs. (26-10 ACI) and (26-11 ACI) shall be as follows:

"1. In members of constant over-all depth, d, equals the effective depth at section of maximum moment, and the . . ."

The section of maximum moment for this beam is at mid-span where

$$d = y_t + e = 8.56 + 17.72 = 26.28 \text{ in.}$$

Substituting in Eq. (4-8)

$$V'_{u,\min} = \frac{2.143 \times 269{,}700}{80}\sqrt{\frac{26.28}{8}} = 13{,}100 \text{ lb}$$
$$\phi V'_{u,\min} = 0.85 \times 13.1 = 11.13 \text{ kips}$$

Calculation 5. Table 4-11. Line 48. Computation of

$$V_u - \phi V'_u = \text{line } 45 - \text{line } 47$$
$$= 42.57 - 11.13 = 31.44$$

Table 4-11. Computation of Remaining Quantities Required to Plot Fig. 4-9.

1	Station	Support	$d/2$	$L/8$	$L/4$	$L/3$	$L/2$
2	Distance from support, in.	0.0	15.8	112.5	225.0	300.0	450.0
45	$V_u = 1.5V_G + 1.5V_S + 1.8V_L$, kips	56.76	54.76	42.57	28.38	18.92	0.0
46	$V'_u \text{ (min)} = \dfrac{A_s f'_s}{80}\sqrt{\dfrac{d}{b'}}$, kips	13.10	13.10	13.10	13.10	13.10	13.10
47	$\phi V'_u \text{ (min)} = 0.85V''_u \text{ (min)}$, kips	11.13	11.13	11.13	11.13	11.13	11.13
48	$V_u - \phi V''_u = \text{line } 45 - \text{line } 47$, kips	45.63	43.63	31.44	17.25	7.79	−11.13
49	ϕV_c, kips	n. a.	73.24	48.92	25.75	21.48	21.48

By definition in ACI 318-63, Sec. 26-10(b), ϕV_c is the lesser of ϕV_{ci} and ϕV_{cw}. From line 36, $\phi V_{ci} = 48.92$, and from line 44, $\phi V_{cw} = 83.03$. Therefore $\phi V_c = 48.92$, which is entered in line 49 of Table 4-11.

The required ultimate shear capacity of the member at location $L/8$ is $V_u = 42.57$ kips from line 45, Table 4-11. Shear carried by minimum stirrups is $\phi V'_u = 11.13$ kips from line 47, Table 4-11. Shear not carried by the minimum stirrups is $V_u - \phi V'_u = 42.57 - 11.13 = 31.44$ kips from line 48, Table 4-11. Since the shear carrying capacity of the concrete, $\phi V_c = 48.92$ kips, is greater than 31.44 kips, no additional stirrups are necessary. The foregoing is a discussion of the application of Eq. (4-9) which can be applied as follows

$$\phi V_c > V_u - \phi V'_{u,\min} \tag{4-9}$$
$$48.92 > 31.44$$

Where Eq. (4-9) is true, minimum stirrups are sufficient and are computed from Eq. (26-11 ACI). Where Eq. (4-9) is not true, stirrups are computed using Eq. (26-10 ACI).

When Fig. 4-9 has been plotted using the values developed in Tables 4-9 through 4-11, it is seen that (for this particular member) $V_u - \phi V'_{u,\min}$ is less than ϕV_c at all stations along the beam and therefore minimum stirrups will be sufficient for the full length of the beam. If ϕV_{ci} or ϕV_{cw} were less than $V_u - \phi V'_{u,\min}$ at any point then Eq. (26-10 ACI) would be used to compute stirrups in that area.

Shear reinforcement as determined from Eq. (26-11 ACI) shall be spaced no farther apart than three-fourths the depth of the member nor more than 24 in., whichever is smaller. In this case the maximum spacing s is 24 in.

$$A_v = \frac{A_s}{80}\frac{f'_s}{f_y}\frac{s}{d}\sqrt{\frac{d}{b'}} \qquad \text{(26-11 ACI)}$$

$$= \frac{2.143}{80} \times \frac{269{,}700}{40{,}000} \times \frac{24}{26.28}\sqrt{\frac{26.28}{8.0}}$$

$$= 0.30 \text{ in.}^2$$

Two #3 bars = $2 \times 0.1105 = 0.221$ in.²

$$24 \times \frac{0.221}{0.30} = 17.65 \text{ in.}$$

Use two #3 bars at 17⅝-in. maximum spacing for the full length of the beam.

Step 9B. Design of shear steel using the simplified method developed by Richard Elstner.

Before using this procedure the designer should be thoroughly familiar with the computations carried out in Step 9A and the relationships between the various parameters as illustrated in Fig. 4-9.

This method is applicable to uniformly loaded beams of constant cross section. Its simplicity is based on the designer's knowledge of the locations where the various parameters are critical (Fig. 4-9) and on a simplified but conservative formula for computing V_{ci} as developed by Elstner. (See Appendix C for derivation of the formulas used here and for values at other stations along the beam.)

For the majority of prestressed concrete building beams three simple calculations will be sufficient to design the shear steel for the entire length of the beam.

Calculation 1. Solve Eq. (26-11 ACI) for A_v. This is the minimum shear steel permitted and will be used for the full length of the beam except at points where subsequent calculations indicate that more is required.

Calculation 2. Solve Eq. (26-13 ACI) for V_{cw} at station $d/2$. (Don't forget to adjust the value used for the prestressing force if $d/2$ is within the transfer length.) Substitute this value of V_{cw} for V_c in Eq. (26-10 ACI) and compute A_v. If this value of A_v is less than that computed from Eq. (26-11 ACI), Calculation 1, Eq. (26-11 ACI) governs and V_{cw} should not be critical at any station.

Calculation 3. Compute V_{ci} at station $L/4$ using Elstner's formula which says that at station $L/4$

$$V_{ci} = 0.6b'd\sqrt{f'_c} + (0.333w_{S+L} + 0.250w_G)L$$

Substitute this value of V_{ci} for V_c in Eq. (26-10 ACI) and solve for A_v. If this value of A_v is less than that computed by Eq. (26-11 ACI) in Calculation 1, then Eq. (26-11 ACI) governs and V_{ci} should not be critical at any station.

In Calculations 2 and 3 we have computed V_{cw} and V_{ci} at the stations where they are usually the most critical and used these values in Eq. (26-10 ACI) to compute A_v. We assumed that A_v from Eq. (26-11 ACI) was greater than A_v from Eq. (26-10 ACI) so that Eq. (26-11 ACI) applies for the full length of the beam. When this is true, as it is in many cases, our shear computations have been cut to a minimum.

When A_v as computed by Eq. (26-10 ACI) is greater than A_v from Eq. (26-11 ACI), additional calculations are necessary.

Calculation 2. If Eq. (26-10 ACI) governs at station $d/2$, use it there and try it at other stations further into the span until the point is reached where Eq. (26-11 ACI) governs. In many cases, Eq. (26-10 ACI) will govern at $d/2$ but Eq. (26-11 ACI) will become governing at the end of the transfer length. (Don't forget that ACI 318-63, Sec. 2610c requires that "shear reinforcement shall be provided for a distance equal to d beyond the point theoretically required.")

Calculation 3. If Eq. (26-10 ACI) governs at station $L/4$, check it at adjacent stations using Elstner's formulas for those stations.

Elstner's formulas for the stations in this example are:

Station	*Formula*
$L/8$	$V_{ci} = 0.6b'd\sqrt{f'_c} + (0.857w_{S+L} + 0.375w_G)L$
$L/4$	$V_{ci} = 0.6b'd\sqrt{f'_c} + (0.333w_{S+L} + 0.250w_G)L$
$L/3$	$V_{ci} = 0.6b'd\sqrt{f'_c} + (0.1875w_{S+L} + 0.167w_G)L$
$L/2$	$V_{ci} = 0.6b'd\sqrt{f'_c} + 0 + 0$

in which L = total span in feet
w_{S+L} = all applied load in kips per linear foot
w_G = weight of beam in kips per linear foot

Substituting numerical values from this example, at $L/8$,

$$V_{ci} = 0.6(8)20.92\sqrt{5{,}000} + (0.857 \times 0.384 + 0.375 \times 0.561)75$$
$$= 7.10 + 40.44 = 47.54 \text{ kips}$$

at $L/4$,

$$V_{ci} = 0.6(8)24.14\sqrt{5{,}000} + (0.333 \times 0.384 + 0.250 \times 0.561)75$$
$$= 8.19 + 20.10 = 28.29 \text{ kips}$$

at $L/3$,

$$V_{ci} = 0.6(8)26.28\sqrt{5{,}000} + (0.1875 \times 0.384 + 0.167 \times 0.561)75$$
$$= 8.92 + 12.43 = 21.35 \text{ kips}$$

at $L/2$, $V_{ci} = 0.6(8)26.28\sqrt{5{,}000} + 0 + 0$

$$= 8.92 + 0 + 0 = 8.92 \text{ kips}$$

Multiplying by the capacity reduction factor and comparing the approximate values with those previously computed by the code formulas,

	Elstner	*ACI 318-63*
$L/8$	$\phi V_{ci} = 40.41$ kips	$\phi V_{ci} = 48.92$ kips
$L/4$	$\phi V_{ci} = 24.05$ kips	$\phi V_{ci} = 25.75$ kips
$L/3$	$\phi V_{ci} = 18.15$ kips	$\phi V_{ci} = 19.05$ kips
$L/2$	$\phi V_{ci} = 7.58$ kips	$\phi V_{ci} = 7.58$ kips

Of course $\phi V_{ci,\min} = \phi 1.7b'd\sqrt{f'_c} = 21.48$ kips still governs near the middle of the span.

Step 10. *Compute camber and deflection.* Camber and deflection are discussed in Art. 10-6. Although we use precise engineering formulas in computing camber and deflection, the results are accurate only within the accuracy with which we can determine the modulus of elasticity of the concrete and estimate the changes which will take place over a period of time due to creep and shrinkage of concrete.

Section 1102(a) of ACI 318-63 gives the generally accepted formula for determining the modulus of elasticity of concrete E_c as a function of its weight and strength. Curves plotted from the formula are shown in Fig. 6-1.

Compute instantaneous camber at the time the strands are released from their external anchors and their prestressing force is applied to the concrete member. The net camber will be the algebraic sum of the upward camber from the prestressing force and the downward deflection from the dead weight of the member. The dead weight becomes effective as the prestress is applied because the prestress causes camber which raises the center part of the beam off the forms and leaves it supported only on its ends.

As computed in Step 6 the prestressing force immediately after release is $F_o = 372{,}500$ lb. At time of release the concrete has a minimum strength of $f'_{ci} = 4{,}000$ psi. From Fig. 6-1 for a concrete weighing 150 lb per cu ft and having a strength of 4,000 psi, $E_c = 3{,}840{,}000$.

Deflection due to dead load is computed by the standard formula

$$\Delta_G = \frac{5wl^4}{384EI}$$

$$= \frac{5 \times 561(75^4)12^3}{384 \times 3{,}840{,}000 \times 49{,}350} = -2.11 \text{ in.}$$

The minus sign indicates a downward deflection which is a negative camber.

As discussed in Chap. 1 in the development of Eq. (1-1), an eccentric prestressing force can be replaced by a force on the c.g.c. and a couple. We do the same thing in computing camber. The force on the c.g.c. can be ignored because it creates a uniform compressive stress over the entire cross section, so that the entire section shortens the same amount and there is no change in camber. The couple produces a bending moment which does cause camber. The bending moment is equal to the prestressing force multiplied by its eccentricity with respect to the c.g.c.

From Fig. 4-4, e at center of support = 9.15 in. and e at mid-span = 17.72. Therefore at center of support

$$M_e = 372{,}500 \times 9.15 = 3{,}410{,}000 \text{ in.-lb}$$

and at mid-span

$$M_e = 372{,}500 \times 17.72 = 6{,}600{,}000 \text{ in.-lb}$$

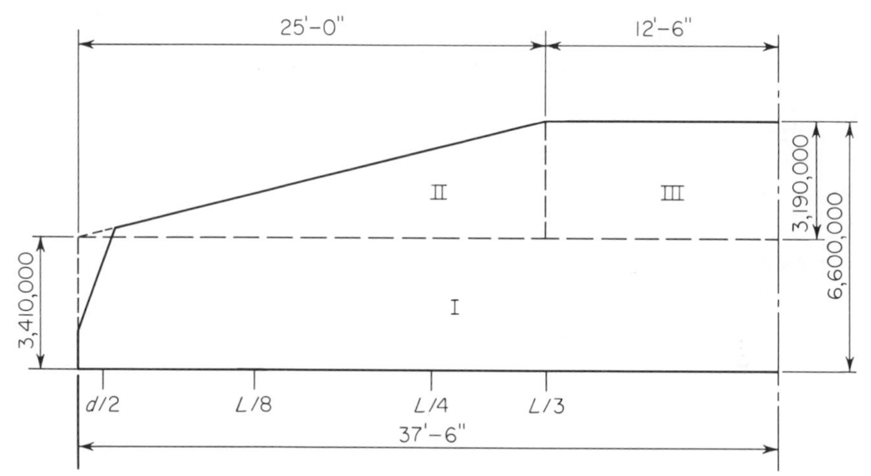

FIG. 4-10. Bending moment due to initial prestress.

The reduction in prestress at the end of the beam due to transfer length will not have an appreciable influence on the magnitude of the moment of the area of the M/EI diagram and has therefore been neglected.

The moment diagram for the foregoing values of M_e is plotted in Fig. 4-10.

Deflection at the center of span due to the moments shown in Fig. 4-10 can be computed by the moment-area method. When applied to this particular problem, the moment-area method says that the deflection at the center of span is equal to the moment of the area of the M/EI diagram about the support. The moment diagram in Fig. 4-10 has been divided into three simple sections, so the moment of the M/EI diagram for each can be easily figured and the total deflection Δ_{F_o} can be obtained by adding the three results.

$$\Delta_{\mathrm{I}} = \frac{3{,}410{,}000(37.5 \times 12)\tfrac{1}{2}(37.5 \times 12)}{EI} = \frac{345.3(10^9)}{EI}$$

$$\Delta_{\mathrm{II}} = \frac{3{,}190{,}000(25 \times 12)\tfrac{1}{2}(25 \times 12)\tfrac{2}{3}}{EI} = \frac{95.7(10^9)}{EI}$$

$$\Delta_{\mathrm{III}} = \frac{3{,}190{,}000(12.5 \times 12)(31.25 \times 12)}{EI} = \frac{179.4(10^9)}{EI}$$

$$\Delta_{F_o} = \Delta_{\mathrm{I}} + \Delta_{\mathrm{II}} + \Delta_{\mathrm{III}} = \frac{620.4(10^9)}{EI}$$

$$= \frac{620.4(10^9)}{3{,}840{,}000 \times 49{,}350} = +3.27 \text{ in.}$$

Net instantaneous camber = 3.27 − 2.11 = +1.16 in.

Accurate computation of final camber will depend upon the designer's knowledge of past experience with similar materials, cross section, and application. This type of information should be available from the local fabricator who will have an empirical value based on past experience with his product.

The reader should study the discussion of camber presented in Art. 10-6 of this book.

Section 909(e) of ACI 318-63 says that the maximum limits for immediate deflection under live load are:

$L/180$ for roofs which do not support plastered ceilings
$L/360$ for roofs which do support plastered ceilings

By the time live load is applied to this member its concrete strength will be at least 5,000 psi. From Fig. 6-1 for 5,000 psi concrete that weighs 150 lb per cu ft, $E_c = 4{,}290{,}000$. From Table 4-3, $w_L = 320$ lb per ft. Using the standard formula for deflection of a uniformly loaded beam,

$$\Delta = \frac{5wL^4}{384EI}$$

$$\Delta_L = \frac{5 \times 320(75^4)12^3}{384 \times 4{,}290{,}000 \times 49{,}350} = -1.08 \text{ in.}$$

$$\frac{L}{360} = \frac{75 \times 12}{360} = 2.50 \text{ in. allowable with plastered ceiling}$$

Since the computed Δ_L is much less than 2.50 in., live-load deflection is not critical.

CHAPTER 5

Design of a Prestressed Concrete Bridge

5-1. General Conditions. The object of this chapter is to present a complete analysis of a typical prestressed concrete bridge. It deals with one of the most popular types, namely, precast pretensioned I section girders supporting a poured-in-place slab.

The bridge is designed to the following conditions:

Live load = AASHO-HS20-44 from Interim 1963 edition of AASHO
Provision for future wearing surface = 20 psf
Span = 80 ft 0 in. center to center of bearings
Width = 28-ft 0-in. roadway

Analysis of the reinforcing steel requirements for the poured-in-place slab is not included, since it is the same as that for a similar slab supported on reinforced concrete stringers.

Details of bearing plates, rockers, expansion joints, etc., are not included, since they will be determined by requirements of customers, which vary from one locality to another.

5-2. Specifications. The structure should be designed and constructed in accordance with the provisions of "Standard Specifications for Highway Bridges" (hereafter referred to as AASHO) published by the American Association of State Highway Officials. References used here are to the 1961 edition and Interim Specifications for 1961 and 1963. When a later edition is available it should be used.

Section 13 of AASHO covers design of prestressed concrete bridges. From Sec. 1.13.7 we will use a final concrete strength of $f'_c = 5{,}000$ psi in the prestressed I beams. From Sec. 1.13.18 the concrete strength at transfer of load from anchors to concrete will be $f'_{ci} = 4{,}000$ psi.

Strength of concrete in the deck slab varies from one designer to another but it is usually from 3,500 to 4,000 psi. For this example we will assume the concrete in the slab has an ultimate strength of 3,850 psi.

5-3. Design Calculations. An AASHO-PCI standard type III beam will be used with pretensioned bonded tendons. Figure 5-1 shows a cross section of the bridge, and Fig. 5-2 shows dimensions of one beam and that portion of the slab which works with it when it is an interior stringer. Calculations are for an interior stringer.

Step 1. Compute properties of the concrete cross section in Fig. 5-2. Properties for AASHO-PCI type III beam listed in Art. 15-2 are:

$$A_c = 560 \text{ sq in.}$$
$$y_b = 20.27 \text{ in.}$$
$$I_c = 125{,}390 \text{ in.}^4$$

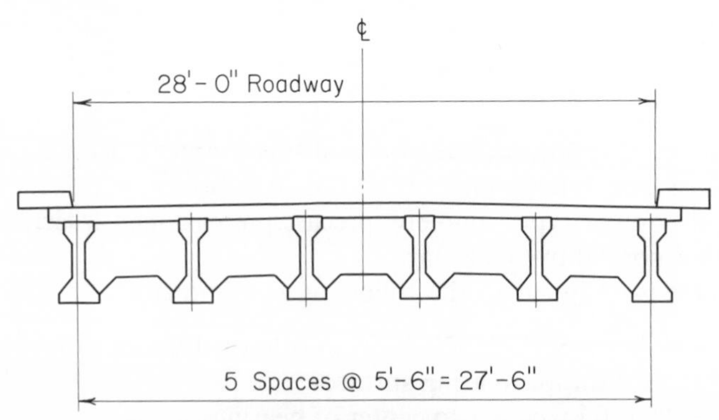

FIG. 5-1. Cross section of bridge at a diaphragm.

From these and Fig. 5-2 we can compute:

$$y_t = 45 - 20.27 = 24.73 \text{ in.}$$
$$Z_t = 125{,}390 \div 24.73 = 5{,}070 \text{ in.}^3$$
$$Z_b = 125{,}390 \div 20.27 = 6{,}186 \text{ in.}^3$$

Weight using regular-weight concrete $= 560 \times {}^{150}\!/_{144} = 583$ lb per ft

The above properties are for the beam alone. We also need the properties of the composite section. The difference in modulus of elasticity between beam and slab should be considered in computing section properties of the composite section.

Table 5-1. Summary of Conditions of Design

Span center to center supports	80 ft 0 in.
Details of structural cross section	Figs. 5-1 and 5-2
Weight of future wearing surface	20 psf
Live load (1963 Interim Specification)	HS20-44
Prestressing strands	7/16-in. type 270K
Initial tension in strands	70% f'_s
28 day concrete strength $= f'_c$	
Precast prestressed I beams	5,000 psi
Poured-in-place slab	3,850 psi
Weight of concrete	150 lb per cu ft
Strength of concrete at transfer $= f'_{ci}$	4,000 psi

From Fig. 6-1,

$$E_c = w^{1.5}33\sqrt{f'_c}$$

For the beam,

$$\begin{aligned} E_c &= 33(150^{1.5})\sqrt{5{,}000} \\ &= 33 \times 1{,}837 \times 70.7 = 4.29(10^6) \end{aligned}$$

For the slab,

$$\begin{aligned} E_c &= 33(150^{1.5})\sqrt{3{,}850} \\ &= 33 \times 1{,}837 \times 62.1 = 3.76(10^6) \\ \text{Ratio} &= 3.76 \div 4.29 = 0.88 \end{aligned}$$

Since the E_c of the slab is less than the E_c of the beam, its contribution to the stiffness, EI, of the composite section is less than if it were the same E_c as the beam. In computing the section properties of the composite

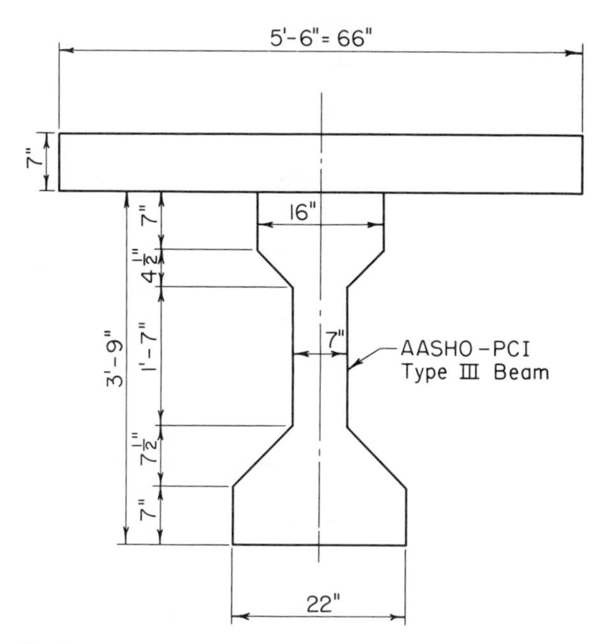

FIG. 5-2. Details of one beam and that portion of the roadway slab which acts with it as a composite section.

section, allowance for the difference in E_c is made by multiplying the area of the slab by 0.88.

Find the section properties of the composite section taking moments about top of slab.

Section	*Area*	y	Ay	Ay^2	I_o
66 × 7 × 0.88	406	3.5	1,421	4,974	1,660
Beam	560	31.73	17,769	563,810	125,390
	966		19,190	568,784	127,050
				127,050	
				695,834	

$$y_t = 19{,}190 \div 966 = 19.86 \text{ in.}$$
$$y_b = 52 - 19.86 = 32.14 \text{ in.}$$
$$I_c = 695{,}834 - 966(19.86^2) = 314{,}825$$
$$Z_t = 314{,}825 \div 19.86 = 15{,}852$$
$$Z_b = 314{,}825 \div 32.14 = 9{,}795$$
$$\text{Weight of slab} = 66 \times 7 \times 150/144 = 480 \text{ lb per ft}$$

Table 5-2. Quantities Computed in Step 1

I beam alone	
$A_c = 560$ in.^{2}	$I_c = 125{,}390$ in.^{4}
$y_t = 24.73$ in.	$Z_t = 5{,}070$ in.^{3}
$y_b = 20.27$ in.	$Z_b = 6{,}186$ in.^{3}
$w_G = 583$ lb per ft	

I beam and slab composite section	
$A_c = 966$ in.^{2}	$I_c = 314{,}825$ in.^{4}
$y_t = 19.86$ in.	$Z_t = 15{,}852$ in.^{3}
$y_b = 32.14$ in.	$Z_b = 9{,}795$ in.^{3}
w_S (weight of slab) $= 480$ lb per ft.	

Step 2. Compute stresses in beam at center of span due to its own dead weight.

$$M_G = \frac{583(80^2) \times 12}{8} = 5{,}600{,}000 \text{ in.-lb}$$
$$f^t_G = 5{,}600{,}000 \div 5{,}070 = +1{,}104$$
$$f^b_G = 5{,}600{,}000 \div 6{,}186 = -905$$

Step 3. Compute stresses in beam at center of span due to applied loads.

The beam alone must support the dead weight of the slab and diaphragms. It does not act as a composite section with the slab until the slab has cured.

Weight of slab $= w_{SS} =$ 480 lb per ft

$$M_{SS} = \frac{480(80^2) \times 12}{8} = 4{,}608{,}000 \text{ in.-lb}$$

The AASHO-PCI standards in Art. 15-2 show a diaphragm at mid-span. The portion carried by one interior stringer will be approximately 8 in. wide by 2 ft 6 in. by 5 ft 0 in. or a concentrated weight of ⅔ × 2.5 × 5 × 150 = 1,250 lb. The moment M_{SD} caused by this load at center of span will be

$$M_{SD} = 1{,}250 \times 80 \times {}^{12}\!/_{4} = 300{,}000 \text{ in.-lb}$$

The total moment carried by the beam alone due to superimposed loads is

$$M_S = M_{SS} + M_{SD} = 4{,}608{,}000 + 300{,}000 = 4{,}908{,}000 \text{ in.-lb}$$
$$f^t{}_S = 4{,}908{,}000 \div 5{,}070 = +968 \text{ psi}$$
$$f^b{}_S = 4{,}908{,}000 \div 6{,}186 = -793 \text{ psi}$$

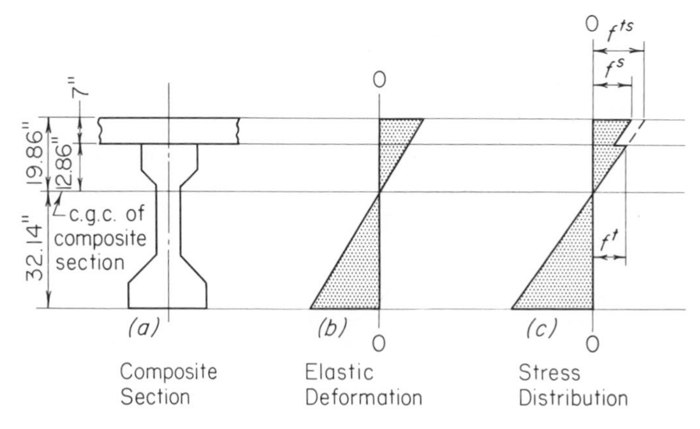

FIG. 5-3. Distribution of strains and stresses in composite section.

The wearing surface and the live load are carried by the composite section. Figure 5-3*b* illustrates the elastic deformations developed when load is applied to the composite section.

Figure 5-3*c* shows the stresses developed in the composite section. When the applied bending moment is divided by the section modulus of the top fiber of the composite section, the resulting stress is that shown as f^{ts} in Fig. 5-3*c*. Because of the difference in E_c between the slab and the beam, the actual unit stress f^s in the top fiber of the slab is less than f^{ts} or

$$f^s = 0.88 f^{ts}$$

The stress in the top fiber of the beam is less than f^{ts} because it is closer to the c.g.c. than the top of the slab:

$$f^t = \frac{12.86}{19.86} f^{ts} = 0.65 f^{ts}$$

Wearing surface weight $w_{WS} = 20 \times 5.5 = 110$ lb per lin ft.

$$M_{WS} = \frac{110(80^2) \times 12}{8} = 1{,}056{,}000 \text{ in.-lb}$$

$$f^{ts}{}_{WS} = 1{,}056{,}000 \div 15{,}852 = +67 \text{ psi}$$
$$f^{b}{}_{WS} = 1{,}056{,}000 \div 9{,}795 = -108 \text{ psi}$$
$$f^{t}{}_{WS} = 67 \times 0.65 = +43 \text{ psi}$$

From AASHO tables the live-load bending moment per lane is 1,164,900 ft-lb.

$$\text{Impact} = \frac{50}{80 + 125} = 24.4\%$$

From AASHO Interim Specifications, 1(61), Sec. 1.3.1, the wheel load per stringer is $S/5.5$, which in this example is $5.5/5.5 = 1.0$ wheel load per stringer. A wheel load is one-half the lane load.

Net live-load moment M_L per stringer is

$$M_L = 1{,}164{,}900 \times 1.244 \times 1.0 \times \tfrac{1}{2} \times 12 = 8{,}695{,}000 \text{ in.-lb}$$
$$f^{ts}{}_L = 8{,}695{,}000 \div 15{,}852 = +548 \text{ psi}$$
$$f^{t}{}_L = 548 \times 0.65 = +356 \text{ psi}$$
$$f^{b}{}_L = 8{,}695{,}000 \div 9{,}795 = -888 \text{ psi}$$

Table 5-3. Quantities Computed in Steps 2 and 3

w_{WS} = Weight of wearing surface = 110 lb per ft			
M_G = 5,600,000 in.-lb	$f^t{}_G = +1{,}104$ psi	$f^b{}_G = -905$ psi	
M_S = 4,908,000 in.-lb	$f^t{}_S = +968$ psi	$f^b{}_S = -793$ psi	
M_{WS} = 1,056,000 in.-lb	$f^t{}_{WS} = +43$ psi	$f^b{}_{WS} = -108$ psi	$f^{ts}{}_{WS} = +67$ psi
M_L = 8,695,000 in.-lb	$f^t{}_L = +356$ psi	$f^b{}_L = -888$ psi	$f^{ts}{}_L = +548$ psi

Step 4. Determine the magnitude of the prestressing force and select the number of tendons.

This computation is based on conditions after all stress losses.

From Sec. 1.13.7(B)(2) maximum allowable compression $= 0.40f'_c = 0.40 \times 5{,}000 = 2{,}000$ psi.

From 1963 Interim Specifications, Sec. 1.13.7(B)(2), the allowable tensile stress (in the precompressed tensile zone) is $(0.40)7.5\sqrt{f'_c}$ but not to exceed 250 psi. On this basis the allowable tensile stress in our beam under full design load after all losses have occurred would be

$$f_{tp} = 0.40 \times 7.5\sqrt{5{,}000} = 212 \text{ psi}$$

At this point we have sufficient information to check stresses in the bottom fiber of the beam and determine whether or not the section chosen is adequate under design-load conditions. Total tensile stresses in the bottom fiber are

$$f^b{}_G + f^b{}_S + f^b{}_{WS} + f^b{}_L = 905 + 793 + 108 + 888 = 2{,}694 \text{ psi}$$

Since the specification permits a tensile stress of 212 psi under design load, the compressive stress which must be developed by the tendons is $2{,}694 - 212 = 2{,}482$ psi. This is in excess of the 2,000 psi permitted and means that some of the strands must be deflected so that the bottom-fiber stress due to prestress at the ends of the beam will be less than 2,000 psi.

At mid-span the dead weight of the beam itself becomes effective as the prestress force is applied and the dead-weight bending-moment stress offsets some of the stress due to prestress. If our strands are selected and located to develop exactly 2,482 psi at mid-span the net stress will be $2{,}482 - f^b{}_G = 2{,}482 - 905 = 1{,}577$ psi, which is well within the permissible 2,000 psi.

The foregoing calculation shows that the section chosen is adequate for the design-load condition. In fact, since only 1,577 psi of an allowable 2,000 psi has been used, the possibility of selecting a more economical section should be investigated. The next smaller standard AASHO-PCI beam is the type II, which would probably be inadequate because its section modulus is only about 50 per cent of that of the type III. Another approach would be to eliminate one beam by spacing them about 7 ft 0 in. center to center instead of 5 ft 6 in. This would increase the moment per beam by about 25 per cent and would probably work out.

We will continue with the section originally chosen because it will demonstrate the design procedure and will also serve to illustrate the progress made in the design of prestressed concrete bridges over the past six years. This same section was used as the design example in chap. 9 of "Practical Prestressed Concrete" published by McGraw-Hill in 1960. Since then the following innovations have taken place:

1. The portion of a wheel load carried by a prestressed concrete stringer as specified by AASHO is now $S/5.5$ instead of $S/5.0$.

2. A tensile stress is now permitted in the bottom fiber of pretensioned beams under design load whereas the previous specification did not permit tensile stress. This change is based on results of the AASHO road-test program which included bridges that were subjected to cyclic loading producing tension in the bottom fiber.

3. Six of the high-strength type 270K strands will give the same final prestressing force as seven of the older strands of the same diameter, and this smaller number of strands can be placed with its center of gravity closer to the bottom of the beam. The increased eccentricity increases the resisting moment developed by the prestressing force.

The magnitude and location of the prestressing force can be determined by solving Eq. (3-1). We have numerical values for all the terms in this equation except F and e.

The prestressing force is applied only to the beam rather than to the composite section, so the section properties of the beam only are used in this computation.

Step 4, Chap. 3, suggests solving Eq. (3-1) by using an approximate value for e and then making an adjustment if necessary. For this approximation the c.g.s. is to be located at 15 per cent of the height of the beam above the bottom of the beam. In this example the beam is 45 in. high and 15 per cent of 45 is 6.75 in. Then

$$e_{\text{approx}} = y_b - 6.75 = 20.27 - 6.75 = 13.52 \text{ in.}$$

We can now substitute numerical values in Eq. (3-1) and get an approximate value for the final prestressing force F.

$$\frac{F}{A} + \frac{Fe}{Z_b} = f^b{}_G + f^b{}_S + f^b{}_L - f_{tp} \tag{3-1}$$

The term $f^b{}_S$ is defined as the stress in the bottom fiber due to additional dead weight. In this example we have two items due to additional dead weight. They are $f^b{}_S$ due to the weight of the slab and carried by the beam alone and $f^b{}_{WS}$ due to the weight of the wearing surface and carried by the composite section. It is therefore necessary to rewrite Eq. (3-1) to read

$$\frac{F}{A} + \frac{Fe}{Z_b} = f^b{}_G + f^b{}_S + f^b{}_{WS} + f^b{}_L - f_{tp} \tag{5-1}$$

Substituting in Eq. (5-1),

$$\frac{F}{560} + \frac{13.52F}{6{,}186} = 905 + 793 + 108 + 888 - 212 = 2{,}482 \text{ psi}$$

Multiplying both sides of the equation by 560,

$$F + 1.224F = 1{,}390{,}000$$
$$F = 625{,}000 \text{ lb}$$

Any size and grade of strand can be selected from Table 7-1. Where a large prestressing force is involved the most economical choice is generally the largest size of type 270K. Of course available casting bed facilities and local regulations must also be considered. We will assume that 7/16-in.-diameter type 270K strands are preferred for this project.

What is the final force (after losses) in a prestressing strand? The maximum allowable stress established by AASHO, Sec. 1.13.7(A)(2) is "$0.60f'_s$ or $0.80f_{sy}$ whichever is smaller." The requirements of ASTM Designation A416 are such that $0.60f'_s$ is less than $0.80f_{sy}$ and is therefore the governing condition. Meeting other AASHO requirements however results in a final stress which is lower than $0.60f'_s$. Section 1.13.7(A)(1) specifies a maximum initial stress of $0.70f'_s$ and Sec. 1.13.8(B) specifies a loss of 35,000 psi for pretensioned members. These values are used in Table 7-1 and indicate a final force of 17,615 lb for each 7/16-in. type 270K strand.

On this basis the total number of strands needed for this member is $625{,}000 \div 17{,}615 = 35.5$. Since this calculation is based on an approximate value of e it seems logical to use 36 strands for a first trial.

Table 5-4. Quantities Computed in Step 4

Allowable stresses in concrete under final conditions

$$\text{Compression} = 0.40f'_c = +2{,}000 \text{ psi}$$
$$\text{Tension} = (0.40)7.5\sqrt{f'_c} = -212 \text{ psi (but not more than 250 psi)}$$

Values of prestressing force, number of strands, etc., computed in Step 4 were first approximation and are not tabulated here to avoid confusion when referring to tables at a later date. Final values of F, etc., will be tabulated in Table 5-5.

Step 5. Locate the strands in the cross section of the beam and compute the eccentricity provided.

The strands should be placed as close to the bottom of the beam as possible to get maximum eccentricity. In the few cases where this causes excessive tensile stress in the top fiber a readjustment will be necessary when the top fiber stress is checked. Where some of the strands must be deflected, as in this example, the strand pattern should be arranged so that about one-third to one-sixth of the total number can be deflected. We will place eight strands in the two center rows so they can slope up in the web.

Most casting yards use templates with holes spaced 2 in. center to center both horizontally and vertically so that they can meet the requirements of AASHO, Sec. 1.13.16, for all sizes of strand up to and including ½ in.

On the basis of the foregoing data we can set up the strand pattern shown in Fig. 5-4.

Compute center of gravity of strand group by taking moments about bottom of beam.

$$30 \times 4 = 120$$
$$6 \times 8 = 48$$
$$36 \qquad 168 \div 36 = 4.67$$
$$e = 20.27 - 4.67 = 15.60 \text{ in.}$$

For thirty-six 7/16-in. type 270K strands, $F = 36 \times 17{,}615 = 634{,}000$ lb.

Compute the compressive stress in the bottom fiber due to the strand pattern in Fig. 5-4 by substituting in Eq. (1-1b).

$$f^b{}_F = \frac{F}{A} + \frac{Fe}{Z_b} \qquad (1\text{-}1b)$$

$$= \frac{634{,}000}{560} + \frac{634{,}000 \times 15.60}{6{,}186}$$

$$= 1{,}132 + 1{,}598 = +2{,}730 \text{ psi}$$

The total final prestress was determined by using an estimated value for e and results in a larger compressive stress than the required 2,482 psi

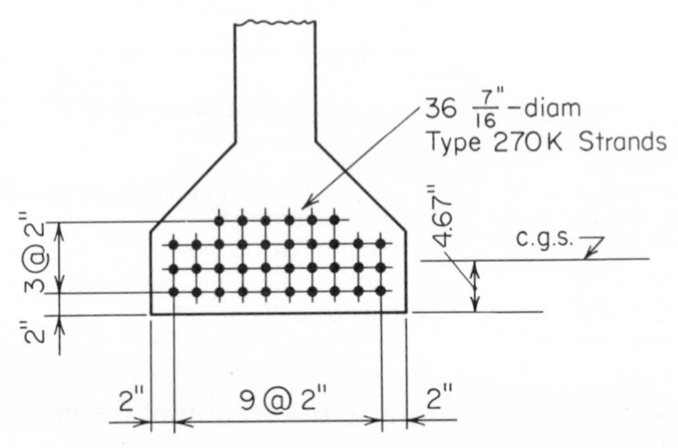

FIG. 5-4. Trial strand pattern at center of span.

determined by previous calculations. We can now make an accurate estimate of the number of strands required by multiplying the number of strands used by the ratio of the stress required over the stress obtained; thus,

$$\frac{2{,}482}{2{,}730} \times 36 = 32.7 \text{ strands}$$

The next larger whole number of strands is 33, but it is not possible to use an odd number of strands in this member and maintain symmetry about the vertical center line. If there were a row of strands on the vertical center line, one could be dropped to give an odd number. For this member, 34 strands will be used in the pattern shown in Fig. 5-5. Compute the center of gravity of the strand group by taking moments about the bottom:

$$\begin{array}{rcl} 30 \times 4 &=& 120 \\ \underline{4} \times 8 &=& \underline{32} \\ 34 && 152 \div 34 = 4.47 \end{array}$$

$$e = 20.27 - 4.47 = 15.80 \text{ in.}$$

For thirty-four 7/16-in. type 270K strands, $F = 34 \times 17{,}615 = 599{,}000$ lb.

Substituting in Eq. (1-1b) the values for the strand pattern in Fig. 5-5,

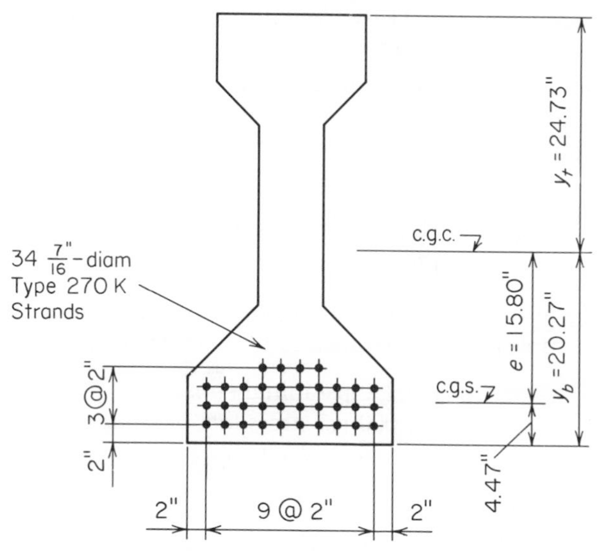

FIG. 5-5. Final strand pattern at center of span.

$$f^b{}_F = \frac{599{,}000}{560} + \frac{599{,}000 \times 15.80}{6{,}186}$$
$$= 1{,}070 + 1{,}530 = +2{,}600 \text{ psi}$$

This is larger than the required 2,482 psi, but a check using 32 strands shows that they would provide slightly less than 2,482 psi, so 34 strands will be used.

Substituting in Eq. (1-1*a*),

$$f^t{}_F = \frac{599{,}000}{560} - \frac{599{,}000 \times 15.80}{5{,}070}$$
$$= 1{,}070 - 1{,}867 = -797 \text{ psi}$$

Table 5-5. Quantities Computed in Step 5

Number of 7/16-in. type 270K strands	34
Strand locations	Fig. 5-5
Final prestressing force F	599,000 lb
Eccentricity e	15.80 in.
$f^b{}_F$ (at mid-span)	+2,600 psi
$f^t{}_F$ (at mid-span)	−797 psi

Step 6. Check the top and bottom stresses at mid-span.

Step 6 in Chap. 3 lists four conditions which should be checked to see that the net stresses are within the allowable. The first two conditions occur immediately after the prestressing force is transferred from external anchors to the concrete. The prestressing force effective at this time is designated as F_o, and Art. 10-5 states that the stress loss in the tendons (for the conditions existing in this example) can be assumed to be 15,000 psi. From Table 7-1 the initial stress in the strands at 70 per cent of ultimate strength is

185,950 psi initial stress
15,000 psi stress loss
170,950 psi after transfer

Article 10-5 also states that the total stress loss (for the conditions existing in this example) can be assumed to be 35,000 psi.

185,950 psi initial stress
35,000 psi stress loss
150,950 psi final stress

From this the ratio between F_o and F is

$$170{,}950 \div 150{,}950 = 1.13$$

Since the stresses in the concrete beam due to the prestressing force vary directly with the magnitude of the prestressing force we can easily find the stresses due to F_o when we already know the stresses due to F.

$$f^t{}_{F_o} = 1.13 f^t{}_F = 1.13(-797) = -900 \text{ psi}$$
$$f^b{}_{F_o} = 1.13 f^b{}_F = 1.13 \times 2{,}600 = +2{,}940 \text{ psi}$$

Maximum allowable temporary stresses in concrete before losses due to creep and shrinkage are given in AASHO, Sec. 1.13.7(B)(1), as

$$\text{Compression: } 0.60 f'_{ci} = 0.60 \times 4{,}000 = +2{,}400 \text{ psi}$$
$$\text{Tension: } 3\sqrt{f'_{ci}} = 3\sqrt{4{,}000} = -190 \text{ psi}$$

Evaluating the conditions which are critical immediately after the prestressing force is transferred to the concrete member we get

$$f^t{}_{F_o+G} = -900 + 1{,}104 = +204$$
$$f^b{}_{F_o+G} = +2{,}940 - 905 = +2{,}035$$

Both these values are within the allowable limits of -190 psi and $+2{,}400$ psi.

Evaluating the conditions which are critical after all stress loss has taken place we get

$$f^t{}_{F+G+S+WS+L} = -797 + 1{,}104 + 968 + 43 + 356 = +1{,}674 \text{ psi}$$

(Note that $f^t{}_{WS}$ has been included. In a noncomposite structure it would be a part of $f^t{}_S$, but in this example it is caused by a dead weight—the wearing surface—which is carried by the composite section and therefore must be computed separately.)

$$f^b{}_{F+G+S+WS+L} = +2{,}600 - 905 - 793 - 108 - 888 = -94 \text{ psi}$$

Both these values are within the allowable limits of -212 psi and $+2{,}000$ psi.

Table 5-6. Quantities Computed in Step 6

Allowable stresses in concrete immediately after transfer of prestressing force to concrete Compression: $0.60f'_{ci} = +2{,}400$ psi Tension: $3\sqrt{f'_{ci}} = -190$ psi	
$F_o/F = 1.13$	$F_o = 677{,}000$ lb
f_{se} (Stress in strand after all losses) $= 150{,}950$ psi	
Stresses in concrete at mid-span	
$f^t{}_{F_o} = -900$ psi	$f^b{}_{F_o} = +2{,}940$ psi
$f^t{}_{F_o+G} = +204$ psi	$f^b{}_{F_o+G} = +2{,}035$ psi
$f^t{}_{F+G+S+WS+L} = +1{,}674$ psi	$f^b{}_{F+G+S+WS+L} = -94$ psi

Step 7. Check ultimate flexural strength to make sure it meets the requirements of AASHO, Sec. 1.13.6. Check percentage of prestressing steel.

AASHO, Sec. 1.13.6 requires that "the computed ultimate load capacity shall not be less than $1.5D + 2.5(L + I)$."

Using values from Table 5-3 the required ultimate load capacity U_M is

$$1.5(5{,}600{,}000 + 4{,}908{,}000 + 1{,}056{,}000) + 2.5(8{,}695{,}000)$$
$$U_M = 17{,}350{,}000 + 21{,}740{,}000 = 39{,}090{,}000 \text{ in.-lb}$$

AASHO, Sec. 1.13.10, gives two formulas for computing the ultimate capacity of a member. Section (A) applies to Rectangular Sections and Sec. (B) applies to Flanged Sections. Section (B) says that the member should be considered a flanged section if the thickness of the flange is less than

$$1.4dp\frac{f_{su}}{f'_c} \tag{5-2}$$

For ultimate flexural capacities the properties of the composite section are used. From Fig. 5-2 the over-all depth of the section is 3 ft 9 in. plus 7 in. = 52 in. From Fig. 5-5 the c.g.s. is 4.47 in. above the bottom of the beam.

$$d = 52.0 - 4.47 = 47.53 \text{ in.}$$

From Table 7-1 the area of one 7⁄16-in. type 270K strand is 0.1167 in.²

$$A_s = 34 \times 0.1167 = 3.97 \text{ in.}^2$$
$$p = A_s \div bd$$
$$p = 3.97 \div (66 \times 47.53) = 0.001265$$

From Table 7-1 the ultimate strength of one 7/16-in. type 270K strand is 31,000 lb and its area is 0.1167 in.^{2}

$$f'_s = 31{,}000 \div 0.1167 = 265{,}500 \text{ psi}$$

From AASHO, Sec. 1.13.10(c), for members with bonded tendons,

$$f_{su} = f'_s\left(1 - \frac{0.5pf'_s}{f'_c}\right)$$

Substituting numerical values,

$$f_{su} = 265{,}500\left(1 - \frac{0.5 \times 0.001265 \times 265{,}500}{5{,}000}\right)$$
$$= 256{,}800 \text{ psi}$$

Substituting numerical values in Eq. (5-2),

$$\frac{1.4(47.53)0.001265(256{,}800)}{5{,}000} = 4.325 \text{ in.}$$

Since this is less than the flange thickness of 7 in. the formula for Rectangular Sections in AASHO, Sec. 1.13.10(A), will apply.

$$M_u = A_s f_{su} d\left(1 - \frac{0.6pf_{su}}{f'_c}\right)$$

Substituting numerical values,

$$M_u = 3.97 \times 256{,}800 \times 47.53\left(1 - \frac{0.6 \times 0.001265 \times 256{,}800}{5{,}000}\right)$$
$$= 46{,}550{,}000 \text{ in.-lb}$$

Since the ultimate flexural capacity M_u is greater than the required capacity U_M, the ultimate flexural capacity is adequate.

AASHO, Sec. 1.13.11, says that, for rectangular sections, pf_{su}/f'_c shall not exceed 0.30. Substituting numerical values in this expression,

$$\frac{0.001265 \times 256{,}800}{5{,}000} = 0.065$$

which is less than 0.30 and therefore satisfactory.

Step 8. Establish the path of the tendons and check critical points along the member under initial and final conditions.

From Tables 5-5 and 5-6 stresses in concrete at mid-span due to prestress only are

$$f^t{}_{F_o} = -900 \text{ psi} \qquad f^t{}_F = -797 \text{ psi}$$
$$f^b{}_{F_o} = +2{,}940 \text{ psi} \qquad f^b{}_F = +2{,}600 \text{ psi}$$

Since this is a simple span structure and there is no dead-load bending moment at the ends, these stresses (which are all in excess of the allowable) would exist at the ends of the beam if the strands were left in a straight line. It will therefore be necessary to slope some of the strands upwards near the ends of the beam.

Table 5-7. Quantities Computed in Step 7

Required ultimate flexural capacity U_M: 39,090,000 in.-lb	
Actual ultimate flexural capacity M_u: 46,550,000 in.-lb	
Properties of composite section at mid-span	
$d = 47.53$ in.	$A_s = 3.97$ in.^{2}
$p = 0.001265$	$f'_s = 265{,}500$ psi
$f_{su} = 256{,}800$ psi	

The maximum permissible value of e is that which brings each of the foregoing stresses within the allowable. The maximum e for each condition can be computed by setting the allowable stress equal to the stress caused by the prestressing force as follows:

1. Top of beam at time of transfer:

$$f^t{}_{F_o} = -190 = \frac{F_o}{A_c} - \frac{F_o e}{Z_t}$$

$$-190 = \frac{677{,}000}{560} - \frac{677{,}000e}{5{,}070} = 1{,}210 - 133.5e$$

$$e = 10.5 \text{ in.}$$

2. Bottom of beam at time of transfer:

$$f^b{}_{F_o} = +2{,}400 = \frac{F_o}{A_c} + \frac{F_o e}{Z_b}$$

$$2{,}400 = \frac{677{,}000}{560} + \frac{677{,}000e}{6{,}186} = 1{,}210 + 109.4e$$

$$e = 10.9 \text{ in.}$$

3. Top of beam under final conditions:
(AASHO is not clear on the allowable tensile stress in the top fiber under final conditions. The writers have assumed an allowable stress of $3\sqrt{f'_c}$ without reinforcing steel and $6\sqrt{f'_c}$ if reinforcing steel is provided to take all of the tension.)

For this example we assume no steel, thus,

$$f^t{}_F = -3\sqrt{5{,}000} = -212 \text{ psi}$$

$$= -212 = \frac{F}{A_c} - \frac{Fe}{Z_t}$$

$$-212 = \frac{599{,}000}{560} - \frac{599{,}000e}{5{,}070} = 1{,}070 - 118.1e$$

$$e = 10.85 \text{ in.}$$

4. Bottom of beam under final conditions:

$$f^b{}_F = +2{,}000 = \frac{F}{A_c} + \frac{Fe}{Z_b}$$

$$+2{,}000 = \frac{599{,}000}{560} + \frac{599{,}000e}{6{,}186} = 1{,}070 + 96.8e$$

$$e = 9.61 \text{ in.}$$

The maximum allowable e at the ends of the beam is 9.61 in., or the minimum distance from bottom of beam to c.g.s. is $y_b - 9.61 = 20.27 - 9.61 = 10.66$ in. Since there are 34 strands, the moment of the entire group about the bottom of the beam must be at least $34 \times 10.66 = 362.4$. The eight strands in the two center rows are to slope up while the others stay in a horizontal line. The moment of those remaining in a straight line is (see Fig. 5-6):

$$\begin{aligned} 24 \times 4 &= 96 \\ 2 \times 8 &= \underline{16} \\ &\ 112 \end{aligned}$$

Therefore the moment of the eight strands which slope up must be at least $362.4 - 112 = 250.4$. The distance from bottom of beam to the c.g.s. of the eight strands must be at least $250.4 \div 8 = 31.30$ in.

The foregoing calculations establish a strand pattern for the maximum allowable e. It is entirely permissible to use a smaller e. Reducing e at the ends of the beam will effect the magnitude of the stress due to prestress at all points along the beam between the end and the hold-down point, but the effect will seldom be critical. In this member the c.g.s. of the eight sloped strands will arbitrarily be set 36 in. above the bottom of the beam as shown in Fig. 5-6. Compute the c.g.s. of the entire group of 34 strands taking moments about the bottom.

$$\begin{aligned} 24 \times 4 &= 96 \\ 2 \times 8 &= 16 \\ \underline{8} \times 36 &= \underline{288} \\ 34 \qquad & \ 400 \div 34 = 11.76 \text{ in.} \end{aligned}$$

$$e = 20.27 - 11.76 = 8.51 \text{ in.}$$

Stresses at ends of beam under final conditions are

$$f^t{}_F = \frac{F}{A_c} - \frac{Fe}{Z_t} = \frac{599{,}000}{560} - \frac{599{,}000 \times 8.51}{5{,}070}$$

$$f^t{}_F = +1{,}070 - 1{,}005 = +65 \text{ psi}$$

$$f^b{}_F = \frac{F}{A_c} - \frac{Fe}{Z_b} = \frac{599{,}000}{560} + \frac{599{,}000 \times 8.51}{6{,}186}$$

$$f^b{}_F = +1{,}070 + 824 = +1{,}894 \text{ psi}$$

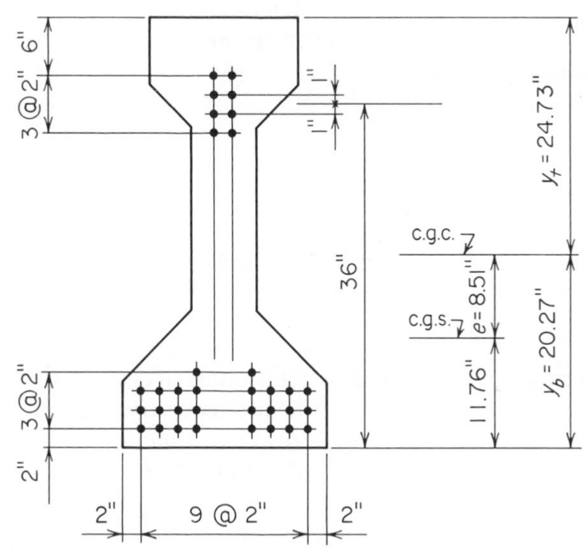

Fig. 5-6. Strand pattern at ends of beam.

Stresses at ends of beam after transfer of prestressing force from external anchors to the concrete beam are

$$f^t{}_{F_o} = 1.13 \times 65 = +73 \text{ psi}$$

$$f^b{}_{F_o} = 1.13 \times 1{,}894 = +2{,}140 \text{ psi}$$

All of these are within the allowables.

The strand pattern is now established at center of span, Fig. 5-5, and ends of span, Fig. 5-6. The next step is to establish the path of the strands along the member. In this example, this means to establish the location of the hold-down points. To do this it is necessary to know the stresses due to the critical combinations of loading conditions at critical points along the beam. After working through a few examples the engineer will recognize the critical conditions and locations and check those stresses only. For this example, to illustrate the net stresses all along the beam under the critical loading conditions, we have tabulated moments and stresses in Table 5-8.

Table 5-8 can be completed in a very short time. All the values at mid-span ($X = 40$) are already available from Table 5-3 except that the moments are converted from inch-pounds to foot-pounds. Since M_G, M_S, and M_{WS} are uniform loads their moment diagrams are parabolas, and their magnitude at points along the span is simply a ratio multiplied by the mid-span moment. In Table 5-8 stresses have been computed at the tenth points along the span in accordance with the following parabolic ratios in which L = span, X = distance from support to point being considered, B = magnitude of moment at mid-span, and Y = magnitude of moment at point being considered. (See Fig. 11-6 for development of these ratios.)

X	Y
0	0
$0.1L$ or $0.9L$	$0.360B$
$0.2L$ or $0.8L$	$0.640B$
$0.3L$ or $0.7L$	$0.840B$
$0.4L$ or $0.6L$	$0.960B$
$0.5L$	B

The same short cut can be used in computing stresses at the tenth points if they are due to a uniform load. It is not necessary to divide the moment by the section modulus. The same result is obtained by multiplying the stress at mid-span by the ratio applicable to the point along the span which is under consideration.

The parabolic ratios are not applicable to the live-load moments and stresses since these are due to truck loadings rather than uniform loading. Live-load moments at points other than mid-span were computed by the simplified formula illustrated in Appendix D. As shown in Step 3, the impact factor for this bridge is 24.4 per cent and each beam carries 0.50 lane load.

The hold-down point must be located far enough from mid-span to eliminate excessive tensile stresses in the bottom fiber under final prestress plus full live load (Fig. 5-7). It must not be too far from mid-span or there will be excessive tensile stresses in the top fiber after transfer of prestress to the member when only its own dead weight is acting (Fig. 5-11). In this example hold-down points have been located 15 ft 0 in. each side of mid-span. Examination of Figs. 5-7, 5-9, 5-10, and 5-11 indicates that they could be moved some distance either way without creating excessive stresses.

We must now check to make sure that stresses are within the allowable at all points along the beam under all conditions of loading and of either initial or final prestress. A study of Table 5-8 indicates that maximum

Table 5-8. Moments* and Stresses from Applied Loads

X†	Precast section						Composite section						$f^t_{G+S+WS+L}$	$f^b_{G+S+WS+L}$
	M_G	f^t_G	f^b_G	M_S	f^t_S	f^b_S	M_{WS}	f^t_{WS}	f^b_{WS}	M_L	f^t_L	f^b_L		
8.0	168,000	+397	−326	147,000	+348	−285	31,700	+15	−39	282,000	+139	−346	+899	−996
16.0	299,000	+707	−579	261,800	+620	−508	56,300	+28	−69	490,000	+241	−600	+1,596	−1,756
24.0	392,000	+927	−760	343,600	+813	−666	73,900	+36	−91	627,000	+308	−768	+2,084	−2,285
32.0	448,000	+1,060	−869	392,600	+929	−761	84,500	+41	−104	706,000	+347	−865	+2,377	−2,599
40.0	467,000	+1,104	−905	409,000	+968	−793	88,000	+43	−108	724,600	+356	−888	+2,471	−2,694

* Moments are in foot-pounds.

† X is the distance from center of support to point being considered.

stresses will exist either under the dead load of the girder only or under all applied loads. There is no need to check the net stresses under the weight of the slab and wearing surface. The critical conditions then are:

I. Final prestress plus
   A. All applied loads
      1. Top fiber—Fig. 5-10
      2. Bottom fiber—Fig. 5-7

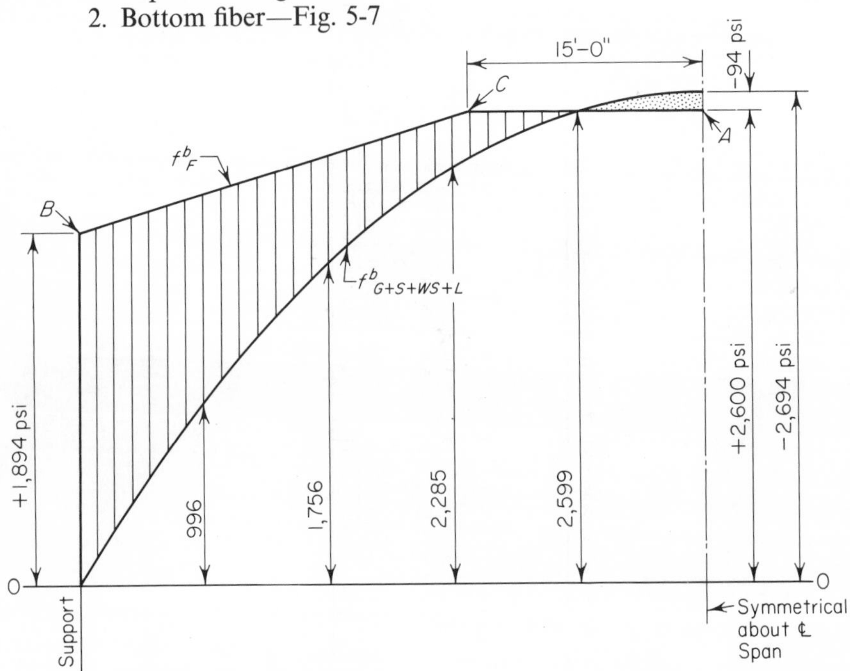

FIG. 5-7. Diagram of stresses in bottom fiber under final prestress plus all applied loads. Shaded area represents net tensile stress and lined area represents net compressive stress.

   B. Dead load of beam only
      1. Top fiber
      2. Bottom fiber

II. Prestress after transfer plus
   A. All applied loads
      1. Top fiber
      2. Bottom fiber
   B. Dead load of beam only
      1. Top fiber—Fig. 5-11
      2. Bottom fiber—Fig. 5-9

Stress diagrams will be plotted for conditions I*A*1, I*A*2, II*B*1, and II*B*2. The conditions not plotted are covered as follows:

IB1 is less critical than IIB1 because the critical stress in these cases is tension and the tension due to F_o is greater than that due to F.

IB2 is less critical than IIB2 because the critical stress in these cases is compression and the compression due to F_o is greater than that due to F.

IIA1 is less critical than IA1 because the critical stress in these cases is compression and the compression due to F_o is less than that due to F.

IIA2 is less critical than IA2 because the critical stress in these cases is tension and the compression due to F is less than that due to F_o.

Diagrams of stresses under critical combinations of prestress and loading are plotted as follows:

Fig 5-7. Stresses in bottom fiber under final prestress plus all applied loads.

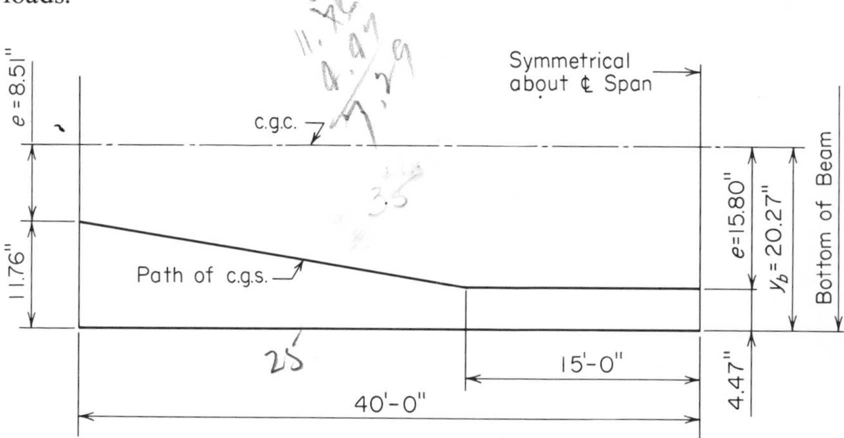

FIG. 5-8. Path of center of gravity of tendons.

Stresses due to final prestress are available from Tables 5-5 and 5-9. These are plotted at A and B. Since the strands are in a straight horizontal line from mid-span to the hold-down point 15 ft 0 in. away, there is no change in $f^b{}_F$ in this region, and the horizontal line AC can be drawn. At the hold-down point the path of the center of gravity of the tendons changes and slopes up to the end of the beam; therefore the magnitude of $f^b{}_F$ decreases steadily, and we can draw the line CB. Now, line ACB represents the magnitude of $f^b{}_F$ at all points along the beam.

Values of the stress due to all applied loads, $f^b{}_{G+S+WS+L}$, from Table 5-8 are plotted and the curve is drawn.

At mid-span $f^b{}_{G+S+WS+L}$ is larger than $f^b{}_F$ so there is a net tensile stress in the concrete which is represented by the shaded area of the diagram. Approximately half-way from A to C the two stresses are equal giving a net stress of zero in the concrete. To the left of this point, $f^b{}_F$ is greater than $f^b{}_{G+S+WS+L}$, giving a net compressive stress which is represented by the lined area of the diagram.

Fig. 5-9. Stresses in bottom fiber under prestress after transfer plus dead weight of beam only.

Applicable stresses are plotted from Tables 5-6, 5-8, and 5-9. Net stress at all points is compressive as represented by the lined area of the

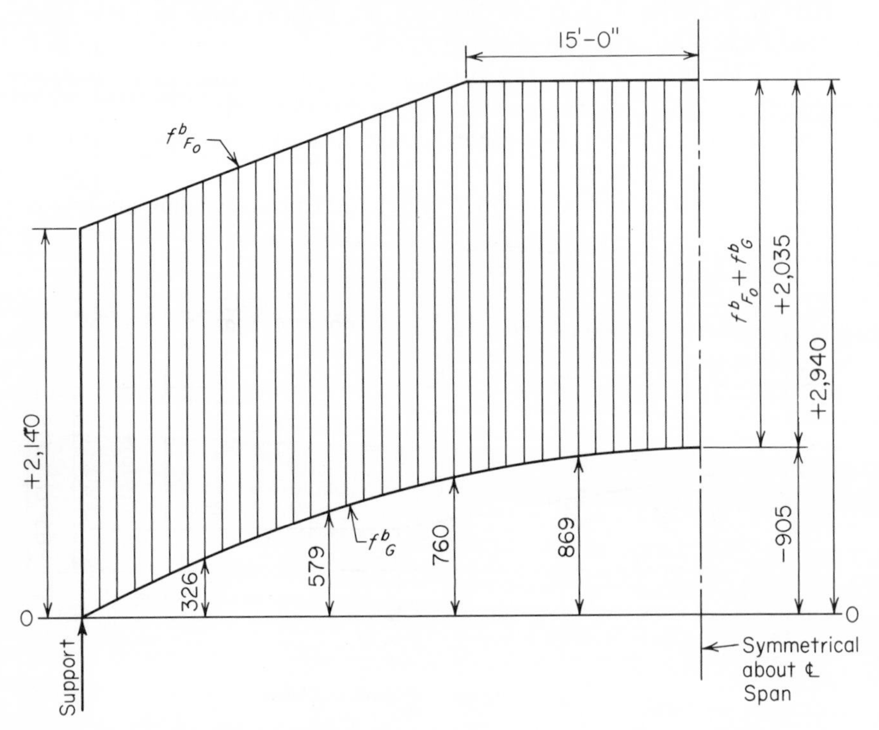

FIG. 5-9. Diagram of stresses in bottom fiber after transfer of prestress plus dead weight of beam only. Lined area represents net compressive stress.

diagram. Maximum net stress is in the vicinity of the hold-down point, but scaling the diagram shows that it is well within the allowable 2,400 psi.

Fig. 5-10. Stresses in top fiber under final prestress plus all applied loads.

The stress in the top fiber due to prestress is a tensile stress for the full length of the beam except for a short distance at the ends. On the diagram its tensile areas are plotted below the base line as negative values. (See Tables 5-5 and 5-9 for values.) Values of $f^t_{G+S+WS+L}$ from Table 5-8 are plotted measuring from the line representing f^t_F. The resulting curve is the algebraic sum of the two values, and the area above the base line (lined area) is the net compressive stress in the top fiber.

Fig. 5-11. Stresses in top fiber under prestress after transfer plus dead weight of beam only.

This diagram is plotted in the same manner as Fig. 5-10, using values from Tables 5-6, 5-8, and 5-9.

Note that the minimum net compressive stress occurs near the hold-down point. If the hold-down were moved much farther from mid-span

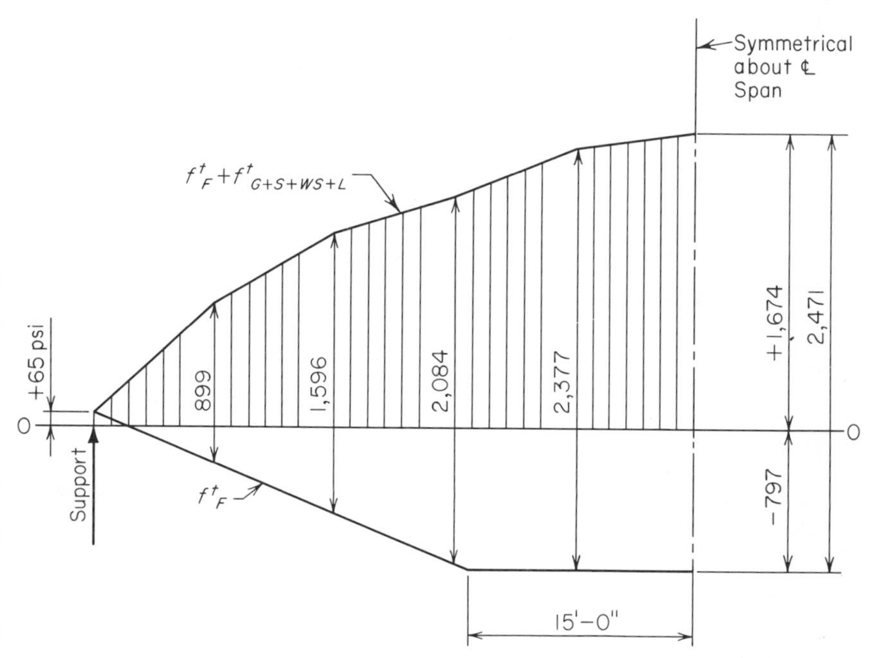

FIG. 5-10. Diagram of stresses in top fiber under final prestress plus all applied loads. Lined area represents net compressive stress.

there would probably be a net tensile stress in its vicinity. From Table 5-6 the maximum allowable tensile stress under the condition plotted here is 190 psi.

Table 5-9. Quantities Computed in Step 8

Stresses at ends of beam due to prestress	
$f^t_{F_0} = +73$ psi $f^b_{F_0} = +2{,}140$ psi	$f^t_F = +65$ psi $f^b_F = +1{,}894$ psi
Location of strands: Figs. 5-5 and 5-6 Path of tendons: Fig. 5-8	

A study of the stress diagrams shows that all stresses are within the allowable.

Step 9. Design of shear steel.

AASHO, Sec. 1.13.13, says:

For the design of web reinforcement in simply supported members carrying moving loads, it is recommended that shear be investigated only in the middle half of the span length. The web reinforcement required at the quarter points should be used throughout the outer quarters of the span.

Dead load is composed of

Beam only	583 lb per ft
Poured slab	480 lb per ft
Wearing surface	110 lb per ft
	1,173 lb per ft

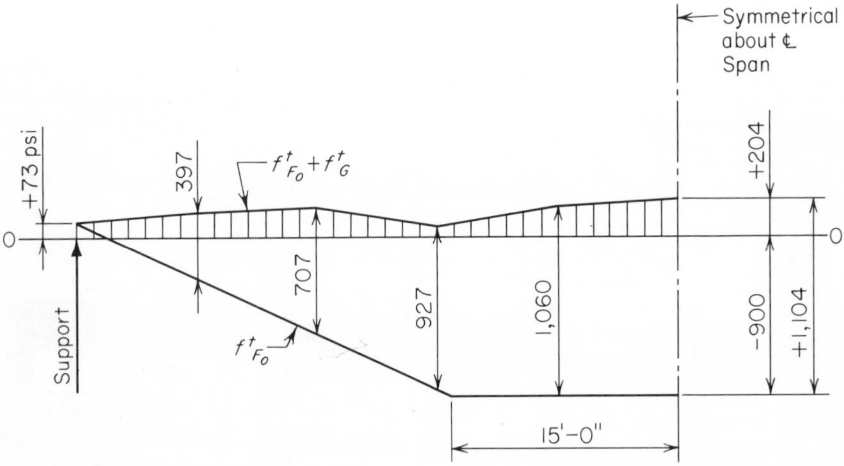

FIG. 5-11. Diagram of stresses in top fiber after transfer of prestress plus dead weight of beam only. Lined area represents net compressive stress.

Dead load shear at the one-quarter point is

$$1{,}173 \times \frac{80}{4} = 23{,}460 \text{ lb}$$

$$\frac{625 \text{ one-half of diaphragm}}{24{,}085 \text{ lb}}$$

From Fig. 5-12 the maximum live-load shear at the quarter point of the span is 45,600 lb per lane. Under this loading the impact factor is

$$I = \frac{50}{60 + 125} = 27\%$$

From previous calculations the beam carries 0.50 lane load. The design live-load shear is therefore

$$45{,}600 \times 1.27 \times 0.50 = 29{,}000 \text{ lb}$$

In prestressed concrete, shear under design-load conditions is never critical in members which meet ultimate load requirements for shear. Therefore only the ultimate condition is checked for shear.

AASHO, Sec. 1.13.6, gives ultimate load requirements as

$$1.5D + 2.5(L + I)$$

Then

$$V_u = (1.5 \times 24{,}085) + 2.5(29{,}000) = 108{,}600 \text{ lb}$$

From Fig. 5-8 the slope of the c.g.s. of the strands is

$$\frac{11.76 - 4.47}{(40 - 15)12} = 0.0243$$

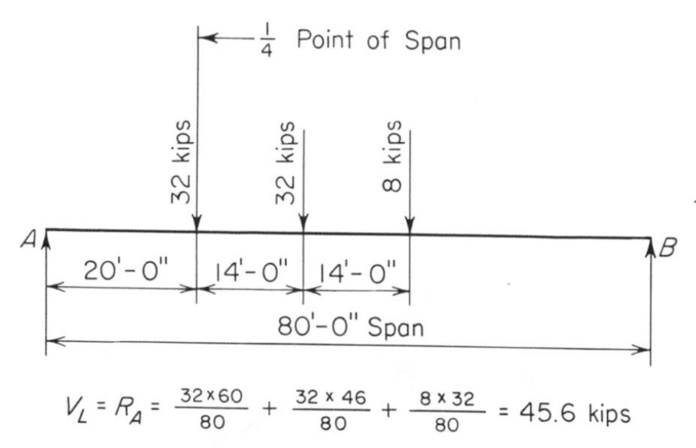

FIG. 5-12. Truck loading to produce maximum live-load shear at one-quarter point of span.

The shear carried by the strands is

$$0.0243F = 0.0243 \times 599{,}000 = 14{,}500 \text{ lb}$$
$$\text{Effective } V_u = 108{,}600 - 14{,}500 = 94{,}100 \text{ lb}$$

The area of web reinforcement required by AASHO, Sec. 1.13.13, is

$$A_v = \frac{(V_u - V_c)s}{2f_y jd} \tag{5-3}$$

AASHO defines V_c as "shear carried by concrete" but does not indicate how to evaluate it. Since the foregoing formula coincides with the formula from Tentative Recommendations for Prestressed Concrete we will use the value of V_c from the Tentative Recommendations which is

$$V_c = 0.06f'_c b'jd \tag{5-4a}$$

but not more than

$$V_c = 180b'jd \tag{5-4b}$$

In this example $0.06f'_c = 300$, so the limiting value of $180b'jd$ is applicable.

For the beam alone, d is equal to $y_t + e$ at the point being considered. e varies in a straight line from 8.51 in. at the end of the beam to 15.80 in.

at the hold-down point 25 ft 0 in. from the end of the beam. (See Fig. 5-8.) Therefore e at the quarter point, 20 ft 0 in. from the end of the beam, will be

$$8.51 + \frac{20}{25}(15.80 - 8.51) = 14.34 \text{ in.}$$

and d for the beam $= 24.73 + 14.34 = 39.07$ in.

Since the shear is resisted by the composite section, d should be measured from the top of the slab and is equal to the slab thickness plus d for the beam or

$$d = 7.00 + 39.07 = 46.07 \text{ in.}$$

AASHO, Sec. 1.13.2, says "j = ratio of distance between centroid of compression and centroid of tension to the depth d."

An equivalent for j is used in the formula in AASHO, Sec. 1.13.10(A), so we can write

$$j = 1 - \frac{0.6pf_{su}}{f'_c} \tag{5-5}$$

$$p = \frac{A_s}{bd} = \frac{3.97}{(66 \times 0.88)46.07} = 0.00148$$

$$f_{su} = f'_s\left(1 - \frac{0.5pf'_s}{f'_c}\right) = 265{,}500\left(1 - \frac{0.5 \times 0.00148 \times 265{,}500}{5{,}000}\right)$$

$$= 255{,}000$$

Substituting in Eq. 5-5,

$$j = 1 - \frac{0.6 \times 0.00148 \times 255{,}000}{5{,}000} = 0.955$$

$$b' = \text{width of web} = 7 \text{ in.}$$

Substituting in Eq. 5-4b,

$$V_c = 180 \times 7 \times 0.955 \times 46.07 = 55{,}400 \text{ lb}$$

AASHO, Sec. 1.13.13, says: "The spacing of web reinforcement shall not exceed three-fourths the depth of the member and *shall provide transverse reinforcement across the bottom flanges.*" Since the beam functions by itself to carry considerable dead weight before becoming part of the composite section, we will establish the maximum spacing on the basis of the beam depth. Then

$$s_{max} = 0.75 \times 45 = 33.75 \text{ in.}$$

The requirements for ties between the beam and composite slab must also be considered. AASHO, Sec. 1.13.14(D), says

All web reinforcement shall extend into cast-in-place decks. The spacing of vertical ties shall not be greater than four times the minimum thickness of either of the composite elements and in any case not greater than 24 in. The total area of vertical ties shall not be less than the area of two #3 bars spaced at 12 in.

If we chose a stirrup spacing near the maximum allowable, say 32 in., we would be required to add vertical ties between each pair of stirrups to keep within the 24-in. requirement. As a trial we will use $s = 24$ in. in the hope that additional ties can be eliminated.

We will assume the use of stirrups having a minimum yield strength of 40,000 psi, so $f'_y = 40{,}000$. Substituting in Eq. 5-3,

$$A_v = \frac{(94{,}100 - 55{,}400)24}{2 \times 40{,}000 \times 0.955 \times 46.07} = 0.264 \text{ in.}^2$$

AASHO, Sec. 1.13.13, also says: "The area of web reinforcement shall not be less than

$$\begin{aligned} A_v &= 0.0025b's \qquad (5\text{-}6) \\ &= 0.0025 \times 7 \times 24 = 0.420 \text{ in.}^2 \end{aligned}$$

This will be the governing factor for the stirrups and will therefore apply for the full length of the beam. If Eq. (5-3) were governing, the stirrups computed from it for the quarter point would be used from the quarter point to the ends of the beam. Values of A_v from Eq. (5-3) would also be computed for points in the middle half of the span length and would apply until they were less than the value obtained from Eq. (5-6).

Since the requirements of Eq. (5-6) and also of AASHO, Sec. 1.13.14(D), relating to vertical ties are based on steel area only without reference to its strength, we may find a saving using a lower strength stirrup. Substitute numerical values in Eq. (5-3) using stirrups with $f'_y =$ 33,000 psi.

$$A_v = \frac{(94{,}100 - 55{,}400)24}{2 \times 33{,}000 \times 0.955 \times 46.07} = 0.320 \text{ in.}^2$$

Thus the minimum area established by Eq. (5-6) is governing, and we might as well use bars with $f'_y = 33{,}000$ psi.

Will it be economical to select stirrups that will provide the total vertical tie area required by AASHO, Sec. 1.13.14(D)? This is two #3 bars at 12 in. or four #3 bars every 24 in. $= 4 \times 0.11 = 0.44$. Obviously, in this particular structure, the stirrups should be designed to the tie requirement rather than add extra bars for ties. We need therefore 0.44 in.² every 24 in. Two #4 bars give $2 \times 0.20 = 0.40$ in.²

$$\frac{0.40}{0.44}(24) = 21.8 \text{ in.}$$

Use two #4 bars at maximum spacing of 21¾ in. for the full length of the beam. Extend all of these bars into the cast-in-place deck as vertical ties.

These beams are being designed without end blocks in accordance with the General Notes for AASHO-PCI Standard Beams printed in Art. 15-2. These notes call for additional stirrups acting at a unit stress of 20,000 psi to resist at least 4 per cent of the total prestressing force to be placed within the distance $d/4$ of the end of the beam.

$$A'_v = \frac{0.04F}{20{,}000} \tag{5-7}$$

in which A'_v = total area of additional stirrups required in end zone.

Substituting numerical values,

$$A'_v = 0.04 \times \frac{599{,}000}{20{,}000} = 1.20 \text{ in.}^2$$

At the ends $d = y_t + e = 24.73 + 8.51 = 33.24$ in.

$$d/4 = 8.3 \text{ in.}$$

Use four #5 bars $= 4 \times 0.31 = 1.24$ in.^{2}

In a composite section it is also necessary to check the shear between the precast section and the cast-in-place section. This check is made under ultimate load conditions. The shearing unit stress is

$$v = \frac{V_u Q}{I_c t'} \tag{5-8}$$

in which t' is the width of contact surface between the two sections. Q is defined as "the statical moment of cross section area, above or below the level being investigated for shear, about the centroid." See Figs. 5-2 and 5-3 for dimensions used in computing Q. The area of the slab must be transformed by the use of the ratio $E_{c,\text{slab}} \div E_{c,\text{beam}} = 0.88$.

$$Q = 7 \times 66 \times 0.88(19.86 - 3.5) = 6{,}650 \text{ in.}^3$$

The shear due to dead weight of beam, slab, and diaphragms is carried by the precast beam alone as discussed in Step 3. The shear carried by the composite section is only that due to the wearing surface and the live load. Shearing stress between slab and beam will be maximum where vertical shear is maximum which is at the end of the span.

Wearing surface shear = 110 lb per ft $\times$ 80/2 = 4,400 lb

Live load shear per lane from AASHO tables is 63,600 lb

$$\text{Impact} = \frac{50}{80 + 125} = 24.4\%$$

Live-load shear on the section is therefore 63,600 × 1.244 × 0.50 = 39,600 lb

(Note that AASHO, Sec. 1.3.1.A, does not permit lateral distribution of the wheel load adjacent to the end of the member. If the portion of the lane load carried by the stringer in bending is less than 0.50, an adjustment must be made to include the full shear of the end wheel of the truck.)

Applying ultimate load factors the total vertical shear carried by the composite section is

$$V_u = (1.5 \times 4{,}400) + (2.5 \times 39{,}600) = 105{,}600 \text{ lb}$$

Substituting numerical values in Eq. (5-8),

$$v = \frac{105{,}600 \times 6{,}650}{314{,}825 \times 16} = 139 \text{ psi}$$

AASHO, Sec. 1.13.14(C), permits a shearing stress up to 150 psi "when the minimum steel tie requirements of (D) are met and the contact surface of the precast element is artificially roughened." We have already met the requirements of (D), so we need only specify the roughening of the top of the beam.

Table 5-10. Quantities Computed in Step 9

Stirrups
$f_y = 33{,}000$ psi
Two #4 bars at maximum spacing of 21¾ in. for full length of beam. Extend bars into poured-in-place slab for vertical ties.
Extra stirrups at each end. Use four #5 bars within 8.3 in. of end. Place two of these as near end face as coverage requirement permits.

Step 10. Compute camber and deflection.

Camber and deflection are discussed in Art. 10-6. In this example computations will use average values for E_c and camber growth, but it should be noted that these values may vary from one locality and casting yard to another. More accurate results can be obtained using the mathematical approach presented here and changing the values for camber growth and/or E_c to match the past experience of the producer who is casting the beams.

In Step 1, E_c for the beam was computed as $4.29(10^6)$ and the properties of the composite section were transformed so that the same E_c applies to it. These values were based on a concrete strength of 5,000 psi in the beam.

When the prestressing force is first applied to the beam the concrete strength is approximately 4,000 psi and

$$E_c = 150^{1.5} \times 33\sqrt{4{,}000} = 3{,}840{,}000$$

Compute the net camber immediately after transfer of the prestressing force from casting bed anchors to concrete beam. At this time the forces acting are the weight of the beam itself, w_G, and F_o. Deflection Δ_G due to w_G is computed by the standard formula:

$$\Delta_G = \frac{5wl^4}{384EI}$$

$$= \frac{5 \times 583(80^4)12^3}{384 \times 3{,}840{,}000 \times 125{,}390} = -1.12 \text{ in.}$$

The minus sign indicates a downward deflection or negative camber.

If the prestressing force were applied at the c.g.c. and the tendons were horizontal for the full length of the beam so that the c.g.s. coincided with the c.g.c. (as shown in Fig. 1-2), the prestressing force would cause a uniform compressive stress for the full depth of the beam and there would be no camber. Camber is a function of the magnitude of the prestressing force and its eccentricity e from the c.g.c. The moment M_e due to the eccentricity of the prestressing force is Fe, or in this particular calculation it is F_oe. Upward deflection (camber) can be computed by the moment-area method in which the deflection at center of span is equal to the moment of the area of the M/EI diagram about the support.

At ends of beam,

$$M_e = F_oe = 677{,}000 \times 8.51 = 5{,}760{,}000 \text{ in.-lb}$$

At center of span,

$$M_e = F_oe = 677{,}000 \times 15.80 = 10{,}700{,}000 \text{ in.-lb}$$

These values are plotted in Fig. 5-13 which is then divided into three simple geometrical shapes so that moments can be taken. Taking moments of the areas in Fig. 5-13, about the support

$$M_{\text{I}} = 5{,}760{,}000(40 \times 12)(20 \times 12) = 664{,}000(10^6)$$

$$M_{\text{II}} = 4{,}940{,}000(25 \times 12)\tfrac{1}{2}(25 \times 12)\tfrac{2}{3} = 148{,}300(10^6)$$

$$M_{\text{III}} = 4{,}940{,}000(15 \times 12)(32.5 \times 12) = 347{,}000(10^6)$$

$$\Delta_{F_o} = \frac{M_{\text{I}} + M_{\text{II}} + M_{\text{III}}}{EI} = \frac{1{,}159{,}300(10^6)}{3.84(10^6) \times 125{,}390} = 2.41 \text{ in.}$$

Net camber is

$$\Delta_{F_o} + \Delta_G = 2.41 - 1.12 = +1.29 \text{ in.}$$

Once the prestressing force has been applied, the concrete in the bottom of the beam will be under constant compression, which causes creep or gradual shortening of the fibers that are under compression. As the bottom fibers shorten, with no corresponding change in the length of the top fibers, the beam gradually arches, producing the effect called "camber growth." Magnitude of camber growth is influenced by a number of things including the humidity. For a prestressed member on which the additional dead weight is small, camber growth can be from one to three

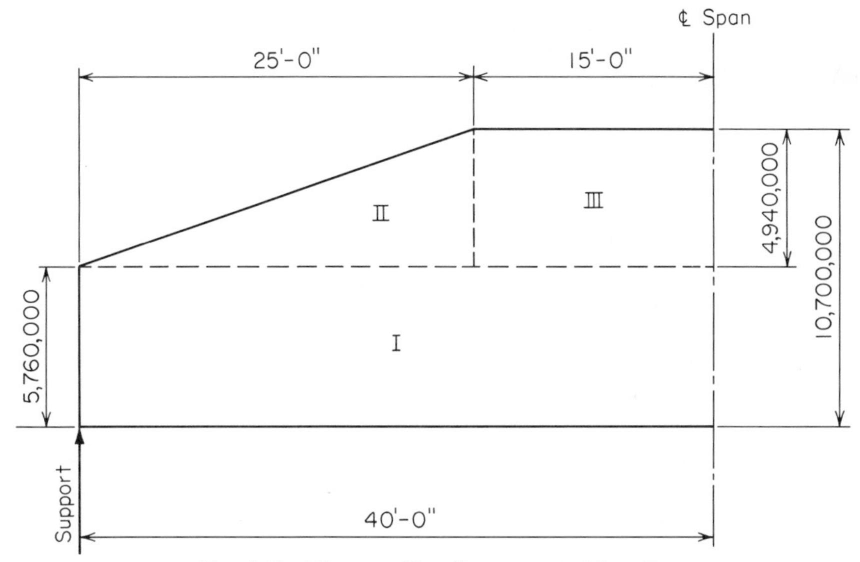

FIG. 5-13. Diagram of bending moments $M_e = F_o e$.

times the initial camber. It is best to try to base an estimate for camber growth on previous experience under similar conditions.

In this example, total camber growth will be affected by the length of time that elapses before the beams are erected and the slab is poured. Immediately after prestressing the net compressive stress in the bottom fiber (Fig. 5-9) is much greater than that in the top fiber (Fig. 5-11). This results in comparatively rapid camber growth. When the weight of the slab is placed on the beam, the compressive stresses decrease in the bottom fiber and increase in the top fiber so that the net difference between the two is less. This results in a slower camber growth and in a smaller total growth.

As already stated, previous experience under similar conditions should be used as a guide in determining the ratio between camber growth and instantaneous camber. For this example we will assume that the camber growth Δ_{CG} has reached a magnitude equal to the instantaneous camber. Then

$$\Delta_{CG} = \Delta_{F_o} + \Delta_G = +1.29$$

In computing beam deflection when the diaphragm and top slab are cast we will assume that the strength of the concrete in the beam has reached 5,000 psi and that the diaphragm and slab are poured in such a manner that the beam carries their entire dead weight before the slab has set sufficiently to create a composite section. Deflection due to slab weight is

$$\Delta_{SS} = \frac{5(480)80^4(12^3)}{384 \times 4.29(10^6) \times 125{,}390} = -0.82 \text{ in.}$$

Deflection at mid-span from the diaphragm is

$$\Delta_{SD} = \frac{1{,}250(80^4 \times 12^3)}{48 \times 4.29(10^6) \times 125{,}390} = -0.04 \text{ in.}$$

After the slab is poured the net camber is

$$(\Delta_{F_o} + \Delta_G) + \Delta_{CG} + \Delta_{SS} + \Delta_{SD} = 1.29 + 1.29 - 0.82 - 0.04 = +1.72 \text{ in.}$$

The composite section will carry the wearing surface so its moment of inertia should be used in computing Δ_{WS}, the deflection due to the wearing surface.

$$\Delta_{WS} = \frac{5(110)80^4(12^3)}{384 \times 4{,}290{,}000 \times 314{,}825} = -0.08$$

Net camber after completion of the bridge is computed as

$$+1.72 - 0.08 = +1.64 \text{ in.}$$

Camber growth will continue but it should not be very large. The difference in stress between top and bottom fiber has been reduced. Only the beam is prestressed and subject to camber growth, but the moment of inertia of the entire cross section is resisting the growth.

Deflection under live load is computed in the same manner as for any elastic member using $E_c = 4{,}290{,}000$ and the moment of inertia of the composite section.

Table 5-11. Quantities Computed in Step 10

Net instantaneous camber = $\Delta_{F_o} + \Delta_G = +1.29$ in.
Net camber under all dead load plus camber growth at completion of bridge = +1.64 in.

CHAPTER 6

Materials

CONCRETE

6-1. Properties. In reinforced concrete structures the usable concrete strength is limited by the behavior of the member. When the tensile stresses in the reinforcing steel exceed 18,000 or 20,000 psi, the member begins to show excessive cracks and deflection. Since 3,000-psi concrete is strong enough to develop these stresses in the reinforcing, higher strengths are seldom specified.

The concrete used in prestressed concrete, although made of the same basic ingredients, is much stronger, since its full strength can be utilized.[1] * In this case the governing factor is economical fabrication. A majority of members are made of 5,000-psi concrete, although some fabricators use appreciably higher strengths and most specifications permit a minimum of 4,000 psi.

High-strength concrete has several advantages:

1. A smaller cross-sectional area can be used to carry a given load. This means less dead weight and in many instances shallower members, giving more clearance.

2. Its higher modulus of elasticity E_c decreases camber, which is frequently a problem in prestressed concrete because of the relatively high negative moment that exists under the dead-load condition. The high E_c also reduces deflection and in pretensioned members cuts down the stress loss due to elastic shortening.

The recognized formula for E_c is the following one taken from Sec. 1102 of ACI 318-63, Building Code Requirements for Reinforced Concrete.

$$E_c = w^{1.5} 33 \sqrt{f'_c} \qquad (6\text{-}1)$$

w = weight of concrete, lb per cu ft

Values of E_c based on Eq. (6-1) can be read from Fig. 6-1.

3. Members can be prestressed and equipment released for the next

* Superscript numbers indicate references listed in the Bibliography at the end of the chapter.

pour in minimum time because the required f'_{ci} is reached sooner. In some cases fabricators of pretensioned members use a mix which gives a higher f'_c than required by specifications in order to reach f'_{ci} in time to maintain their pouring schedule.

4. In bridges and other members exposed to freezing or salt spray the combination of high strength and density plus freedom from cracks makes prestressed concrete virtually maintenance-free.

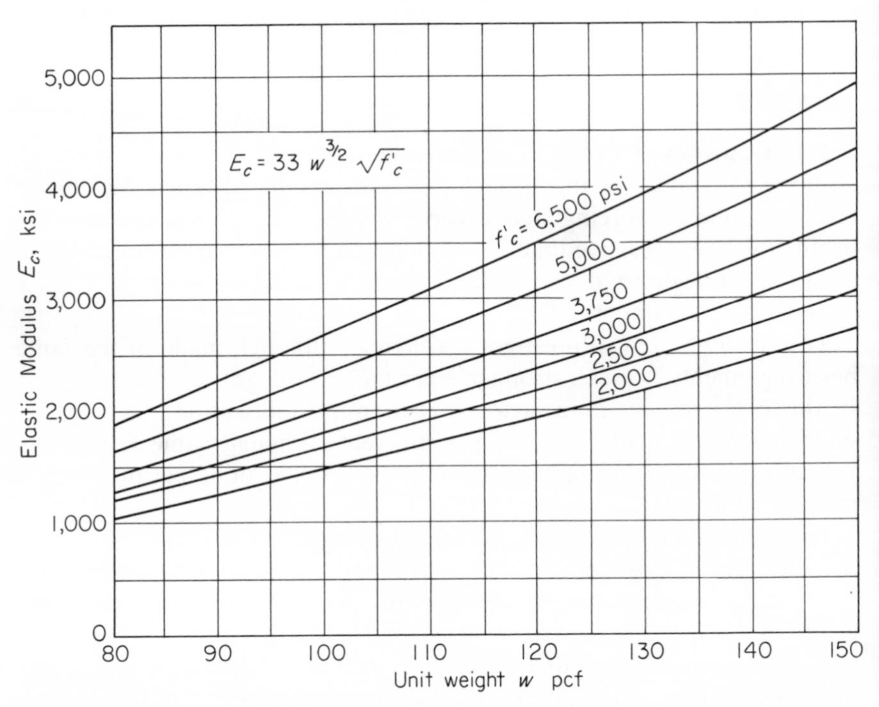

FIG. 6-1. Modulus of elasticity of concrete as a function of its strength and weight. (This chart first appeared in Adrian Pauw, Static Modulus of Elasticity of Concrete as Affected by Density, *J. Am. Concrete Inst.*, December, 1960, pp. 679–687.)

6-2. Placing. Prestressed concrete fabricators employ several aids for handling the high-strength low-slump concrete to achieve satisfactory placement and rapid curing.

Vibrators are standard equipment for all work of this type. Vibrating screeds are frequently used on area-type members such as double T's and channels.

Air entraining cement and admixtures such as Plastiment are employed to increase the workability of the mix. Use of additional cement to increase workability while maintaining the water-cement ratio and ultimate strength is avoided because the extra cement paste in the rich mix is a source of additional shrinkage.

6-3. Curing. Rapid curing to permit frequent reuse of equipment is an important factor in the economical production of prestressed concrete and especially of the pretensioned type. Figure 6-2 illustrates the curing speed obtained by steam curing on a typical pretensioned project.[2,3] Some casting yards accelerate curing by circulating hot oil through pipes

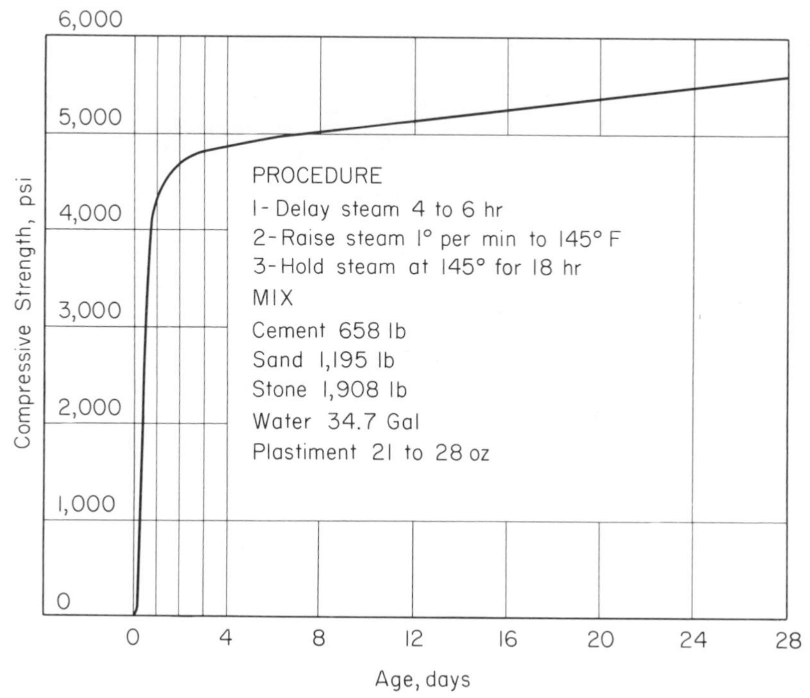

FIG. 6-2. Time-strength curve for steam cured concrete. (*By permission of Sika Chemical Corporation.*)

located under and beside the curing concrete. Many use high-early-strength cement.

Calcium chloride is not used to accelerate curing because existing test data indicate that its action is damaging to the properties of the high-strength tendons.

6-4. Lightweight Concrete. Prestressed concrete members both pretensioned and post-tensioned can be fabricated satisfactorily from lightweight concrete.[4-6]

The chief differences between regular-weight concrete and lightweight concrete of the same ultimate strength are in the elastic properties. As indicated by Eq. (6-1), E_c is a function of the weight of the concrete. For concretes of equal strength the lighter weight concrete has the smaller E_c.

Shortening of lightweight concrete due to creep plus shrinkage is usually greater than the shortening of regular-weight concrete under similar conditions, but with properly designed mixes it is not excessive and can be provided for in the design. One report based on a large number of tests found that the average creep plus shrinkage of all the lightweight mixes tested was about 1.5 times the average of the regular-weight mixes tested.[7] The minimum ratio of any one lightweight mix to any one regular-weight mix was one-to-one and the maximum ratio was two-to-one.

Properties of many lightweight aggregates are such that it is difficult to measure their water content, thus making it difficult to know how much water to add to the mix. A slump test is sometimes used as a check. In order to obtain a uniform product it is often desirable to use regular sand as the fine aggregate even though it does increase the weight of the finished product.

Producers of lightweight aggregates know their own products. They can recommend mixes and handling methods to give concretes with the desired properties. They should also have test reports to indicate what creep and shrinkage can be expected in concretes made from their product.

In addition to ASTM and ACI specifications a Guide Specification for Structural Lightweight Concrete is available from the Expanded Shale Clay and Slate Institute in Washington, D.C. The Bureau of Public Roads has issued a Proposed Specification for Lightweight Aggregates which is available from their headquarters in Washington, D.C.

Prestressed concrete members made of lightweight concrete have a longer fire endurance than identical members made of regular-weight concrete.[8] One possible exception is thin slabs (less than 3 in.) in which fire rating is determined by heat transfer through the slab.

6-5. Grout. This section refers to the grout used around post-tensioned tendons to protect them and to bond them to the concrete of the member. It does not apply to grout used to connect precast units to each other.

The process of grouting is not complicated, but a thorough job is essential and there are several features of design, detail, and application that should be understood and given consideration.

After a tendon has been post-tensioned and anchored, grout is pumped into the tube around the tendon to fill the tube and eliminate all air spaces and water pockets. The inside diameter of the tube should be as small as is feasible without interfering with an easy flow of the grout. Abrupt changes in the path or cross section of the tube should be avoided. A tendon whose path moves up or down should have a vent at each crest to prevent an air pocket and a vent at each valley to drain off any accumulation of water. Maximum spacing of vents regardless of curvature should be 50 to 60 ft.

Grout should be injected at the lowest vent or anchor fitting by using a positive displacement type of pump. Flow must be continuous and uninterrupted from the start of the operation to its completion. Individual vents can be plugged when grout of the same consistency as that from the pump is flowing smoothly from the vent. When all vents have been plugged, a pressure of approximately 100 psi should be applied by the pump for at least 1 min.

Numerous variations of mix proportions and ingredients have been used successfully. Flyash and/or admixtures are popular with some users. Sand is desirable in the mix if large voids are to be filled. Calcium chloride is not permitted because of its tendency to cause corrosion of the tendons. Best results are obtained with colloidal-type mixers. Grout should go through a strainer as it passes from mixer to pump. Typical specifications call for an ultimate cylinder strength (3 × 6 in.) of 3,000 psi at 7 days and 4,500 psi at 28 days.

One mix which should give satisfactory results consists of

1 bag of type II portland cement
1 lb of Interplast "C" (Sika Chemical Corporation)
5 gal of water

Section 2619 of ACI 318-63 sets forth requirements for grout in building members. It is reproduced in Appendix A.

TENDONS

6-6. General Properties. The need for high-strength tendons to minimize the drop in prestressing force due to stress losses was discussed at the end of Chap. 2. Satisfactory tendons are made from wire or combinations of wires and from high-strength bars, all of which are fabricated with special prestressed concrete properties. These tendons differ from most other steels in several ways in order to meet the particular requirements of prestressed concrete.

Uniform elongation to initial tension is essential in obtaining an accurate prestressing force. Figure 6-3 is the stress-strain curve of a 0.196-in.-diameter uncoated stress-relieved prestressed concrete wire. (Its general shape is typical of all types of prestressed concrete tendons.)

For all practical purposes this curve is a straight line to a point above the initial tension of 165,000 to 175,000 psi normally used for 0.196-in. wire. Tendons do not have a yield point like that of ordinary steel where elongation continues without increase of stress. They do reach a point where the ratio of strain to stress increases rapidly and remains large until the tendon fails. This change in slope of the curve is also a necessary property. When a beam is overloaded and the stress in a tendon approaches ultimate, the tendon elongates, causing a visible deflection which warns of impending failure.

In addition to high strength and special elastic properties, a tendon must have a low coefficient of relaxation at the high stresses used.

Tables giving the physical properties of the various types of tendons are given in Chap. 7 and 8, where their application is explained.

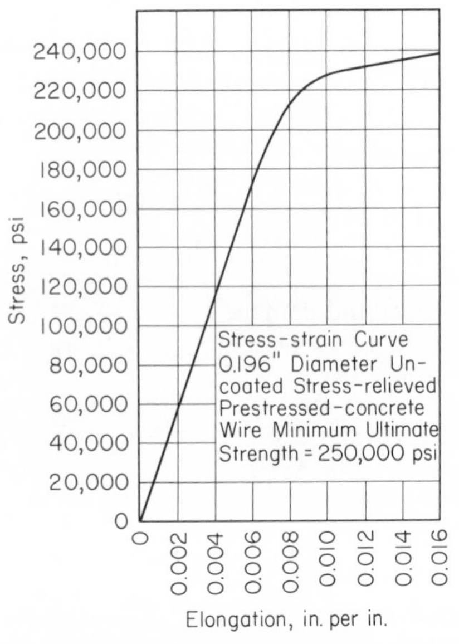

FIG. 6-3. Stress-strain curve for 0.196-in.-diameter uncoated stress-relieved prestressed concrete wire.

6-7. Wire. The chief use of wire is for making post-tensioned cables which are composed of a number of parallel wires assembled in a flexible metal hose or inserted in a cavity left in the concrete.

Figure 6-4 is a load-elongation curve for 0.196-in.-diameter uncoated stress-relieved prestressed concrete wire. This curve differs from Fig. 6-3 in that load is plotted against elongation instead of stress against elongation. This is done for the convenience of the fabricator so that, knowing the elongation, he can read the load directly from the curve without going through a calculation involving area and unit stress. Curves similar to this and plotted to a large scale for easy reading are available from the steel fabricators for each size and type of tendon.

Prestressed concrete wire is made by drawing a high-carbon hot-rolled round steel rod through several conical dies. As the rod passes through each die, its diameter is decreased and its length is correspondingly increased. In the process the crystals that made up the rod are drawn out into the long parallel fibers which give wire its extremely high tensile strength. Figure 6-5 is a microphotograph showing the crystalline structure of the hot-rolled rod. Figure 6-6 is a microphotograph showing the fibers in a completed wire. After the wire has passed through the last die, it is called "cold drawn" wire.

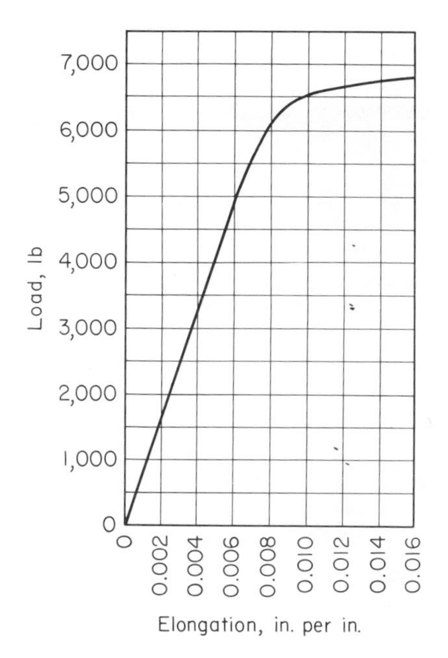

FIG. 6-4. Load-elongation curve for 0.196-in.-diameter uncoated stress-relieved prestressed concrete wire. (*By permission of CF&I Steel Corporation.*)

Cold drawn wire does not have the uniform elastic properties needed in prestressed concrete. As the crystals elongate into fibers, they develop stresses in excess of their yield point. The resulting wire is made up of fibers which are full of internal stresses. When a tensile load is applied to the wire, each fiber elongates the same amount. The stress due to this elongation is added to the internal stress, and soon the fiber with the highest internal stress reaches the point where its ratio of strain to stress increases rapidly. From here on this fiber will elongate with the others but its stress will not increase so rapidly. Since the internal stresses in the various fibers are not uniform, first one fiber, then another, then another will reach the yield point. The resulting load-elongation curve of the wire is not uniform.

The internal stresses in the cold drawn wire are eliminated by subjecting it to a stress-relieving process in which it is raised to a moderate temperature (750 to 850°F) for a short period of time (15 to 30 sec.). This is not a heat-treatment and does not change the ultimate strength of the wire to any noticeable degree. It does create a wire which has the desired uniform elastic properties because it is free of internal stresses.

At stresses up to 70 per cent of its ultimate strength the modulus of elasticity E_s of uncoated stress-relieved prestressed concrete wire is approximately 29,000,000 psi. While this value is sufficiently accurate for

FIG. 6-5. Microphotograph showing crystalline structure of hot-rolled rod. (*By permission of CF&I Steel Corporation.*)

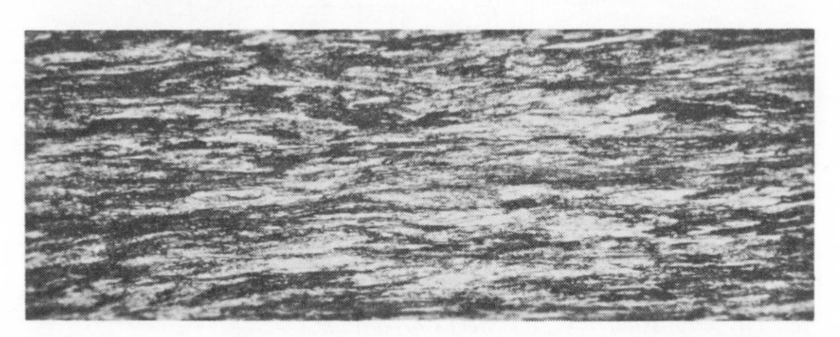

FIG. 6-6. Microphotograph showing fibers in a completed wire. (*By permission of CF&I Steel Corporation.*)

design calculations, it is recommended that the steel fabricator's load-elongation curve for the particular wire be used in computing the elongation required in jacking wires to a specified tension.

Although oil-tempered wire has many of the properties desired in a prestressed concrete tendon, it is extremely susceptible to stress corrosion and is therefore not used.

6-8. Seven-wire Strand. Seven-wire uncoated stress-relieved prestressed concrete strands conforming to ASTM Designation A 416 are the tendons normally used in pretensioned bonded members. Figure 6-7 is a photograph of a seven-wire strand and Fig. 6-8 is the load-elongation curve.

Each strand is made up of seven cold drawn wires. The center wire is straight, and the six outside wires are laid helically around it. All six outside wires are the same diameter, and the center wire is slightly larger. The difference in diameter is sufficient to guarantee that each of the outside wires will bear on and grip the center wire. This is an important detail in members where the tension in the strand is developed through

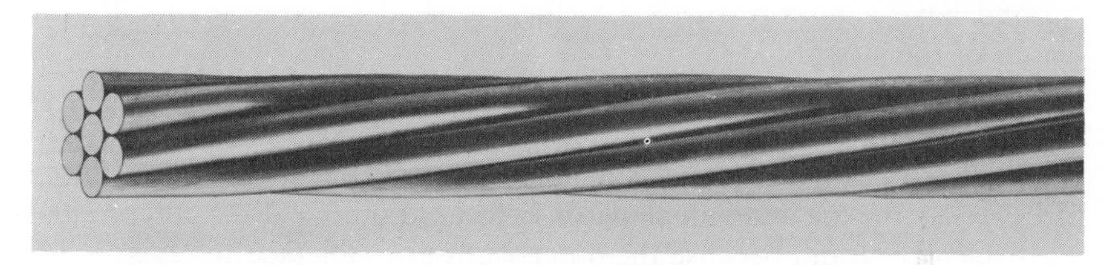

FIG. 6-7. Seven-wire strand. (*By permission of CF&I Steel Corporation.*)

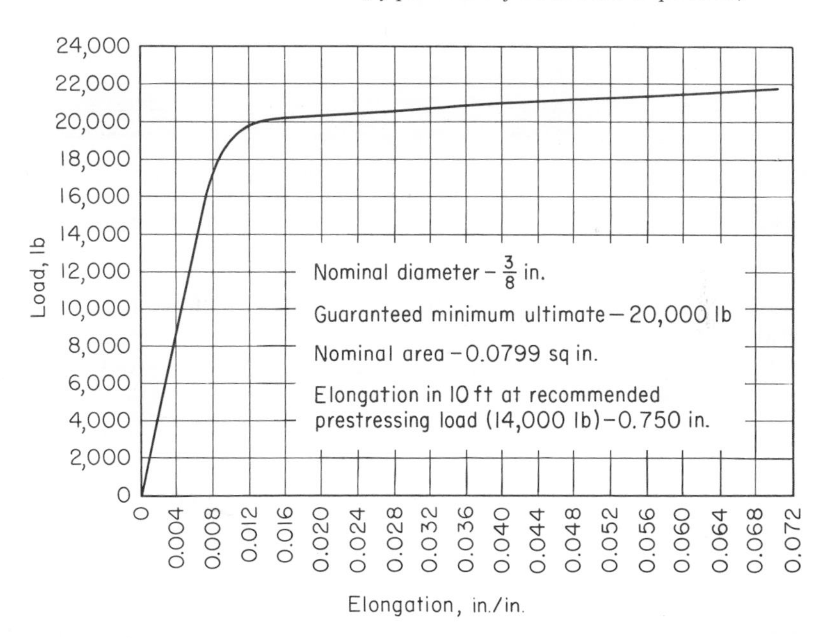

FIG. 6-8. Load-elongation curve for seven-wire uncoated stress-relieved prestressed concrete strand. (*By permission of CF&I Steel Corporation.*)

bond. The six outer wires are bonded to the concrete both by adhesive bond and by mechanical bond of the concrete in the valleys between the wires. Under tension each of the outer wires bears on the center wire, and the resulting friction makes the center wire elongate with the outer wires. If the center wire were too small, the outer wires would bear against each other to form a pipe through which the center wire would slip without carrying any load.

The "lay" of the outer wires is the distance along the strand in which an outer wire makes one complete turn around the center wire. Each strand fabricator chooses the lay best suited to his production and within the limits of not less than twelve and not more than sixteen times the strand diameter. Since a strand with a long lay will have an appreciably higher E_s than a strand with a short lay, specifications require that the "strand have a uniform lay"; i.e., the lay shall be constant for the full length of the strand.

After the seven wires are formed into a strand, the strand is subjected to a stress-relieving process like that described in Art. 6-7 for wire. The result is a strand with ideal elastic properties and free of internal stresses. Stress-relieving the individual wires before stranding does not produce a satisfactory strand because the outer wires are subjected to severe twisting in the standing operation, which creates internal stresses. These are

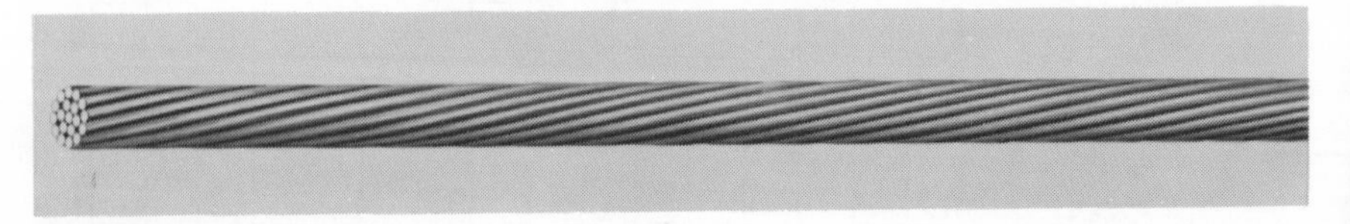

FIG. 6-9. One-inch-diameter prestressed concrete strand. (*By permission of CF&I Steel Corporation.*)

removed only when the stress-relieving operation follows the stranding operation.

6-9. Large Strands. Large-diameter galvanized prestressed concrete strands are used only for post-tensioned work. Their cost per pound of prestressing force is appreciably higher than that of other types of post-tensioned tendons and they are used only when their special properties are needed. The anchor fittings on these strands are so designed that the tendon can resist millions of cycles of repeated loads without failure of the fitting or the strand.

A galvanized prestressed concrete strand is composed of seven or more hot-dip galvanized wires. Here again the cold drawn wire is the basic unit. In this case the cold drawn wire is passed through a bath of molten zinc and emerges as a galvanized wire with an outer coating of pure zinc plus a hard zinc-steel alloy between the zinc and the steel. The temperature of the zinc bath and the speed of the wire passing through it are regulated so that the wire is stress-relieved as it is galvanized.

The smallest galvanized strand is 0.600 in. diameter. This strand is composed of a center wire plus six outer wires assembled in the same manner as the strands described in Art. 6-8. Larger strands are made by adding extra layers of wire. A 1-in.-diameter strand is a 0.600-in.-diameter 7-wire strand with one more layer composed of 12 wires for a total of 19 wires.

A 1⅜-in.-diameter strand is a 1-in.-diameter 19-wire strand with one more layer composed of 18 wires for a total of 37 wires. Intermediate sizes are made by varying either the number or size of the wires or both to produce the most economical strand.

Strands well over 3 in. in diameter have been produced on existing equipment.

When the ultimate strengths are divided by the areas, it is found that the stresses at ultimate load are only 202,000 to 214,000 psi instead of the much higher stresses given for uncoated wire and strand. This is because it is standard practice to list the gross area of the strand including the zinc, even though the zinc carries practically no load.

In Art. 6-8 it was stated that the seven-wire strands are stress-relieved after stranding to eliminate internal stresses due to stranding. It is not practical to stress-relieve the larger strands because the outer wires would reach the critical temperature and revert to a crystalline structure before the inside wires were stress-relieved. To achieve uniform elongation characteristics the large strands are prestretched. In this operation a full shop length (usually 3,600 ft) is stretched in a tensioning rig to approximately 60 per cent of its ultimate and held at that load a short time. During this period of high stress the fibers which are stressed beyond their yield point because of the applied load added to internal stresses will continue to elongate until they reach a point of stability. After the load is released, the strand has good elastic properties up to its prestretching load. Figure 6-10 is the load-elongation curve of a 1¹¹⁄₁₆-in.-diameter strand after the prestretching operation.

Ungalvanized strands having the same areas as the galvanized strands can be made to order. Since their entire area is steel, their ultimate strength is approximately 10 per cent greater than that of the galvanized strands. Ungalvanized strands of this type are seldom used because they do not offer enough advantages to justify their additional cost over other tendons.

6-10. High-strength Steel Bars. The high strength needed for prestressed concrete tendons is achieved by using a specially selected hot-rolled alloy steel. In their hot-rolled state the alloy bars have neither the uniform elastic properties nor the ductility required. These properties are obtained by subjecting the bars to a stress-relieving treatment at 600°F for 8 hr in a furnace. They are then left closed in the furnace until cooled to approximately room temperature.

After stress-relieving, the bar has all the desirable properties except the stress-strain curve and high yield point typical of prestressed concrete tendons. These are developed by cold-stretching each bar to at least 90 per cent of its ultimate strength. The cold-stretching process also serves as a proof-stressing which eliminates any bars having surface imperfections or metallurgical defects.

Figure 6-11 is the stress-strain curve for a high-strength steel bar of the type used in prestressed concrete in the United States.

6-11. Splices. Manufacturing and/or shipping conditions are often such that some tendons must be spliced. All splices have certain disadvantages, but they cannot be eliminated, so an understanding of the reasons for and conditions of their use is important.

Splices are not permitted in the wires of parallel-wire cables. There

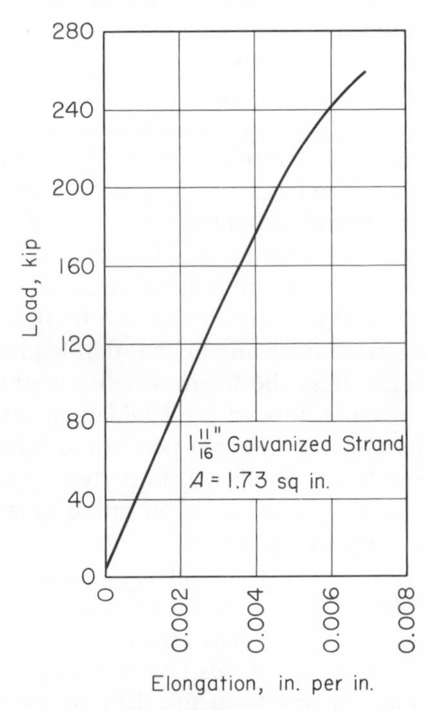

FIG. 6-10. Load-elongation curve for $1\frac{11}{16}$-in.-diameter galvanized prestressed concrete strand. (*By permission of CF&I Corporation.*)

are two reasons for this. Since these wires are manufactured in lengths of 1,800 to over 7,000 ft, it is a simple matter to make a cable without splices. When wires are spliced, they are spliced by welding, which damages the fibers and therefore lowers the strength of the wire. In a parallel-wire cable each individual wire must carry its full share of the tension at all times. It cannot distribute part of its load to adjacent wires at a point where it is weak. As a result if a welded wire is used in a parallel-wire cable, it usually fails during the tensioning operation.

Seven-wire strands are produced in endless lengths and are shipped on reels or in packs containing 8,000 to 22,000 ft. In use the full length required is cut from the reel in one length and no splice is needed. When the last length on the reel is not long enough to reach the full length of the casting bed, some operators find it desirable to salvage this length by splicing it to another short length. Splicing two short lengths of strand together in this manner is permissible if the coupling used will develop the full strength of the strand, will not cause failure of the strand under the fatigue loadings which will be applied to the structure being fabricated, and does not weaken the member by replacing too much of the concrete

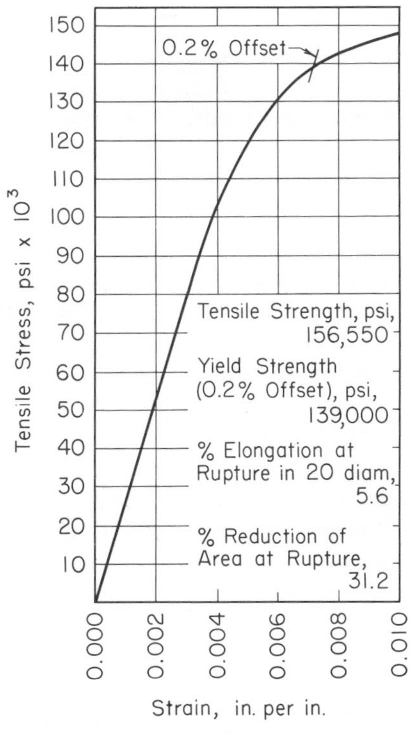

FIG. 6-11. Stress-strain curve of Stressteel bar. (*By permission of Stressteel Corp.*)

in the cross section. The engineer must determine whether or not a particular splice has the necessary qualifications. Information on this subject is also available from some of the fabricators of seven-wire strands.

Splices are used in the individual wires of seven-wire strands during shop fabrication. ASTM Specification A 416 says "During fabrication of the strand, butt-welded joints may be made in the individual wires, provided there is not more than one such joint in any 150 foot section of the completed strand." Although the strength of the welded wire is reduced in the vicinity of the weld, the total strength of the strand is not seriously reduced. The welded wire is treated in the vicinity of the weld so that part of its load is transferred to the six other wires.

High-strength bars are spliced at the job site with couplers when necessary. Bars are fabricated in maximum lengths of about 80 ft and are sometimes furnished in shorter lengths with splices because of shipping difficulties. The couplers are larger in diameter than the bar and are tapped to take a threaded bar in each end. They develop the full strength of the bar. The disadvantage in the use of couplers is the large hole which must be cored in the concrete at the location of the coupler. If concrete stresses are high, the net section should be checked with the area of the hole deducted.

Large strands are never spliced, since they can be supplied in any length required without splices. They do have occasional splices in individual wires similar to the seven-wire strands.

6-12. Anchorages for Post-tensioned Tendons. The various types of anchor fittings used in the United States are illustrated in Chap. 8 in conjunction with the discussion of the tendons used with them.

Analysis of the stresses in anchor fittings for most post-tensioned tendons is a very specialized procedure and varies with the type of fitting. Responsibility for proper design of a fitting rests with the fabricator of that fitting.

In addition to anchoring the tendon, the anchor fitting must spread the load to the concrete or it must be anchored against a bearing plate which will spread the load. Allowable bearing stresses and other requirements are presented in both the specifications for bridges and for buildings. Some universally approved fittings cause stresses in excess of those allowed by the specifications but are used successfully in any case. The concrete beneath them is confined by a reinforcing spiral or other device. Where adequate test data indicate that such fittings are safe, their use should be permitted.

6-13. Flexible Metal Hose. Post-tensioned tendons are frequently encased in flexible metal hose. For large strands and bars the inside diameter of the hose is ¼ to ⅜ in. larger than the diameter of the tendon. For parallel-wire cables the diameter is selected to provide sufficient room for grouting the wires in the cable.

The purpose of the metal hose is to keep the tendon from bonding to the concrete until the concrete has cured and the tendon has been tensioned. After tensioning, the metal hose is pumped full of grout, which bonds the tendon to the hose. Since the hose is already bonded to the concrete, the grout actually bonds the tendon to the concrete.

Flexible metal hose is available in various thicknesses, profiles, degrees of watertightness, etc. Experience indicates that it should be made of metal having a minimum thickness of 0.011 in. Hose made of lighter metal is easily damaged. Special packings to make the hose absolutely watertight are not necessary. Properly formed hose without packing will be tight enough to prevent any leakage from the low-slump concrete into the hose.

Choice of uncoated metal hose or galvanized hose depends upon its use. Uncoated hose is satisfactory when it is to be used in a reasonably short time and will not be left exposed to the weather for more than a few days. Galvanized hose is used when exposure is expected to be severe enough to rust the light metal seriously in an uncoated hose. Galvanized hose is also used when friction between tendon and hose is a problem, because the galvanized hose gives a lower coefficient of friction.

The profile of the hose is an important factor. Profile refers to the shape into which the flat wire is bent as it is formed into the hose. The crushing strength of the hose, its ability to resist tearing as it is handled, and the radius to which it can be bent without damage are all functions of the profile.

Many fabricators of flexible metal hose have developed hoses especially for use with prestressed concrete and can refer designers to previous users of this material if necessary. They can also furnish hose with attachments for grout connections at specified locations.

6-14. Standard Materials. Cement, aggregate, water, reinforcing bars, etc., are of the same types and specifications as those used in ordinary reinforced concrete.

BIBLIOGRAPHY

1. Klieger, Paul: Early High-Strength Concrete for Prestressing, *Research and Develop. Labs., Portland Cement Assoc. Research Dept. Bull.* 91, Chicago, Ill., March, 1958.
2. Schmid, Emil, and Raymond J. Schutz: Steam Curing, *J. Prestressed Concrete Inst.,* September, 1957, pp. 37–42.
3. Hanson, J. A.: Optimum Steam Curing Procedure in Precasting Plants, *J. Am. Concrete Inst.,* January, 1963, pp. 75–100.
4. Schideler, J. J.: Lightweight-Aggregate Concrete for Structural Use, *J. Am. Concrete Inst.,* October, 1957, pp. 299–328.
5. Nordby, Gene M., and William J. Venuti: Fatigue and Static Tests of Steel Strand Prestressed Beams of Expanded Shale Concrete and Conventional Concrete, *J. Am. Concrete Inst.,* August, 1957, pp. 141–160.
6. Gray, Warren H., John F. McLaughlin, and John D. Antrim: Fatigue Properties of Lightweight Aggregate Concrete, *J. Am. Concrete Inst.,* August, 1961, pp. 149–162.
7. Creep and Drying Shrinkage of Light-Weight and Normal-Weight Concretes: *Natl. Bur. Std. Monograph* 74.
8. Selvaggio, S. L., and C. C. Carlson: Fire Resistance of Prestressed Concrete Beams Study B. Influence of Aggregate and Load Intensity, *J. PCA Research and Develop. Labs,* January, 1964.

CHAPTER 7

Pretensioned Method

7-1. Basic Operation. Pretensioning is defined by the ACI-ASCE Committee on Prestressed Concrete as "a method of prestressing reinforced concrete in which the reinforcement is tensioned before the concrete has hardened." Applied to standard practice in the United States, a more specific definition would be "a method of prestressing reinforced concrete in which the reinforcement is tensioned before the concrete is placed."

Basically, one complete cycle on a casting bed has five steps:

1. Tendons are placed on the bed in the specified pattern. They are tensioned to full load and attached to anchors at each end of the bed so that the load is maintained.

2. Forms, reinforcing bars, wire mesh, etc., are assembled around the tendons.

3. Concrete is placed and allowed to cure. In many cases curing is accelerated by the use of steam or other similar methods.

4. When the concrete reaches the strength specified for f'_{ci}, the load in the tendons is released from the anchors. Since the tendons are now bonded to the concrete, they cannot move independently of the concrete. As they try to shorten, their load is transferred to the concrete by bond. This load is the prestressing force in the concrete member.

5. The tendons are cut at each end of each prestressed concrete member, and the members are moved to a storage area so the bed can be prepared for the next cycle.

7-2. Tendons. The standard tendon for pretensioned members in the United States is the seven-wire uncoated stress-relieved prestressed concrete strand illustrated in Fig. 6-7 and covered by ASTM Specification A 416. These strands have proved superior to single wires because they have much better bonding properties, they take up less space for a given prestressing force, and they can be placed and tensioned with less labor.

An improved seven-wire strand meeting all the requirements of ASTM Specification A 416 and having 15 per cent greater strength was placed on the market in 1962. The physical properties of both the ASTM grade and the high-strength grade are given in Table 7-1. For structural analysis (but not for computing exact elongation in a casting bed) the modulus of elasticity of these strands can be taken as 28 million psi.

In a pretensioned member the load in the tendon is transferred to the concrete by bond. This transfer of load takes place in a short distance, called the transfer length, at each end of the member. For example, test data show that the full initial tension in a 7/16-in.-diameter strand is transferred to the concrete in a distance which varies from 15 to 30 in. depending upon the conditions in the particular member. Between the two transfer lengths the only transfer of load from the strand to the concrete is the small amount due to changes in bending moment. If a member is loaded to failure, the bond between strand and concrete must be sufficient to develop the ultimate strength of the strand. Prestressed concrete members

Table 7-1. Physical Properties of Seven-wire Uncoated Stress-Relieved Prestressed Concrete Strands.*

	3/8-in. diameter		7/16-in. diameter		1/2-in. diameter	
	ASTM grade	Type 270K	ASTM grade	Type 270K	ASTM grade	Type 270K
Minimum ultimate strength....	20,000 lb	23,000 lb	27,000 lb	31,000 lb	36,000 lb	41,300 lb
Area, sq in..................	0.0799	0.0854	0.1089	0.1167	0.1438	0.1531
Initial tension—70% ult.......	14,000 lb	16,100 lb	18,900 lb	21,700 lb	25,200 lb	28,910 lb
Initial stress................	175,220 psi	188,520 psi	173,550 psi	185,950 psi	175,240 psi	188,830 psi
Stress loss..................	35,000 psi	35,000 psi	35,000 psi	35,000 psi	35,000 psi	35,000 psi
Final stress	140,220 psi	153,520 psi	138,550 psi	150,950 psi	140,240 psi	153,830 psi
Final tension	11,205 lb	13,110 lb	15,090 lb	17,615 lb	20,165 lb	23,550 lb
Final tension, 6-270K strands.	....	78,660 lb	....	105,690 lb	....	141,300 lb
Final tension, 7-ASTM Grade Strands..................	78,435 lb	....	105,630 lb	....	141,155 lb	

* The 35,000 psi stress loss used herein is specified in Sec. 208.3.2 of Tentative Recommendations for Prestressed Concrete and in Sec. 1.13.8(B) of AASHO Standard Specifications for Highway Bridges for 1961. (Table courtesy of CF&I Steel Corporation.)

fabricated in accordance with standard requirements have the necessary bond properties. Both the ACI-318 Building Code and the AASHO Specification for Highway Bridges include criteria on the embedment length necessary to develop adequate bond.

Seven-wire strands develop both adhesive and mechanical bond. The heat of the stress-relieving process burns off oil, grease, and other foreign materials which might impair the adhesion between concrete and the wires in the strand. In many cases "on-the-job" conditions make it difficult to keep the strand entirely free of rust. Rust is not harmful as long as it is within the limits permitted by ASTM A 416; in fact a small amount of rust actually improves bond. Mechanical bond is developed by the concrete in the valleys between the wires of the strand.

All requirements affecting the bond of the seven-wire strand in prestressed concrete are based on test results. Numerous comprehensive tests have been conducted to establish these bonding properties. They include everything from simple pull-out tests to full-size bridge girders subjected to millions of cycles of repeated loadings.[1-6]* Pull-out tests seldom give accurate results because the full advantage of the mechanical bond is not obtained. As shown by Fig. 6-7 the seven-wire strand looks something like a stud on which the threads have a very long pitch. When subjected to a pull-out test, the strand rotates or unscrews so that only the adhesive bond is effective. The mechanical bond is effective in a typical beam or girder because the strand is encased in concrete for its full length and cannot unscrew.

Single wires are used for pretensioned structures in Europe and were tried in this country when prestressed concrete fabricators were looking for the most efficient combinations of materials and production methods. Tests comparing the bonding properties of wires and seven-wire strands soon demonstrated the superiority of the strands. The full ultimate strength of 0.196-in.-diameter or smaller stress-relieved wires is developed in prestressed concrete beams statically loaded to failure. However, after such beams have been loaded to design load a million or more times, they fail under static test at loads as low as 60 per cent of their computed ultimate strength, and the failure is due to failure of bond between the concrete and the wires. Beams pretensioned with seven-wire strands have just as high an ultimate strength after millions of cycles of loading as they have when loaded only once. This leads to the conclusion that smooth wires undergo a progressive loss of bond, whereas strands with the additional advantage of mechanical bond are not subject to this type of bond failure. Since a typical highway bridge can be subjected in its lifetime to over a million cycles of full load plus numerous severe overloads, use of single wires anchored only by bond is not recommended.

For members not subjected to repeated loads, development of satisfactory bond limits wire sizes to about 0.196-in. maximum diameter. A single 0.196-in.-diameter wire in a pretensioned member has a final tension of 4,000 lb. Comparing this with 17,615 lb and 23,550 lb for 7⁄16- and ½-in. high-strength strands we see that the number of wires needed is between four and six times the number of strands needed. In many members the space between strands is the minimum permitted by the size of the aggregate being used. If the number of tendons were increased four to six times, a much larger concrete section would be needed to accommodate them. Since the labor required to place and tension a single wire is practically the same as that for a seven-wire strand, use of strands saves labor.

The first seven-wire strands used in pretensioned bonded work were ¼ in. diameter. These were followed closely by 5⁄16 and ⅜ in. diameters before the question of bond was raised. Comprehensive fatigue tests of concrete members pretensioned with ⅜-in. strands proved that these

*Superscript numbers indicate references listed in the Bibliography at the end of the chapter.

strands had the necessary bonding properties. As larger strands were introduced, their bonding properties were checked by similar tests. At this writing the strongest strand generally approved as having adequate bonding properties is the ½-in. high-strength strand (type 270K) listed in Table 7-1.

7-3. The Casting Bed. Figure 7-1 is a line drawing showing the basic elements of a casting bed on which three pretensioned members are curing.[7,8]

The anchor posts must carry the full load of the tendons until the concrete has cured to the specified strength, f'_{ci}. Since the tendons are placed above the surface of the concrete slab, they create a bending moment as well as a horizontal force which must be resisted by the anchor posts. Details of the anchor posts depend upon soil conditions. Some are prestressed concrete piles like those shown in Fig. 7-1, some are gravity blocks

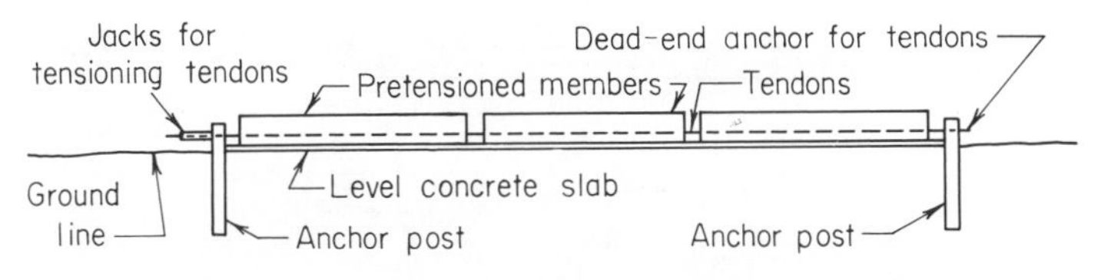

FIG. 7-1. Basic elements of a casting bed.

similar to those used to anchor suspension bridge cables, and in some cases the concrete slab is made heavy enough to carry the load in compression without anchor posts. Anchor posts strong enough to hold the tendons of a large bridge girder are often one of the most expensive items in a casting bed. For this reason casting yards with more than one bed usually have just enough beds with heavy anchor posts to meet their requirements for production of heavy members. They have additional beds with smaller anchor posts for the lighter members.

The concrete slab between the anchor posts serves as the pallet on which the members are cast. It is usually equipped with insets to which the forms can be fastened. When a bed is to be used for casting members with deflected strands as discussed in Art. 7-6, the slab is also equipped with inserts for holding the strands down at deflection points.

Casting beds have been built in lengths from under 100 to over 600 ft, but the efficient length for most installations seems to be in the neighborhood of 300 to 400 ft. No matter what bed length is chosen, there will be occasions when the total length of members to be cast is appreciably less than the distance between end anchor posts. In these cases the strand between the end of the last member cast and the end anchor post is wasted. Casting yards deal with this waste in various manners.

When the quantity is small, the waste strand can be cast into concrete members in loops that serve as lifting hooks or it can be used in the unstressed condition to replace small reinforcing bars. Some casting yards have various arrangements of strands or long bars that can be used as tag lines between the anchor post and the end of the last member.

Lengths of strand too short to reach the full distance between anchor posts are frequently salvaged by splicing to a length long enough to complete the distance. A popular splice for this purpose is the PLP splice for prestressed concrete strands made by Preformed Line Products Co. of Cleveland, Ohio. These splices shown in Fig. 7-4 and 7-5 not only develop the full ultimate strength of the strand, they are small in diameter and can be cast into the concrete member.

FIG. 7-2. Straight and deflected strands in place for a precast pretensioned I beam at plant of Capital Prestress Co., Inc., Jacksonville, Florida.

The details of a casting bed are determined by the members it is to produce. Even the largest casting yards have at least one universal-type bed. A universal bed can be used to fabricate any pretensioned member that has a prestressing force within the capacity of the anchor posts. It is composed of anchor posts, a level concrete slab between the posts, and jacking equipment. In addition it usually has inserts in the slab for supporting forms, steam, or other curing facilities and inserts for strand deflection. Forms for the members to be cast are placed on the concrete slab. Templates for spacing and holding strands at the anchor posts have uniform spacings for strand patterns, and every effort is made to design members with strand patterns that can be worked into the standard templates. Jacking equipment is designed to accommodate any strand pattern within the capacity of the bed. Universal beds permit small yards to produce all types of members, and they are essential to any yard in the production

FIG. 7-3. Prestressing plant of Material Service Division of General Dynamics Corp. Reel rack in lower left-hand corner contains about 40 reels permitting 40 lengths of strand to be pulled into bed at one time. To right of rack are three casting beds in operation. Bed on right is complete with tensioned strands and reinforcing cages ready for side forms. Beams on center bed are being steam cured under temporary covers. Last bed has just been cleared of finished beams. Stored members are I beams for bridges.

of a complete line of prestressed concrete members. They are not quite so efficient in the production of standard members as a standard bed.

A standard bed is designed to produce a specific member in permanent forms. One of the most versatile and popular standard forms is the double T. Figure 7-6 shows the various sections which can be made in a double-T bed. The deepest double T is cast in the permanent form. Shallower double T's are cast by placing fillers in the legs of the form to the desired height. Channels are formed by using fillers which block out the over-

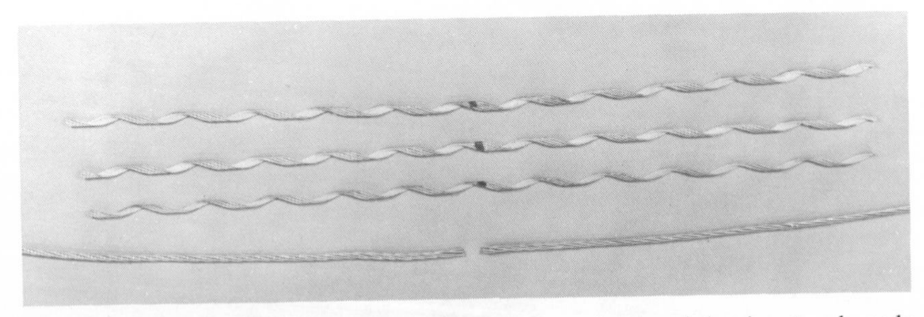

FIG. 7-4. Preformed Line Products Co. splice for seven-wire strand showing strands ready for splicing and the splice subsets to be used. The splice subsets are made of high-strength wires which have their contact surfaces with the strand coated with a gripping compound.

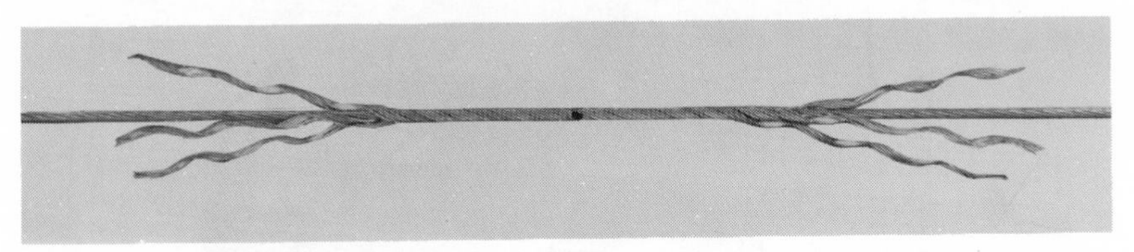

FIG. 7-5. Preformed Line Products Co. splice partially assembled on seven-wire strands.

hanging portions of the top slab. Two rows of joists can be cast at one time by placing a divider of the desired width along the center line of the slab portion of the form. Depth of channels and joists can be varied by using fillers in the legs of the forms.

Some projects are large enough to justify the construction of custom-made beds for the one job. These beds are very efficient because each detail can be chosen to suit the members to be made. For instance, the length between anchor posts can be set to leave a minimum of strand waste.

A large casting plant installed at the job site to turn out members for just one structure is illustrated in Chap. 15 in connection with the A & P project.

7-4. Handling Strands. As a seven-wire strand emerges from the stress-relieving process it is wound onto reels and prepared for storage or shipment. It is shipped either on a wooden reel or in a "reel-less pak."

The typical wooden reel is 50 in. in diameter with an over-all width of 36 to 37 in. Each reel carries approximately 4,000 lb of strand. At a casting bed the reels of strand are set in a reel rack (see Fig. 7-3) and the strand is pulled into the bed. When the reels are empty they are dis-

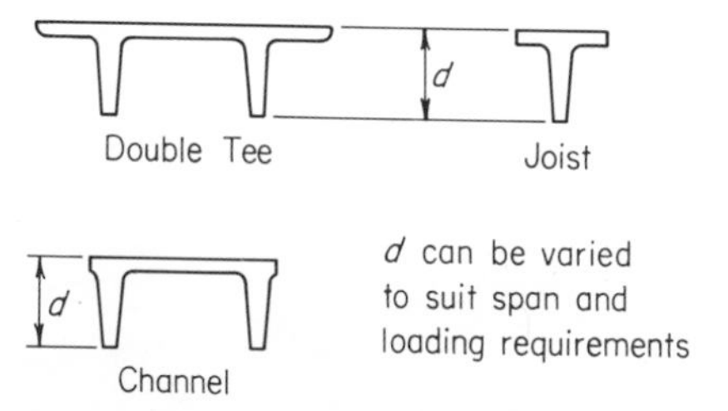

FIG. 7-6. Sections which can be cast in a standard double-T form.

FIG. 7-7. Typical reel-less pak 50-in. outside diameter, 27-in. inside diameter, 30-in. wide, contains approximately 6,000 lb of strand. Pak illustrated here is wrapped on a ½-in. thick cardboard core, has corrugated cardboard sides and waterproof burlap outer wrap. (*Courtesy of CF&I Steel Corporation.*)

carded because the cost of return freight and repairs is greater than the cost of new reels.

The reel-less pak, illustrated in Fig. 7-7, was developed for use in conjunction with steel take-apart reels to eliminate the expensive wooden reels which are discarded after one use. At the casting yard a reel-less pak is placed on a take-apart reel (Fig. 7-8), steel bands on the pak are cut, the outer paper is removed, and the strand can then be pulled off in the same manner as with a wooden reel. Except for six steel bands and a cardboard core there is no waste involved in this procedure. When the reel is empty, another pak of strand is placed on it.

As the strand is pulled from a reel into a casting bed the entire mass of strand and reel is started rotating, and at the end of the pull the rotation must be stopped. This procedure involves a considerable waste of energy and requires power equipment or at least two men. An ingenious casting-yard man eliminated this problem. He built a cage around a reel-less pak, cut the bands, and pulled the strand from a hole in the side of the cage as if it were a ball of twine.

FIG. 7-8. Steel take-apart reel. One piece consists of one side and the drum permanently attached to each other. After the reel-less pak is placed on the drum the other side is set in place and attached by tightening three bolts. (*Courtesy of CF&I Steel Corporation.*)

Figure 7-9 shows a pak rack, which has become the most popular method of dispensing strand. A reel-less pak, made without a cardboard core when used in this manner, is placed in the rack, the bands are cut, and the strand is pulled through the large-diameter ring in the side of the rack. One man can easily pull out strand in this manner. Only the weight of the strand going into the bed is involved; there is no inertia to overcome at the beginning and end of the operation.

A length of strand is left outside the steel bands of a reel-less pak as shown in Fig. 7-7. This tag which is about 12 ft long and continuous with the strand in the pak is available for test samples. Many state highway departments and some other organizations require samples taken by their own representatives after the material is delivered to the casting yard. The length outside the band is cut for these samples. If there were no tag it would be necessary to place the pak on a reel or in a rack and cut the bands to get a sample. In this case the take-apart reels or pak racks would be out of use until the strand was tested and approved for use.

Reasonable care should be exercised in handling and storing strand to avoid mechanical damage, corrosion, excessive rust, and injurious temperatures. When strand must be kept for any length of time, it should be stored indoors if possible. Strand stored outdoors should be supported a few inches above ground on timbers and covered with tarpaulins.

Strands, being made of high-carbon steel, are susceptible to mechanical damage. A nick or kink can be enough of a stress raiser to cause failure

FIG. 7-9. Reel-less pak in pak rack. Bands will be cut and strand pulled through large-diameter ring in side of rack. Note that paks intended for pulling from center are shipped without cardboard core. (*Courtesy of CF&I Steel Corporation.*)

when the strand is tensioned to the high stresses normally used. Mechanical damage occurs frequently when the strand has been pulled into the bed but not tensioned. It is caused by the edges of carelessly handled steel tools, wheelbarrow wheels, etc. Damage from vibrators during placing of concrete does not seem to be a problem.

Improper handling of welding equipment can be a cause of strand failure. A single drop of molten weld metal on a strand will raise the temperature of several wires to the point where the fibers return to crystals, thus losing more than half their strength. The same change in structure will result when an electric arc jumps to or from the strand. If the strand is under tension when the high temperature occurs, failure will be instantaneous. If it is not under tension, the change in structure will still take place and failure will occur during the tensioning operation.

The foregoing comments may give the impression that prestressed concrete strands are extremely delicate. This is not the case. Properly organized casting yards seldom have trouble with strand damage, but the problems connected with handling these strands must be pointed out for the benefit of new operators.

7-5. Placing and Tensioning Strands. Reels or paks of strand are placed at one end of the casting bed, and strands are pulled to the far end of the bed. Some stands have as many reels or paks as there are strands in the member to be cast so that all the strands can be pulled into the bed in one operation. With smaller stands it is necessary to make two or more passes to get the required number of strands into the bed. Fig. 7-3 shows one type of reel stand. When a casting yard has several beds, the stand is sometimes set up so that it can serve more than one bed.

When the proper length has been pulled into the bed, the strand is cut from that remaining on the reel. The cut is made with a cutting torch or with a shear. The high temperature needed to sever the strand with welding equipment returns the fibers to a crystalline structure, but only a few inches of strand are damaged, and the damaged portion will not be under tension.

A thick steel plate which serves as a combination anchor plate and template is provided at each end of the bed. The plate has holes, slightly larger than the strand diameter, which are spaced in the pattern the strands are to have in the concrete member. At one end of the bed each strand is threaded through its hole in the template and anchored on the far side by a temporary grip. Typical grips are shown in Figs. 7-10 and 7-11. When the concrete member is completed and removed from the bed, the grips are recovered for reuse.

The object of the tensioning operation is to stretch the strands and anchor them to end posts at the specified initial tension which is usually 70 per cent of their guaranteed minimum ultimate strength. Load is measured by two methods: elongation of the strand during tensioning; and load on the jacking unit as measured by its hydraulic or electric gage at the end of the tensioning operation. Most specifications are satisfied if the results of measurements by the two methods check each other within 5 per cent.

Pulling a strand hand-tight in a casting bed before starting the tensioning operation will not eliminate all of its slack. For this reason many specifications require that each strand be tensioned individually to a low tension—1,000, 1,500, or 2,000 lb—as measured by a dynamometer and that its elongation be measured from this point to full tension. Elongation from low tension to full tension is determined from the typical curve (Fig. 6-8) for the size and grade of strand being used. Be sure to use the typical curve prepared by the strand manufacturer who furnished the strand being used.

Since a single-strand jack must be attached to the strand to bring it to the specified low tension, many casting yards simply stop a few seconds at this point to set a reference mark and then continue to tension each strand individually to full load. Some single-strand jacks have a built-in scale that measures in inches and a pointer that moves along the scale measuring strand elongation as the jack is extended. Most jacks are hydraulic and are subjected to periodic calibrations in conjunction with

FIG. 7-10. Supreme Chuck (anchor grip) in place on a seven-wire strand. Made by Supreme Products Corp., Chicago, Illinois.

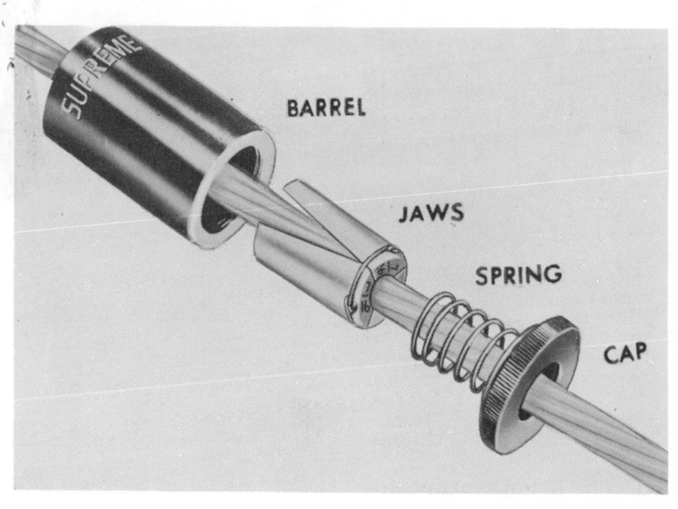

FIG. 7-11. Exploded view of Supreme Chuck showing case (barrel), wedges (jaws), spring, and cap.

their pressure gages so that the gage reading can be translated into pounds of tension on the strand. It is the experience of the writers that determining tension by means of elongation is more consistently accurate than is determining it by oil pressure in a jack.

As the load in the strand is taken up by the anchor chuck (Figs. 7-10 and 7-11) the jaws seat themselves in the barrel, and there is a motion of the strand with respect to the chuck barrel and the anchor plate. This motion is small—never more than ⅛ in. per chuck—compared with the total elongation which will be 22.5 to 24.5 in. for a 300-ft-long bed. It can be compensated for by measuring the motion of strand with respect to the chuck at each end of the bed for several tensioning operations and then increasing the required elongation by the amount determined.

Some casting yards pull each strand individually to the specified low tension and anchor it to a multiple-strand cross head. This cross head (Fig. 7-12) is then jacked to the required elongation tensioning all strands simultaneously.

FIG. 7-12. Jacking end of a universal casting bed in plant of Schuylkill Products, Inc., Cresona, Pennsylvania.

The jacking arrangement shown in Fig. 7-12 is well designed for a universal bed. On a universal bed the elevation of the center of gravity of the strand group will vary as different members are fabricated. With some jacking units the jacks would have to be raised or lowered to match the c.g.s. in each new member. With this unit the top jack is well above the highest possible elevation of the c.g.s., and the bottom jack is well below the lowest possible elevation. When the c.g.s. is not exactly halfway between the two jacks, there will be a difference in the oil pressure required in each, but it will not be too great because the space between the jacks is large in comparison with the amount the c.g.s. can move. The problem of the difference in oil pressure is eliminated by using a special pump which supplies an equal volume of oil to each of the four jacks regardless of pressure. Since each jack gets the same volume of oil and oil is noncompressible, all the jacks have equal runout.

Some designers feel that overtensioning is necessary to reduce stress losses from relaxation. Overtensioning is tensioning to a load greater than the specified initial tension and holding for a few minutes before relaxing to initial tension. This procedure was originally developed in Europe for wire which had a 12 per cent stress loss when tensioned to 57 per cent of its ultimate. When this wire was overtensioned to 63.5 per cent of its ultimate for 2 min and then relaxed to 57 per cent of ultimate as an initial tension, its stress loss was cut to 4 per cent. Such an operation was obviously worth the trouble. Stress-relieving replaces and improves upon overtensioning and makes it feasible to use strand at initial tensions up to 70 per cent of ultimate strength without excessive stress loss.

Infrequently a single wire in a seven-wire strand will break during or after the pretensioning operation. Unless other wires are damaged too, by exposure of the strand to excessive heat or electric arc, it is seldom necessary to delay production by replacing the strand. In many members the designer does not need all the strands shown in the details. If his calculations call for 9.63 strands, he will, of course, show 10 strands on the drawing. When one wire breaks in a group of 10 strands, it represents one-seventh of one strand, which leaves $9\frac{6}{7}$, or 9.86, strands intact. Since the operation of replacing a strand often represents a considerable expense, it should not be required unless necessary for the stability of the structure. Current specifications recognize the possibility of occasional single-wire breaks and permit them as long as they do not exceed one wire per strand or 2 per cent of the total area of prestressing steel in the member. Specific references are Sec. 2621(d) of ACI 318-63, Building Code Requirements for Reinforced Concrete, and "Wire Failures in Prestressing Tendons" on page 16 of Tentative Standards for Prestressed Concrete Piles, Slabs, I-Beams and Box Beams for Bridges published by the AASHO Committee on Bridges and Structures.

The engineer may find it desirable to use his judgment in establishing initial tension when elongation and jack readings do not coincide within the accuracy given in the specification. In general the elongation of the seven-wire strands gives the most consistent results. One quick check that often definitely locates the source of error is to compute the E_s of the strand using its measured elongation and the corresponding load read from the jack. If E_s is greater than 29,000,000, we know that the jack reading is too high because the wire from which the strand is made has an E of 29,000,000 and the E of a seven-wire strand is necessarily less than that of the wire from which it is made. The converse is not necessarily true, because improper stranding techniques can produce a strand with a low modulus.

7-6. Deflected Strands. In Chap. 1 it was shown that a member with deflected tendons could be designed to carry a larger live-load moment than a like member with straight tendons and that the magnitude of the additional moment capacity was equal to the dead-load moment of the prestressed member. The advantages of deflected tendons are twofold. If a given member with straight tendons needs added capacity equal to its own dead-weight moment, this capacity can be gained by using deflected tendons without changing the original concrete section. If the additional capacity must be gained using straight tendons, a larger concrete section must be used. This not only uses more concrete, it creates additional dead weight requiring a still larger section to carry the extra

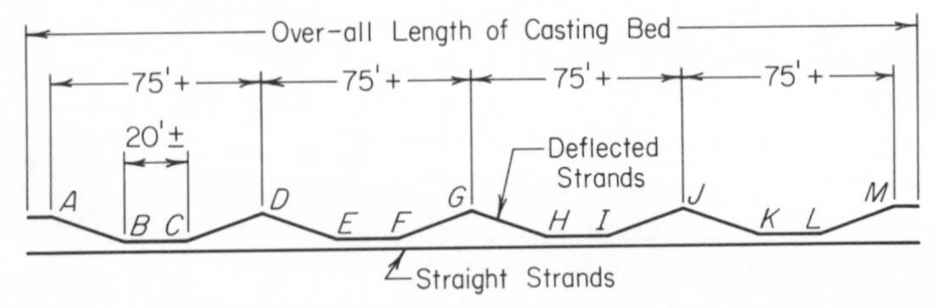

FIG. 7-13. Typical strand pattern for bed casting girders with deflected strands.

dead-weight moment. It is apparent, therefore, that even though deflected strands are more expensive to install than straight ones, they can show a net saving in the finished member when the dead-weight moment is appreciable.

Placing and tensioning deflected strands present problems which are not met in working with straight strands. On a typical 300-ft bed with four 75-ft members the strands would be held down at eight points and held up at three intermediate points and at each end. Figure 7-13 shows the approximate path the strands would follow. The desired location of the c.g.s. can usually be attained by tensioning 50 to 70 per cent of the total number of strands in a straight line and deflecting the remaining strands.

There are two methods of installing deflected strands, and each has its drawbacks. One is to place the strands in their final path under no tension and then tension them. The other is to tension the strands in a straight line and then deflect them up or down to their final position.

When the strands are placed in their final position and then tensioned, the problem is friction. If the deflected strands in Fig. 7-13 are tensioned from one end, they must be pulled through 12 deflection points each of which is a source of loss in tension due to friction. Even if the strands are tensioned from both ends at one time, the tension at the middle of the bed is less than that at the jacks by the friction losses at 6 deflection points. In Fig. 7-13 points *A*, *D*, *G*, *J*, and *M* at which the strands are supported are located beyond the ends of the concrete members. Since these points are outside the concrete, the supports are reusable and it is feasible to use permanent roller bearings under the strands to minimize friction losses. The hold-down devices at points *B*, *C*, *E*, *F*, *H*, *I*, *K*, and *L* become a permanent part of the precast member. In many cases the device shown in Fig. 7-14, although not frictionless, gives usable performance at reasonable cost. When friction must be almost completely eliminated, roller bearings can be used at all hold-down as well as support points. After tensioning and before pouring concrete, the roller bearings at the hold-down points can be replaced with less expensive devices like those in Fig. 7-14. Figure 7-15 shows details of tendons and reinforcing at the end of a bridge girder with deflected strands.

When the strands are tensioned in a straight line and then deflected, losses due to friction are greatly reduced. One method is to tension the strands in a straight line at the lowest elevation they will have in the finished girder. Hold-downs similar to those shown in Fig. 7-14 are assembled around the strands and attached to hold-down bolts. The strands are then raised at all support points to their final elevation. Raising the strands increases their length, which, in turn, increases their tension. The amount of this increase in tension is computed before the tensioning operation and deducted from the specified initial tension to establish the tension placed in the strands in their straight-line position.

Another approach is to tension the strands in a straight line at their highest elevation and then pull them down at deflection points. Figure 7-16 illustrates a method of deflecting strands by pulling them down.

7-7. Forms. Forms for pretensioned work are usually made of steel, which provides the durability needed to maintain accurate dimensions during constant reuse. Forms for standard shapes are available from fabricators who specialize in forms for prestressed concrete beds. Standard cross sections include I beams and box beams for bridges, plus single T's, double T's, ledger beams, and slabs for buildings. A single standard form can often be used to produce an entire series of sections. By the addition of fillers a standard form for an 8-ft-wide by 36-in.-deep single T can turn out single T's varying in depth from 12 to 36 in. in 4-in. increments. It can also produce a 6-ft-wide single T in any of these depths.

In rectangular prestressed concrete sections, concrete near the neutral axis of the section is undesirable because it not only adds unnecessarily to the volume of concrete used and to the dead weight of the member, it absorbs a lot of the prestressing force without appreciably increasing the load-carrying capacity of the member. This undesirable concrete is

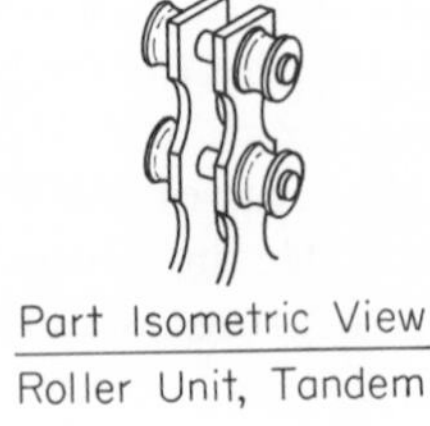

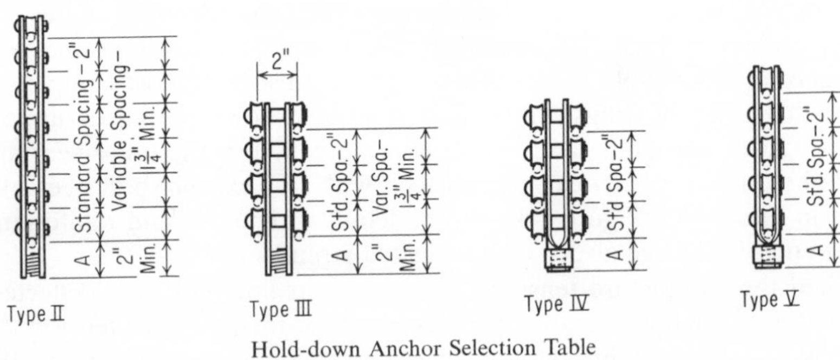

Hold-down Anchor Selection Table

Design data	Type II	Type III	Type IV	Type V
Recommended working loads (vertical components of force from deflected strands):				
Per strand—not to exceed, lb	3,000	3,000	3,000	6,500
Per complete unit—not to exceed, lb	18,000	18,000	30,000	30,000
Ultimate capacity of unit or attachment bolt, approximately, lb	27,000	27,000	45,000	45,000
Number of strands per unit: Normal maximum (at maximum recommended working loads)*	6	6	10	4
Strand spacing:				
Horizontal strand spacing, in.	.....	2	2	.....
Minimum vertical spacing:				
1. Center line of bottom strand to base of unit, in., see dimension *A*	2	2	2	2
2. Strands above bottom strand, in.	1¾	1¾	1¾	1¾
Normal vertical spacing, in.	2	2	2	2
Maximum vertical spacing	As required	As required	As required	As required
Attachment bolt diameter: contour-threaded high-strength coil bolt—length as required, in.	¾	¾	1 or 1¼, specify	1 or 1¼, specify

* If actual working loads per strand are less than maximum recommended working loads, divide working capacity of unit by actual load per strand to determine maximum permissible number of strands per unit.

FIG. 7-14. One type of hold-down for groups of seven-wire strands as made by Superior Concrete Accessories, Inc., Franklin Park, Illinois. A bolt projects up through the bottom of the form to hold the unit down.

eliminated by coring holes in the section. Figures 7-17 and 7-18 show a hollow-box bridge beam in which the hole was cored by casting in a paper tube. When ordered in quantity, paper tubes can be obtained in practically any size and shape.

Round holes can be cored with inflated tubes of heavy rubber. After the concrete has set, the tubes are deflated and withdrawn.

FIG. 7-15. Details at end of bridge girder with deflected strands at plant of Consumers Company, Chicago, Illinois. The sloping strands were placed in their deflected path and then tensioned. Support with roller bearings will be outside concrete beam on left of end form plate. Tin cans core holes in web for transverse steel in diaphragm. Six loops of seven-wire strand form lifting hook. Note that this member is an I section for its full length. There will be no end block.

Cored flat slabs cast in one continuous piece for the full length of the bed are made by the extrusion process. Strands are placed and tensioned and then the concrete placing equipment moves slowly along the bed while extruding concrete around them. The finished slab has smooth top and bottom surfaces, cored holes and sides which are usually grooved to take a dry pack material to provide shear transfer between the slabs. After the concrete has cured, the slab is saw-cut into desired lengths.

7-8. Completing the Member. When the strands are in place and tensioned, reinforcing bar cages, wire mesh, lifting hooks, inserts, etc., and forms are assembled in their places. Concrete is placed by using vibrators, often both internal and external, to ensure proper compaction of the low slump mix.

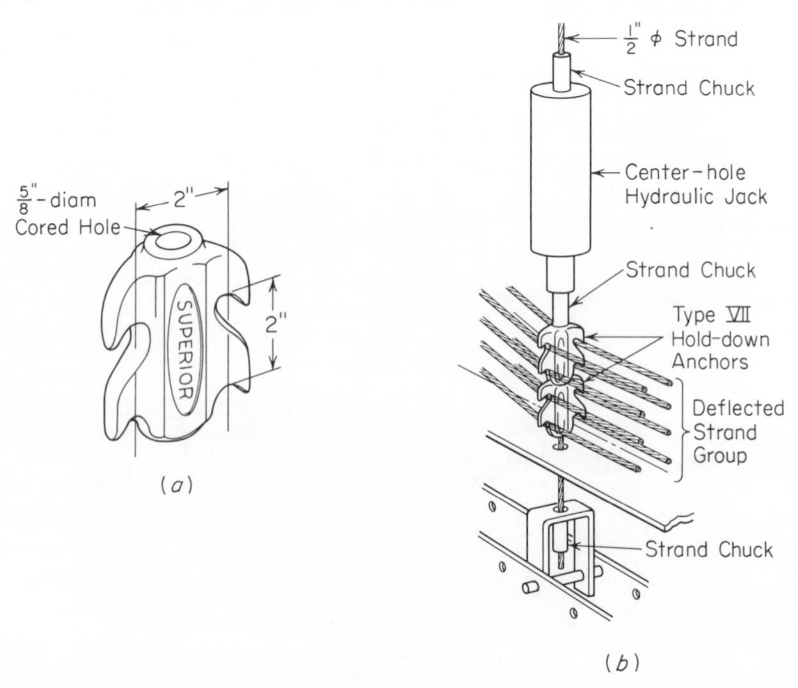

FIG. 7-16. (*a*) Malleable iron casting for holding down four strands. Curvature of deflecting surface on casting is equivalent to that provided by a 2-in.-diameter pin. Safe load vertical component is 3,000 lb per strand. Total load in the vertical hold-down strand should not exceed 50 per cent of its ultimate strength. (*b*) A short length of strand is anchored to the bed and threaded through one or more hold-down castings, a strand chuck, a center-hole hydraulic jack, and another strand chuck. As the jack is operated it reacts against the top chuck and forces downward the center-strand chuck, the hold-down anchors, and the strands. When the strands reach proper elevation, jack pressure is released and the hold-down castings are held in place by the center-strand chuck. The jack and top-strand chuck are removed. After concrete has been poured and cured, the hold-down strand is burned through at the under side of the beam and the bottom chuck reclaimed. Materials left in the concrete are one strand chuck, the anchor castings, and a short length of strand. (*Courtesy Superior Concrete Accessories, Inc.*)

Where live steam is used for accelerated curing, it is applied to the members through one or more pipes running the full length of the bed. An enclosure is placed around the concrete member and the steam pipe. Steam escapes through holes spaced along the pipe, filling the enclosure around the member, both raising the temperature and keeping it moist. On some beds pipes are placed in or around the forms to carry hot oil, hot water, or steam. These pipes raise the temperature of the concrete members through radiation. A steam pipe can be seen in Fig. 7-18. Rate of strength gain obtained by steam curing is shown by the chart in Fig. 6-2.

Since steam curing has been used for years to accelerate the curing of concrete and standard procedures for its application have been established, it will not be discussed at length here. Steam curing is not a simple matter of allowing some steam to flow around a concrete member. Time lapse between pouring concrete and turning on of steam, rate of temperature increase, maximum temperature reached, etc., must be determined to suit the conditions and then properly controlled. Improperly applied steam curing can damage the concrete in a member so that it will never reach its design strength. Reference to specifications and literature on this subject is recommended for those not already familiar with it.[9]

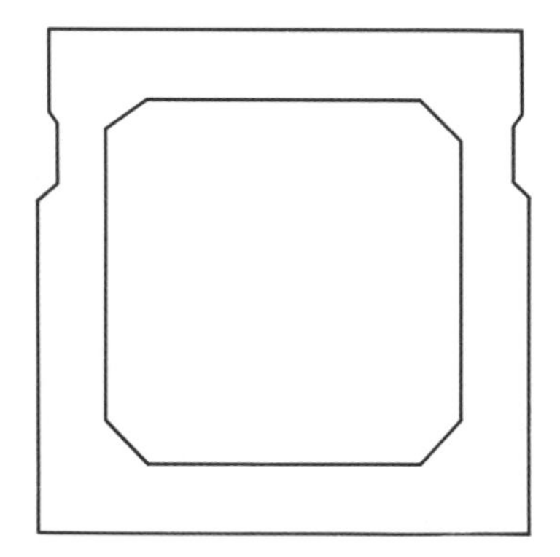

FIG. 7-17. Rectangular section with hole cored by specially shaped corrugated-paper tube.

One of the big factors which makes pretensioned bonded prestressed concrete members so economical is the fact that a well-organized casting yard can be run on a production-line schedule. A bed casting small standard members such as double T's, joists, channels, or piles can be operated on a 24-hr cycle when curing is accelerated by steam or radiant heat. Larger and more complicated members are usually produced in a 48-hr cycle or more.

7-9. Removal from Bed. When the concrete has cured to the required strength and load in the strands is transferred from anchor posts to the concrete, elastic shortening of the members takes place. This means that there is longitudinal motion of the member with respect to the bed. Wherever possible the cross section of the member is designed so that the forms will not restrict this motion. Other types of forms are usually removed before the strand load is transferred.

There are two methods of releasing strand load from anchor posts into the concrete: unjacking the strands and cutting the strands. Unjacking is practical only on beds which are equipped for tensioning, or detensioning, the entire group of strands in one operation. It is not practical for beds which tension one strand at a time.

Cutting strands is the most prevalent method of detensioning a bed. It is usually done with an acetylene torch, although a portable power cutter has also been used. The sequence in which strands are cut is important.

FIG. 7-18. Fabricating rectangular section at plant of Shockey Bros., Inc., Winchester, Va. Concrete has been poured to level of bottom of paper tube. Tube is being placed and will be held down while sides and top are poured. Pipe near bottom of form will supply steam for curing.

Cutting strands in improper sequence can develop critical stresses in the concrete. When a group of strands is cut between an end beam and an anchor plate, the same group of strands should be cut between all beams and between the other end beam and its anchor plate.

Consider first the procedure for cutting strands in a member which has all straight strands. Strands are cut in small groups, each group usually representing 20 to 30 per cent of the total number of strands. The strands which make up the first group, and each succeeding group, are selected so that cutting them will not cause excessive stresses at any point. It is desirable for the strands of any one group to make a symmetrical pattern about the vertical axis of the concrete member so that no warping will be induced.

Since each strand is under high tension when it is cut, it parts with a snap and its entire load is applied suddenly to the concrete. For a number of years it was assumed that this sudden application was detrimental to bond; therefore, some specifications required a gradual release rather than cutting. Comprehensive tests have now shown that the pattern of load transfer at a cut end is different from that at an end where release was gradual but complete load transfer was achieved in the same distance on each end. As a result cutting of strands is universally accepted and is the most common procedure for detensioning.

When each group of strands is cut its load is applied to the concrete member which undergoes additional elastic shortening. As the concrete members shorten, the spaces between the members grow, and the remaining strands between the members are stretched to appreciably higher tensions. When these last, highly tensioned strands are cut they apply the greatest impact load to the concrete. Experience indicates that this does not have an injurious effect on bond, but if the strand is near a concrete face it can sometimes cause spalling. For this reason strands near the sides or top surface should be cut in one of the first groups, and the last group to be cut should be made up of strands that are away from the edges.

If a beam has some of its strands deflected, other factors also must be considered in establishing the cutting procedure. In addition to cutting the strands which prestress the beam, it is necessary to cut or otherwise release the members which tie the strands down to the bed at deflection points. When the tie downs are released, the vertical load which was resisted by the bed is transferred to the concrete beam as a concentrated upward force. This force causes a negative moment that will crack the top flange of the beam if it is applied before the beam is equipped to resist it. If strands are cut at one end of a beam only or at one end of a line of beams there will be horizontal motion of the beam with respect to the bed. If such horizontal motion occurs before the tie-downs are released there will be a heavy shearing force in the tie-down, which can damage it or the beam and also make it difficult to release the tie down.

In summary, the strand-cutting procedure should be designed to minimize horizontal motion of the concrete member with respect to the bed, delay release of tie-downs until the beam can withstand the vertical loads caused by their release, and eliminate the chance of cracking and spalling by cutting the strands nearest the edges of the beam first.

One procedure which has been found to give a minimum amount of difficulty in releasing a member that has both straight and deflected strands is presented herewith. Strands listed for cutting in one step should be cut at each end of each beam before proceeding to the next step.

Step 1. Cut one-half of the deflected strands
Step 2. Cut one-third of the straight strands
Step 3. Cut one-fourth of the remaining deflected strands
Step 4. Cut one-fourth of the remaining straight strands
Step 5. Release hold-down
Step 6. Cut next quarter of deflected strands
Step 7. Cut next quarter of straight strands
Step 8. Cut remaining deflected and straight strands one-fourth at a time.

After the strands have been detensioned and cut, the members are removed from the casting bed to a storage area. In all members designed to resist bending such as girders, beams, and double T's the strands are located eccentrically to produce the maximum allowable negative bending moment under the dead load of the member. When these members are handled, they must be picked at or near their normal support points to avoid creating additional negative moment. They are often transported on equipment like that used for poles. Symmetrically stressed members such as piles are lifted at points 21 per cent of their length from each end so that the positive moment at the center of the length is equal to the negative moment at the pickup points. Scrap lengths of seven-wire strand can be used for lifting loops, as shown in Fig. 7-15.

7-10. Economy. Pretensioned bonded prestressed concrete members are the most economical prestressed concrete members available when existing conditions are reasonably suited to their use. Ideal conditions for pretensioned members are:

1. The casting plant is located within economical hauling distance.
2. Members used are standard with the casting plant, or enough similar members are needed to justify a setup.
3. Size and weight of members are within capacity of hauling equipment.
4. The design is adapted to make use of the properties of prestressed concrete members rather than simply trying to substitute prestressed concrete for another type of member.

Prestressed concrete members show a lower first cost for many structures than other materials. In addition they offer greater durability, less maintenance, and fire resistance.

BIBLIOGRAPHY

1. Slutter, Roger G., and Carl E. Ekberg, Jr.: Static and Fatigue Tests on Prestressed Concrete Railway Slabs, *AREA Bull.* 544, June–July, 1958. Also printed as Special Report 6, Fritz Engineering Laboratory, Structural Concrete Division, Lehigh University, Bethlehem, Pa.
2. Ozell, A. M., and J. F. Diniz: Fatigue Tests of Prestressed Concrete Beams Pretensioned with ½ Inch Strands, *J. Prestressed Concrete Inst.,* June, 1958, pp. 79–88.
3. Ekberg, C. E., Jr.: The Characteristics of Prestressed Concrete under Repetitive Loading, *J. Prestressed Concrete Inst.,* December, 1956, pp. 7–16.
4. Ozell, A. M., and E. Ardaman: Fatigue Tests of Pretensioned Prestressed Beams, *J. Am. Concrete Inst.,* October, 1956, pp. 413–424.
5. Kaar, Paul H., Robert W. LaFraugh, and Mark A. Mass: Influence of Concrete Strength on Strand Transfer Length, *J. Prestressed Concrete Inst.,* October, 1963, pp. 47–67.
6. Hanson, Norman W., and Paul H. Karr: Flexural Bond Tests of Pretensioned Prestressed Beams, *J. Am. Concrete Inst.,* January, 1959, pp. 783–802.
7. PCI Standards for Prestressed Concrete Plants, *J. Prestressed Concrete Inst.,* September, 1956, pp. 36–45, or *PCI-STD-103-58T.*
8. "Concrete Industries Yearbook," Pit & Quarry, Chicago, Ill., 1957, 1958, or later edition.
9. Schmid, Emil, and Raymond J. Schutz: Steam Curing, *J. Prestressed Concrete Inst.,* September, 1957, pp. 37–42.

CHAPTER 8

Post-tensioned Method

8-1. Basic Operation. Post-tensioning is defined as "a method of prestressing reinforced concrete in which the reinforcement is tensioned after the concrete has hardened." Basically, the complete operation has six steps:

1. The tendon is assembled in a flexible metal hose, and anchor fittings are attached to the ends of the tendon.
2. The tendon assembly is placed in the form and tied in place in the same manner as a reinforcing bar. Reinforcing bars, wire mesh, etc., are placed.
3. Concrete is poured and allowed to cure to the strength specified for tensioning.
4. Tendons are elongated by hydraulic jacks, and the anchor fittings are adjusted to hold the load in the tendons.
5. The space around the tendon is pumped full of cement grout under pressure.
6. Anchor fittings are covered with a protective coating.

Although the foregoing procedure is the most common, others are used to suit various conditions. In some cases a hole is cored in the concrete and the tendon is threaded through the hole just before it is to be tensioned. Holes can be cored by casting in a rubber tube of the desired shape and then withdrawing it after the concrete has set. Holes can also be cored by casting in a flexible metal hose. The hose becomes a permanent part of the structure. Since the hose is not stiff enough to maintain its position while the concrete is placed, one or more steel bars are placed inside the hose and are withdrawn after the concrete has set.

Another departure from the foregoing is the use of ungrouted tendons. These tendons are coated with a thick corrosion-preventing greaselike material and then wrapped with waterproof paper or encased in plastic tubing. The coating minimizes friction during the tensioning operation. Since these tendons are not bonded to the concrete after tensioning, the members in which they are used must be designed in accordance with the provisions that the specifications set forth for unbonded tendons.

In large hollow structures such as hollow-box bridges, the tendons are threaded through the hollow spaces and tensioned against anchor plates cast in the end block of the structure. Galvanized strands are used in these structures, and grouting is not required.

8-2. Combination of Pretensioned and Post-tensioned Methods. When the two methods are combined, pretensioned strands are tensioned in a straight line to provide as much of the prestressing force as possible and post-tensioned tendons are used in a deflected path to provide the remaining force and vary the location of the c.g.s. as required.

The pretensioned post-tensioned combination is used where some deflected tendons are needed and lack of facilities or other reasons prevent the economical use of deflected pretensioned strands. Under most conditions the combination of methods is more economical then an all-post-tensioned structure.

8-3. Systems. Several different systems or types of post-tensioned tendons are used in the United States. Procedure for fabricating a post-tensioned member is essentially the same with all systems except for the details of the tendons and their anchorages.

Most systems are patented to some degree, but there are seldom any royalty fees. Purchase of materials for a particular system from the patent holders who fabricate the parts includes permission to use the system.

Jacking equipment, grouting equipment, technical advice on use of the system, and any necessary field supervision are available from the suppliers of materials for the various systems.

8-4. The Post-tensioned Member. When the cross-sectional dimensions of a beam or girder are being established, they must provide adequate space for the tendons as well as sufficient section modulus to carry the design load.

In comparison with the load they carry, post-tensioned tendons are small in cross section and usually fit easily into the cross section of the concrete member. However, an end block is required at each end of the beam or girder so that the anchor fitting (or bearing plate under the fitting) can distribute its load into the concrete without causing excessive stresses.

Drawings of post-tensioned girders that are being put out for bid usually show details of the concrete and unprestressed reinforcing plus the magnitude and location of the prestressing force which must remain after all stress losses have taken place. The bidder is then permitted to offer whichever recognized system he prefers. Of course the designer should make sure that the girder and end blocks he uses provide adequate room for one and preferably two or more systems to encourage competitive bidding and keep his costs down. Data on minimum spacing between anchors, distance from anchorage to edge of beam, etc., are found in the catalogues of most tendon suppliers.

Unless the members are especially large or heavy, post-tensioning one or two girders on a job is seldom economical. Jacking and grouting equipment must be brought to the job site; this represents an appreciable expense unless it can be spread over a number of members.

8-5. Parallel-strand Cables. Each of these cables is composed of one to twelve seven-wire strands meeting ASTM Designation A 416. Both grades of strand are used, with ½-in. diameter being most popular.

Anchor fittings consist of two main parts, an external steel socket with a tapered hole in its center and a grooved tapered plug. The plug has one groove for each strand in the cable. The cable is tensioned by attaching the strands to grips in a hydraulic jacking unit which stretches them until they reach the specified load. A 100-ft cable will elongate 7.5 to 8 in. When the proper tension is reached, a separate plunger on the jacking unit thrusts the tapered plug into place and the load on the jack is released. Since each strand is wedged between the plug and the socket, it is held in place and its load is transferred to the socket and from the socket into the concrete. The projecting ends of the strands are burned off a short distance from the socket; the cable is pumped full of grout; and the ends and socket are encased in concrete, coated with bitumen or otherwise protected.

In some areas cables can be purchased with the strands cut to length, inserted in flexible metal hose, and coiled for shipment. For jobs of any size, however, it is usually cheaper for the contractor to purchase the component parts and fabricate his own cables at the job site. Anchor fittings and jacking equipment for the Anderson System are available from Concrete Technology Corporation in Tacoma, Washington, and for the Freyssinet System from Freyssinet Company, Inc., in New York, New York. Seven-wire strand is available from numerous manufacturers who supply it to pretensioned fabricators. Flexible metal hose is also available from numerous sources, and a hose specifically for use in prestressed concrete has been developed and is marketed by Flexico Products, Inc., of Metuchen, New Jersey.

Figures 8-1 to 8-9 illustrate details and use of parallel-strand cables.

8-6. Parallel-wire Cables. Each of these cables is composed of a number of single wires meeting ASTM Specification A 421. Anchor fittings are of two types, the button-head and the wedge type. In other respects, such as assembling in metal hose, tensioning, anchoring, and grouting, these cables are similar to the parallel-strand cables described in Sec. 8-5.

In the United States the standard wire for button-head cables is ¼ in. diameter and has a minimum ultimate strength of 240,000 psi. It is specified as ASTM A 421 type BA. With this system it is quite easy to make fittings for any number of wires, and cable sizes have varied from two wires per cable to over forty wires per cable.

The anchor fitting is a piece of steel with one hole through it for each

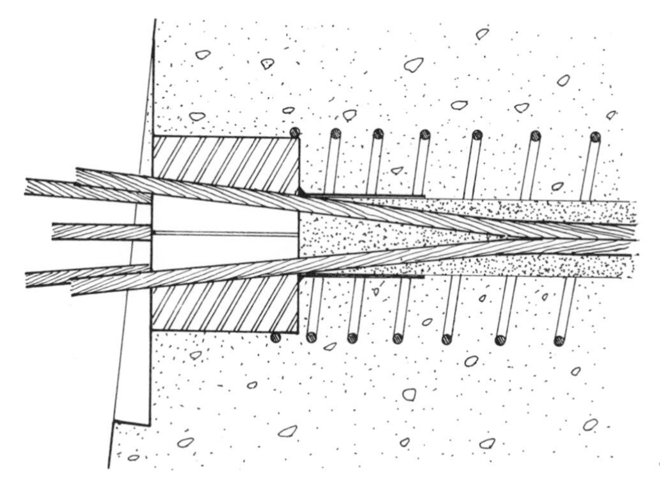

FIG. 8-1. Cross section through Anderson fitting for anchoring parallel seven-wire strands of a post-tensioned cable. Note the reinforcing spiral used to confine the concrete in the zone of high bearing pressure under the socket. (*Courtesy Concrete Technology Corporation.*)

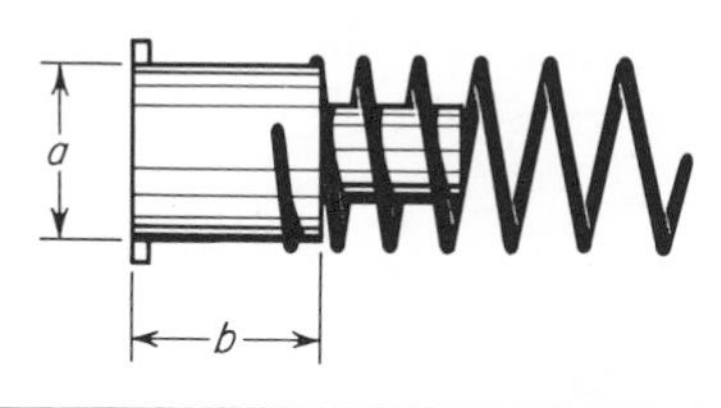

Dimensions *a*/*b*	2¾ in./3½ in.	3⅝ in./4 in.	5¼ in./4 in.	6¼ in./5 in.
Spiral	8 turns of ¼-in. ⌀ bar	12 turns of ¼-in. ⌀ bar	12 turns of ¼-in. ⌀ bar	12 turns of ⅜-in. ⌀ bar
Weight	5.25 lb	9.5 lb	24.5 lb	45 lb
Maximum design load	80 kips	170 kips	230 kips	300 kips

FIG. 8-2. Socket dimensions, Anderson System.

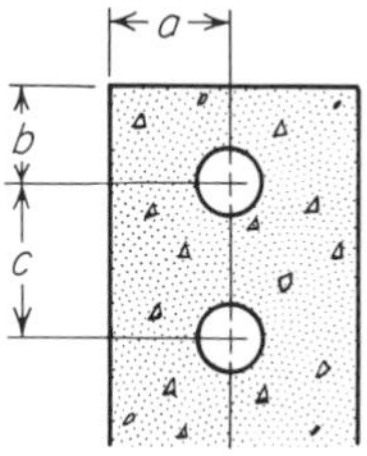

	80 K	170 K	230 K	300 K
a	2"	3"	4"	6"
b	3"	4"	4"	6"
c	8"	8"	8"	10"
d	8"	8"	8"	10"

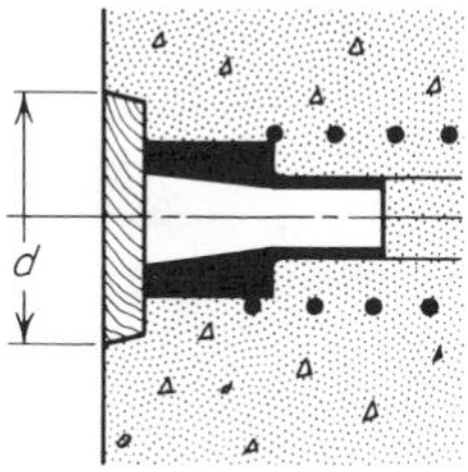

A tapered block should be provided to recess the anchor. A patch will seal the anchor to prevent corrosion and will give a clean, even appearance.

FIG. 8-3. End details recommended by the Anderson System.

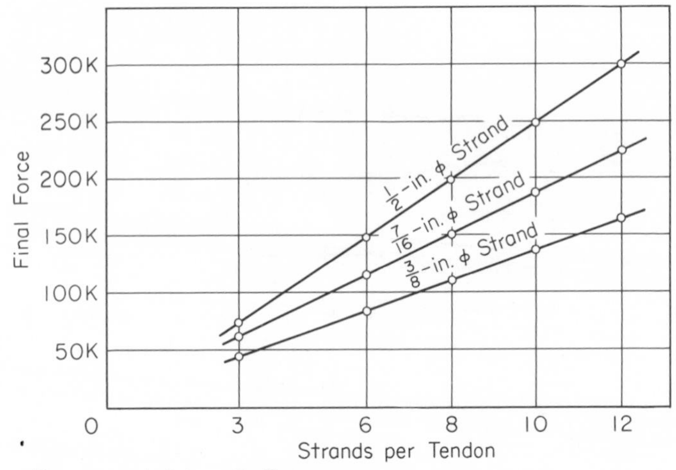

Recommended Length Ranges:
$\frac{3}{8}$-in. ϕ strand, 30 to 60 ft; $\frac{7}{16}$-in. ϕ strand, 40 to 100 ft;
$\frac{1}{2}$-in. ϕ strand, 60 ft. up.

FIG. 8-4. Tendon selection chart, Anderson System. (Losses due to friction between strands and enclosures during tensioning have not been subtracted from these values.)

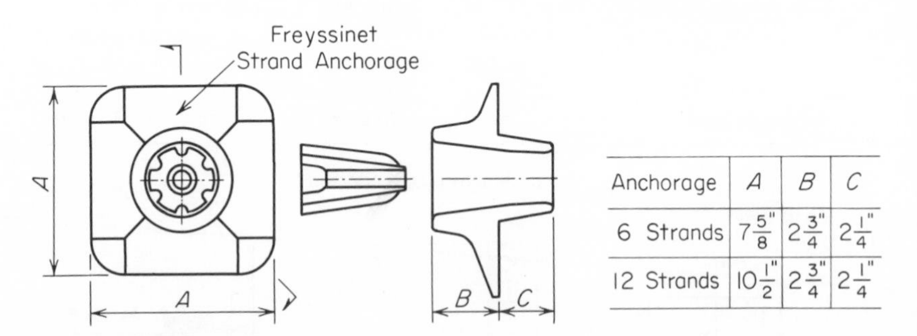

Anchorage	A	B	C
6 Strands	$7\frac{5}{8}''$	$2\frac{3}{4}''$	$2\frac{1}{4}''$
12 Strands	$10\frac{1}{2}''$	$2\frac{3}{4}''$	$2\frac{1}{4}''$

FIG. 8-5. Detail of Freyssinet anchor for cable composed of seven-wire strands. Plug shown will accommodate six strands.

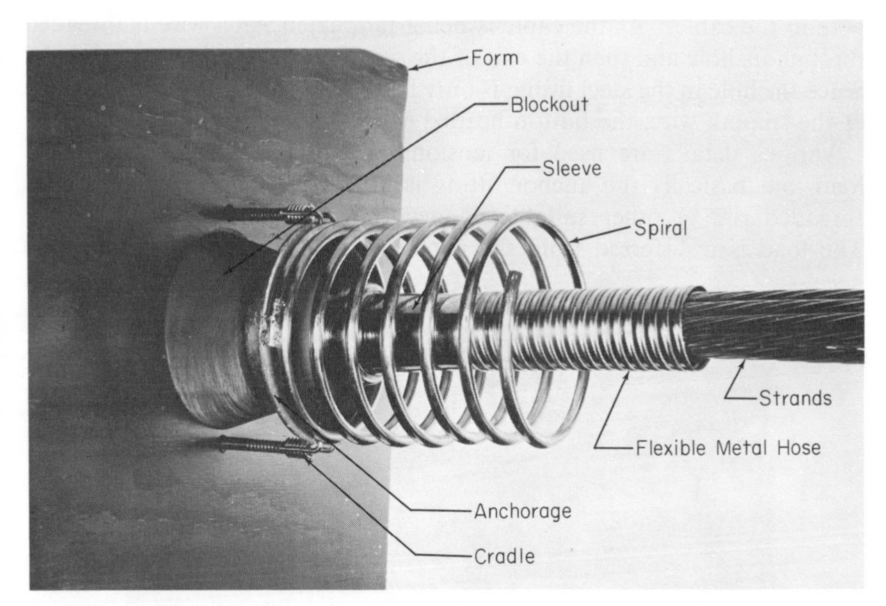

FIG. 8-6. Freyssinet anchorage assembly is held against form by cradle composed of two bolts and connectors. Sleeve from metal hose to anchorage keeps concrete from seeping through to strands during pouring of girder.

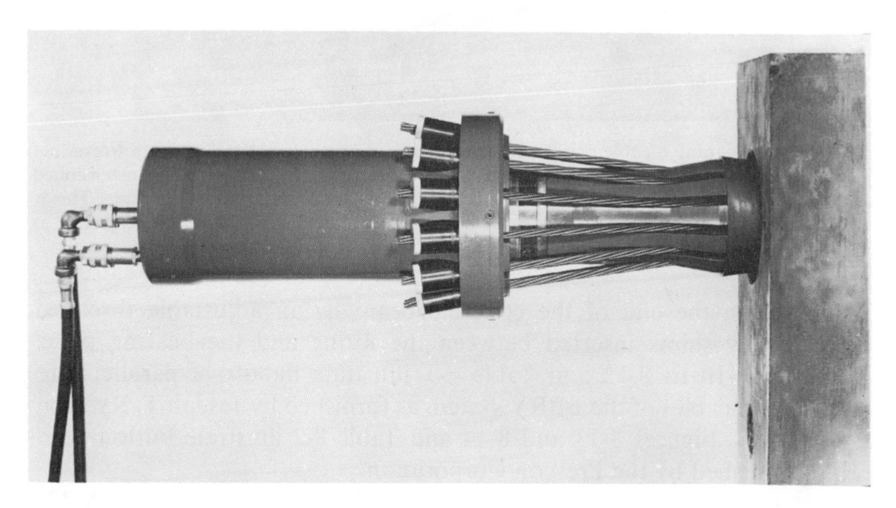

FIG. 8-7. Freyssinet jack for tensioning twelve ½-in.-diameter strands. End of jack bears directly on anchor fitting.

wire in the cable. As the cable is being fabricated, each wire is threaded through its hole and then the end of the wire is upset or "button-headed." Since the hole in the steel fitting is only large enough to permit the passage of the smooth wire, the button-headed end cannot pull through.

Various details are used for tensioning the cable and anchoring the load, but basically the anchor fitting is attached to a jacking unit by a threaded rod or other suitable connection and the cable is tensioned. The load is transferred from the anchor fitting to a steel bearing plate

FIG. 8-8. Tensioning a cable composed of twelve ½-in.-diameter strands using a Freyssinet jacking unit in plant of Nebraska Prestressed Concrete Co. Top cable has been tensioned and anchored. Strands in center cable attached to jack are nearing full elongation. This is a combination pretensioned-post-tensioned girder with the pretensioned strands projecting at the bottom of the girder.

embedded in the end of the concrete beam by an adjustable threaded device or by shims inserted between the fitting and the bearing plate.

Figures 8-10 to 8-12 and Table 8-1 illustrate details of parallel-wire button-head cables of the BBRV system as furnished by Joseph T. Ryerson & Son, Inc. Figures 8-13 and 8-14 and Table 8-2 illustrate button-head cables furnished by the Prescon Corporation.

Wire for cables with wedge-type anchors is specified as ASTM A 421 type WA. These cables and fittings are similar to the parallel-strand cables described in Sec. 8-5 except that they are made of a number of

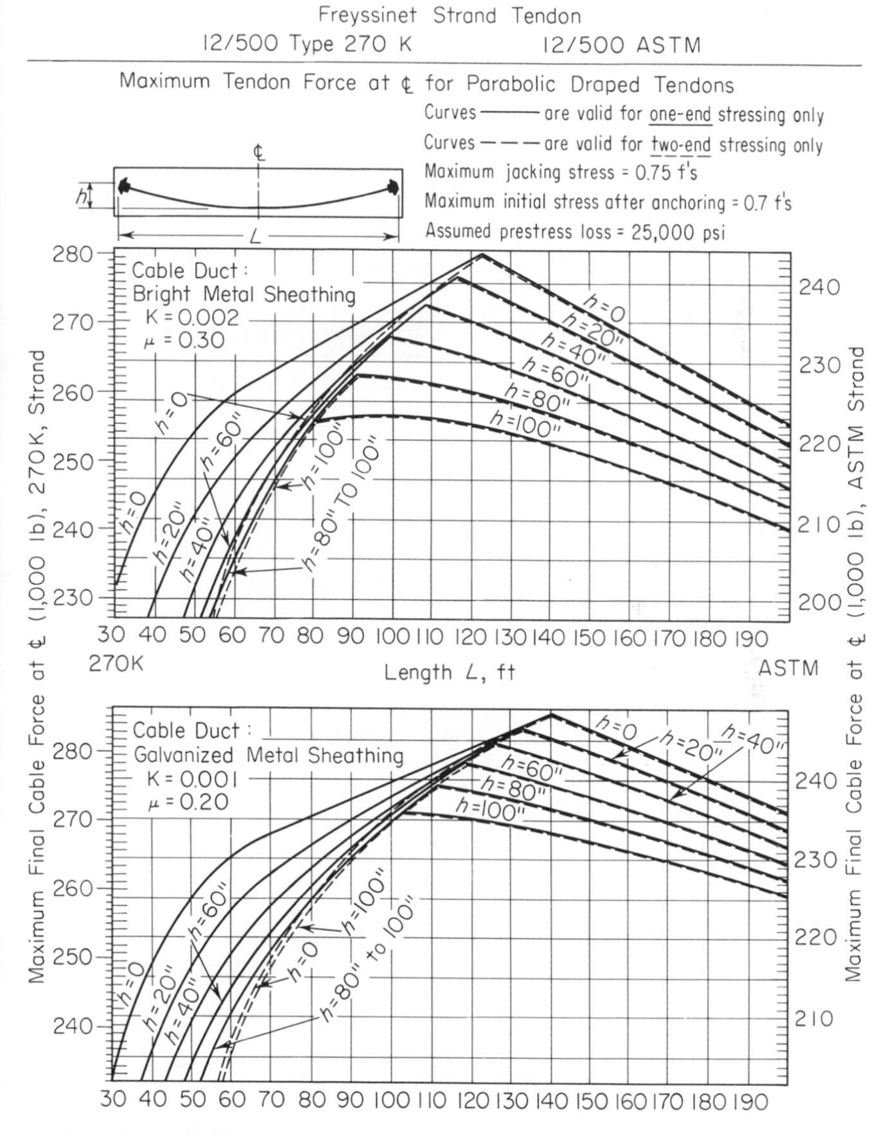

FIG. 8-9. Prestressing forces obtained with Freyssinet cables of twelve ½-in.-diameter strands, including effect of friction and seating of plugs. Forces obtained with fewer strands are in direct proportion to the number of strands. The force obtained with six strands would be exactly one-half of that shown here.

single wires instead of a number of strands. Anchor fittings and jacking equipment are furnished by Freyssinet Company, Inc. Up to 18 wires 0.196 in. diameter (ultimate strength, 250,000 psi) per cable and up to 12 wires 0.276 in. diameter (ultimate strength, 236,000 psi) per cable are standard sizes.

8-7. High-strength Bars. The familiar alloy, heat-treated, and other high-strength bars available from many warehouses are not suitable for prestressed concrete. High yield strength, followed by high elongation before failure (discussed in Sec. 6-6), is required in addition to high ultimate strength. Bars made by Stressteel Corporation are subjected to cold-working and stress-relieving treatments to give them these specific properties. Ultimate strength is about 145,000 psi. At present there is no ASTM specification; however, good specifications are available from the bar manufacturers, and the subject is also covered in Section 405(f) of ACI 318-63.

Table 8-1. Dimensions and Properties of BBRV Cables

(Forces shown are in accordance with maximum allowable stresses specified by Bureau of Public Roads and PCI Building Code Requirements)

Dimension data for bonded (grout-type) tendons

Type of tendon	14MM* or 14MF	28MM or 28MF	40MM or 40MF
Number of wires, diameter	14(¼ in.)	28(¼ in.)	40(¼ in.)
Base-plate size, in.	6¾ × 6¾	9¼ × 9¼	11 × 11
Trumpet OD, in.	4	5	5¾
Conduit OD, in.	1⅝	2⅛	2½

Stressing data for bonded (grout-type) tendons

Number of wires, diameter	1(¼ in.)	14(¼ in.)	28(¼ in.)	40(¼ in.)
Section of wires, sq in.	0.04909	0.687	1.3744	1.963
Final force—after losses, lb	7,070	98,980	197,960	283,000
Initial force—before losses, lb	8,250	115,500	231,000	330,000
Overstressing force, lb	9,420	131,880	263,760	377,000
Ultimate force of tendon, lb	11,780	164,920	329,840	471,260

* M—movable or stressing-end anchor; F—fixed-end anchor.
MM—tendon that can be stressed from both ends.
MF—tendon that can be stressed from one end only.

Bars are available in diameters of ¾ to 1⅜ in. Maximum length is slightly over 80 ft, but longer length can be achieved through the use of couplers. Common procedure is to encase the bar in a flexible metal hose which is pumped full of grout after the bar has been tensioned. It can also be threaded through a cored hole.

A steel bearing plate is provided at each end of the concrete member,

FIG. 8-10. Button-head wires are shown in lower right-hand corner of picture. Wires cut with square ends ready for button-heading are shown in upper left-hand corner of picture. (*Courtesy of Joseph T. Ryerson & Son, Inc.*)

and anchor fittings of either wedge type or threaded type transfer the load in the bar to the bearing plate which distributes it to the concrete.

Table 8-3 and Figs. 8-15 to 8-17 illustrate details of Stressteel high-strength bars.

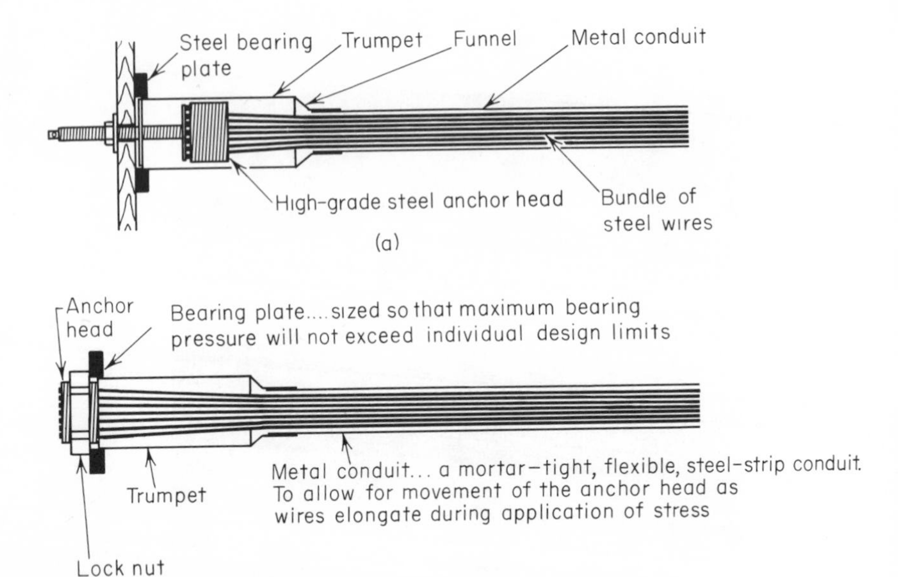

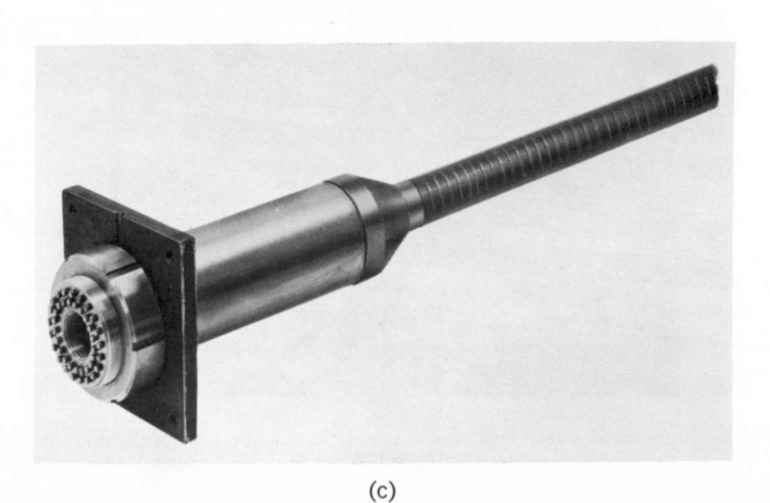

(c)

FIG. 8-11. BBRV movable anchor-head assembly for cable composed of button-head wires. (*a*) Before tensioning; (*b*) and (*c*) after tensioning. (*Courtesy of Joseph T. Ryerson & Son, Inc.*)

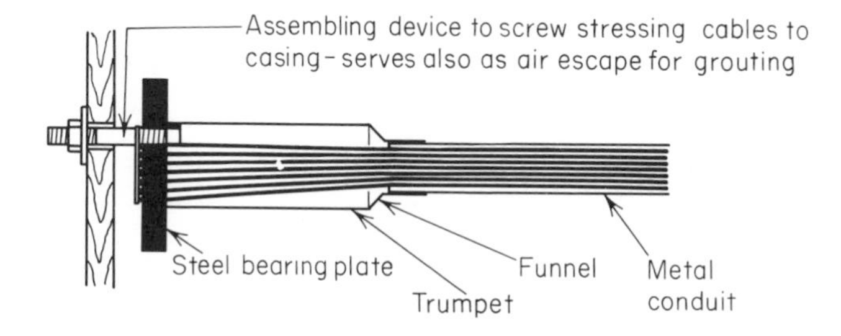

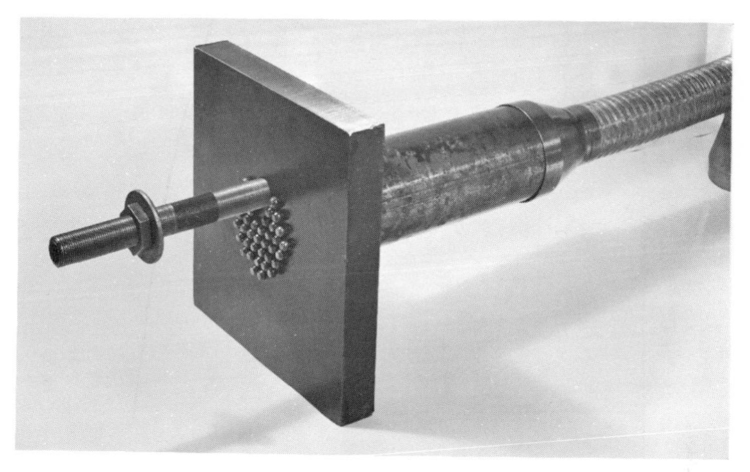

FIG. 8-12. BBRV fixed anchor-head assembly. (*Courtesy of Joseph T. Ryerson & Son, Inc.*)

FIG. 8-13. Standard Prescon tendon showing button-headed wires, stressing washer, shims, and bearing plate. (*Courtesy of the Prescon Corporation.*)

Table 8-2. Tendon Size Chart, Prescon System. (See Fig. 8-14 for meaning of letters.)

NOTE: Prescon prestressing tendons are identified as to number of wires; coated or grouted; and types of end anchorages

EXAMPLE: 6MSP; 6—Number of ¼-in. wires per tendon; M—mastic-coated, paper wrapped; SP—1 stressing end, 1 "dead-end" plate

10GSS; 10—Number of ¼-in. wires per tendon; G—encased in flexible metal hose; SS—2 stressing ends

No. 0.250 in. wires either M or G	Prestress force, kips 0.6f's	A	B	C	D	E*	F†	G Diameter coated tendon	H‡ Cond. size OD grouted tendon	J§	T¶	f'_c§
2	14.1	3	4	2	1¼	3	3¾	½	1	¾	⅝	3,000
3	21.2	3	4	2	1¼	3	4½	½	1	¾	⅝	3,000
4	28.3	3	4½	2	1¼	3½	5	⅝	1⅜	1¼	⅝	4,000
5	35.3	3	5½	2	1¼	4½	5½	¾	1⅜	1¼	⅝	4,000
6	42.4	4	6	2½	1¼	5	5½	¾	1⅜	1¼	¾	4,000
7	49.5	4	6	2½	1¼	5	6½	¾	1⅜	1¼	¾	4,000
8	56.6	4	7	2½	1¼	6	6¾	⅞	1⅝	1¼	¾	4,000
9	63.6	4	7½	3	1¼	6	6¾	1	1⅝	1¼	¾	4,000
10	70.7	4	8½	3	1¼	7	6¾	1	1⅝	1¼	¾	4,000
11	77.8	5	7½	3	1¼	6½	7	1⅛	1⅝	1¼	1	4,000
12	84.8	5	8	3	1¼	7	7	1⅛	1⅝	1½	1	4,000
13	91.9	6	7½	3	1¼	6½	7	1⅛	1⅝	1½	1	4,000
14	98.9	6	8½	3½	1¼	7	8¼	1⅛	1⅞	1½	1	4,000
15	106.0	6	8½	3½	1¼	7	8¼	1⅛	1⅞	1½	1	4,000
16	113.1	6	9½	3½	1¼	7	8¼	1¼	1⅞	1½	1	4,000
17	120.2	6	9½	3½	1¼	7	8¼	1¼	1⅞	1½	1	4,000
18	127.2	6	10	3½	1½	8½	8¼	1¼	2	1½	1	4,000
19	134.3	6	11	3½	1½	8½	8¼	1⅜	2	1½	1	4,000
20	141.4	6	11	3½	1½	8½	8¼	1⅜	2	1½	1	4,000

21	148.5	6	11½	4	1⅞	9	8¼	1⅜	2	1½	1	4,000
22	155.5	6	11½	4	1⅞	9	8¼	1⅜	2	1½	1	4,000
23	162.6	6	12	4	1⅞	9	8¼	1½	2	1½	1	4,000
24	169.7	6	12	4	1⅞	9	8¼	1½	2	1½	1	4,000
25	176.8	7	11½	4	1⅞	9	10	1⅝	2⅛	1½	1¼	5,000
26	183.8	7	11½	4	1⅞	9	10	1⅝	2⅛	1½	1¼	5,000
27	190.9	7	12	4	1⅞	9	10	1⅝	2⅛	1½	1¼	5,000
28	197.9	7	12	4	1⅞	9	10	1⅝	2⅛	1½	1¼	5,000
29	205.0	7	12½	4	1⅞	9	10	1¾	2⅛	1½	1¼	5,000
30	212.0	8	12	5¼	2¼	9	10¾	1¾	2⅛	1½	1¼	5,000

Tendons are readily available to 42 wires with larger tendons available on special order. Due to varying applications of larger tendons, terminal hardware will be custom designed to fit the application.

* Anchor holes are ⅝ in. for 14-wire tendons and larger, ½ in. for all others.

† Clear dimension required on one side only for inserting shims, measured in the direction of the long dimension of the plate.

‡ Inside diameter of tubing is ⅛ in. less than this value.

§ If either J or f'_c is less than the values shown on table, then bearing plates must be increased in size to prevent excess bearing stresses in concrete.

¶ 3-10 wire—mild steel; yield 33,000, tensile 60,000; 11-30 wire—high tensile steel; yield 59,000, tensile 90,000.

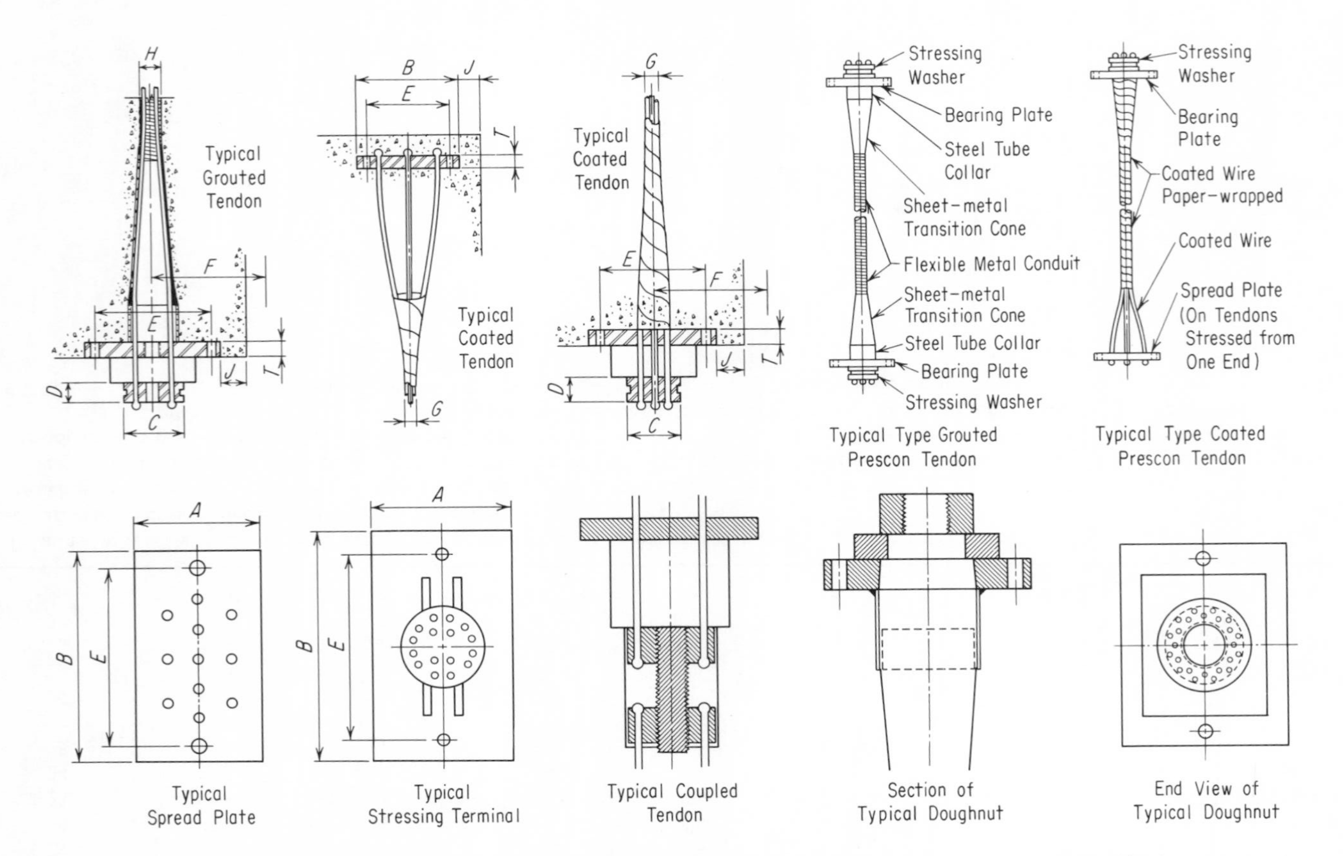

FIG. 8-14. Details of end hardware, Prescon System. See Table 8-2 for dimensions for various cable sizes.

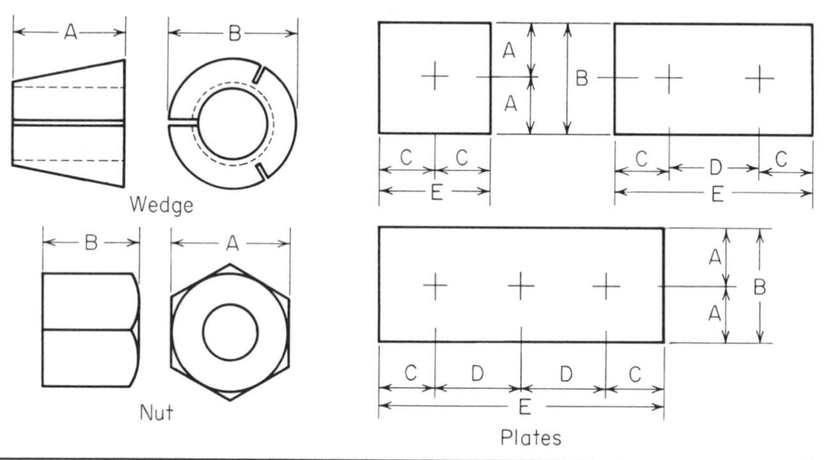

Wedges					Nuts			
Bar diameter, in.	Part No.	Dimensions, in.		Weight, lb each	Part No.	Dimensions, in.		Weight, lb each
		A	*B*			*A*	*B*	
¾	W-6	1³⁄₁₆	1¼	0.22	HN6	1⅜	1¼	0.5
⅞	W-7	1½	1½	0.32	HN7	1⅝	1⁷⁄₁₆	0.7
1	W-8	1½	1¾	0.50	HN8	1⅞	1⅝	1.0
1⅛	W-9	1¾	2	0.70	HN9	2⅛	1¹³⁄₁₆	1.5
1¼	W-10	2	2¼	0.80	HN10	2⅜	2	2.0
1⅜	W-11	2³⁄₁₆	2½	1.10	HN11	2⅜	2	2.0

PLATES

Bar diam, in.	Part No. WP, TP, or P*	No. of holes	Dimensions, in.						Weight lb each
			A	*B*	*C*	*D*	*E*	Thickness, in.	
¾	6	1	2	4	2	....	4	1	4.5
⅞	7	1	2¼	4½	2½	....	5	1½	9.5
1	8	1	2½	5	2¾	....	5½	1½	11.7
1⅛	9	1	3	6	3	....	6	1¾	17.8
1¼	10	1	3	6	3½	....	7	1¾	20.8
1⅜	11	1	3½	7	3¾	....	7½	2	29.7
2@ ¾	6-2	2	2	4	2	4	8	1	9.1
2@ ⅞	7-2	2	2½	5	2½	4	9	1½	19.0
2@1	8-2	2	2½	5	3	5	11	1½	23.4
2@1⅛	9-2	2	3	6	3¼	5	11½	1¾	34.2
2@1¼	10-2	2	3½	7	3¾	5	12½	1¾	43.4
2@1⅜	11-2	2	3½	7	4¼	6	14½	2	57.5
3@ ¾	6-3	3	2	4	2	4	12	1	13.6
3@ ⅞	7-3	3	2½	5	2½	4	13	1½	27.6
3@1	8-3	3	2½	5	3	5	16	1½	34.0
3@1⅛	9-3	3	3	6	3¾	5	17½	1¾	52.0
3@1¼	10-3	3	3½	7	3¾	5½	18½	1¾	64.2

* WP = wedge plate; TP = threaded plate; P = plate with drilled hole.

FIG. 8-15. Dimensions of wedges, nuts, and bearing plates for Stressteel bars.

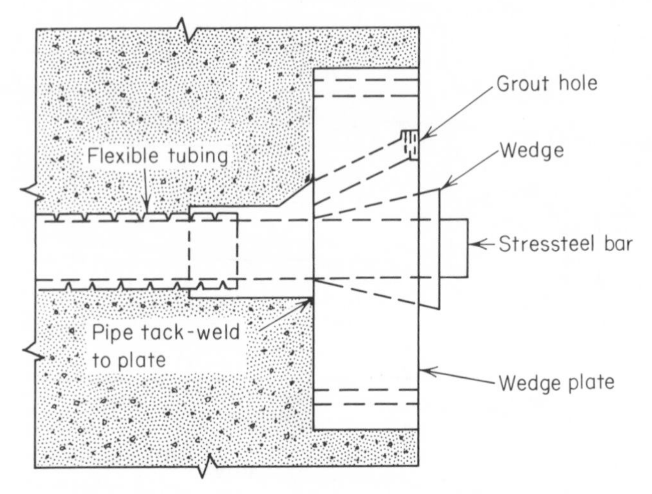

FIG. 8-16. Details of Stressteel bearing plate for wedge anchor.

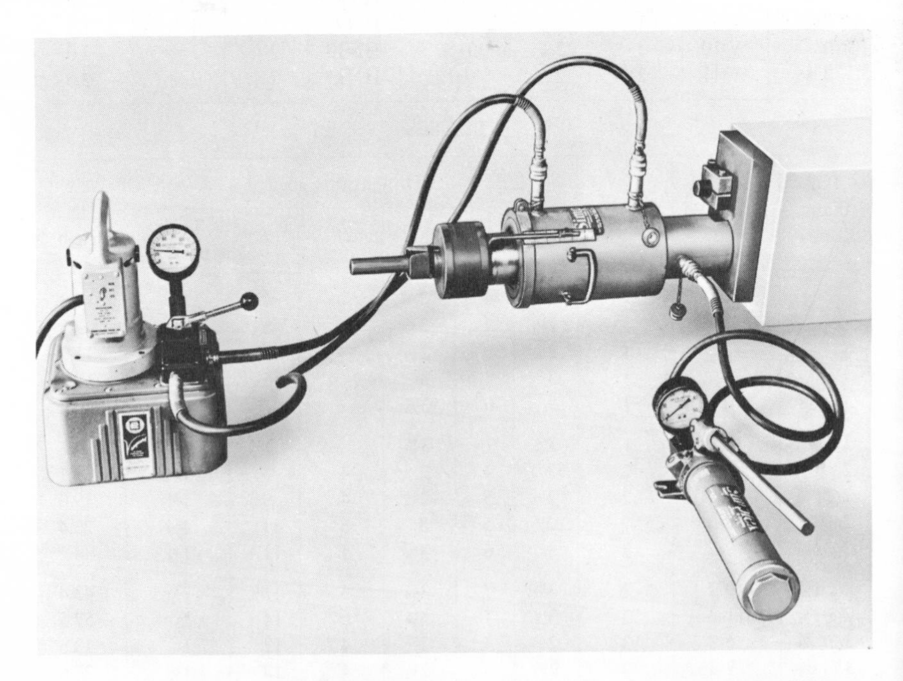

FIG. 8-17. Stressteel tensioning equipment. Power pump at left includes hydraulic gauge which measures tension in bar. Jack has built-on scale which measures elongation of bar during jacking operation. Hand pump operates plunger which sets wedge after bar is stretched to full load.

Table 8-3. Properties of Stressteel Bars

Bar size diameter, in.*	Weight, lb per lin ft	Area, sq in.	Minimum guaranteed ultimate strength		Initial tensioning load, 0.7 f'_s†		Final design load, 0.6 f'_s‡	
			Regular	Special	Regular	Special	Regular	Special
			(All values in units of 1,000 lb)					
¾	1.50	0.442	64.1	70.7	44.9	49.5	38.5	42.4
⅞	2.04	0.601	87.1	96.2	61.0	67.3	52.3	57.7
1	2.67	0.785	113.8	125.6	79.7	87.9	68.3	75.4
1⅛	3.38	0.994	144.1	159.0	100.9	111.3	86.5	95.4
1¼	4.17	1.227	177.9	196.3	124.5	137.4	106.7	117.8
1⅜	5.05	1.485	215.3	237.6	150.7	166.3	129.2	142.6

* ½- and ⅝-in.-diameter bars are available on special request.

† Losses due to creep, shrinkage, and plastic flow of concrete and steel relaxation should be deducted from this value. Overtension to $0.8f'_s$ is permitted to account for friction loss and/or wedge seating loss.

‡ Working stress in the steel should not exceed this value.

8-8. Large-diameter Strands. Galvanized strands are used for structures where the prestressing tendons are not encased in concrete. A typical example is a large hollow-box girder in which the tendons pass through the open area of the box and bear only against the diaphragms in the box and against the end blocks of the girder.

Each strand is composed of seven or more galvanized wires. Sizes run from ⅝ in. diameter to over 2 in. diameter. For example, the 1$\frac{11}{16}$-in.-diameter galvanized strand has over 60 wires and has an ultimate strength of 352,000 lb.

The strand is cut to required length in the shop, where threaded anchor fittings are attached to each end. When the strand is placed in the girder

Table 8-4. Properties of Galvanized Strands for Post-tensioning

Diameter in.	Weight per foot, lb	Area, sq in.	Minimum guaranteed ultimate strength, lb	Recommended final design load, lb
0.600	0.737	0.215	46,000	26,000
1	2.00	0.577	122,000	69,000
1⅛	2.61	0.751	156,000	90,000
1¼	3.22	0.931	192,000	112,000
1⅜	3.89	1.12	232,000	134,000
1½	4.70	1.36	276,000	163,000
1$\frac{9}{16}$	5.11	1.48	300,000	177,000
1⅝	5.52	1.60	324,000	192,000
1$\frac{11}{16}$	5.98	1.73	352,000	208,000

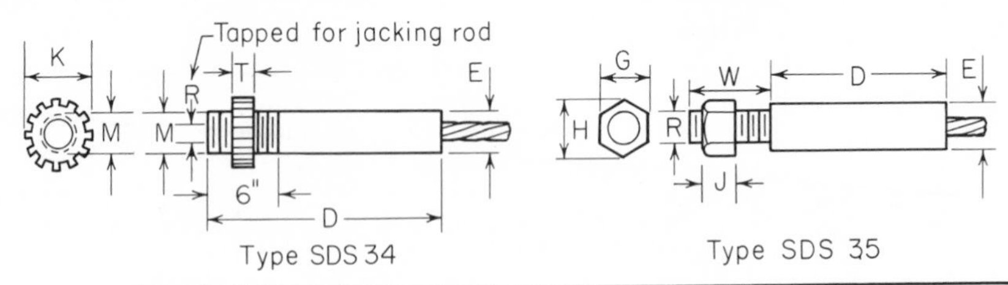

Strand diameter, in.	Measurements, in.										Total weight, lb	
	D	*W*	*E*	*M*	*R*	*G*	*H*	*J*	*K*	*T*	Type SDS34	Type SDS35
0.600	9½	8	1¹¹⁄₁₆	....	1¼–12N	1¹³⁄₁₆	2⅙	1³⁄₁₆	....	....	....	9¼
0.835	12½	10	2¼	12N	1⅝–12N	2¼	2⁹⁄₁₆	1⁵⁄₁₆	3¾	1	16½	21
1	13	11	2¾	8N	2 - 8N	3	3⁷⁄₁₆	1¾	4⅜	1¼	24¾	32½
1⅛	16½	11	3	8N	2¼- 8N	3	3⁷⁄₁₆	1¾	4¾	1½	39½	50

All SDS 34 and 35 fittings are proofloaded to a stress in excess of the recommended design stress after being attached to the strand.

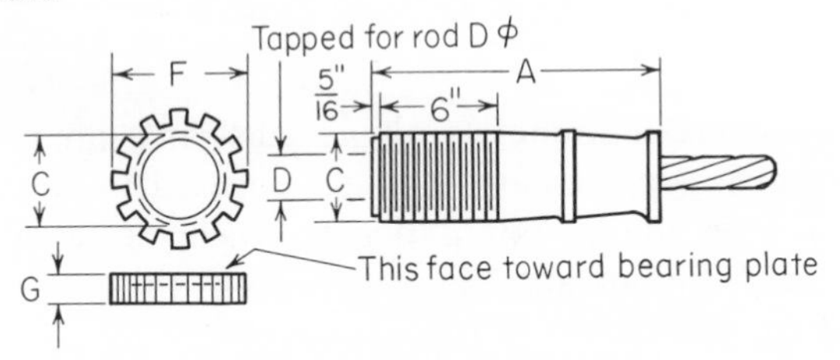

Strand diameter, in.	Measurements, in.					Total weight, lb
	A	*C*	*D*	*F*	*G*	
1	10	3½–4N	2 –4½NC	5⅜	1¼	29
1⅛	10¾	4 –4N	2¼–4½NC	6	1⅜	38
1¼	11⅜	4⅜–4N	2½–4NC	6½	1⅝	45
1⅜	11¾	4⅞–4N	3 –4NC	7	1⅝	54
1½	12⅜	5¼–4N	3 –4NC	7⅝	1¾	74
1⁹⁄₁₆	12¾	5½–4N	3½–4NC	7⅞	1¾	77
1⅝	13	5⅝–4N	3½–4NC	8	1¾	84
1¹¹⁄₁₆	13¼	5⅞–4N	3½–4NC	8⅜	1¾	88

Fittings Type SS–2 can also be supplied with permanent studs and no external threads when necessary.

FIG. 8-18. Anchor fittings for galvanized strands. (*Courtesy of CF&I Steel Corporation.*)

and tensioned, a nut is turned down on the threaded fitting to transfer the load to a steel bearing plate.

Table 8-4 and Fig. 8-18 illustrate details of large-diameter galvanized strands.

CHAPTER 9

Specifications

9-1. Development. The first specification for prestressed concrete published in the United States was Criteria for Prestressed Concrete Bridges issued by the Bureau of Public Roads in 1954. From this beginning there have been numerous new and revised codes and specifications.

The first really complete document, which is still a good reference, was Tentative Recommendations for Prestressed Concrete prepared by the Joint ACI-ASCE Committee 323 on Prestressed Reinforced Concrete. These recommendations are printed in the *Journal of the American Concrete Institute* for January, 1958 and the *Journal of the Structural Division, American Society of Civil Engineers,* Paper 1519, January, 1958.

Since prestressed concrete is a relatively new structural material new developments occur frequently. Data from research programs are making our design procedures more accurate so that we do not have to overdesign to insure an adequate factor of safety. The codes and specifications covered in the following sections are current at this writing. The reader should be sure that he always follows the latest issue of any code as time goes on.

9-2. Materials. As pointed out in Chap. 6, desirable properties for prestressed concrete tendons are in a class by themselves. We cannot use any of the high-strength steels previously available.

At present there are two specifications prepared specifically for prestressed concrete tendons by the American Society for Testing Materials. They are ASTM A 416, Uncoated Seven-Wire Stress-Relieved Strand for Prestressed Concrete and ASTM A 421, Uncoated Stress-Relieved Wire for Prestressed Concrete. It should be noted that ACI 318-63, Sec. 405(e) calls for these two specifications and then goes on to say: "Wires used in making strands for post-tensioning shall be cold-drawn and either stress-relieved in the case of uncoated strands, or hot-dip galvanized in the case of galvanized strands." There is no mention of permitting the use of either oil-tempered or unstress-relieved wire.

A brief specification for high strength steel bars is given in Sec. 405(f) of ACI 318-63.

Except for tendons the materials that go into prestressed concrete are the same as those that go into reinforced concrete and the same specifications apply for reinforcing steel, cement, aggregate, etc. These materials are merely used in different proportions to give higher strength concrete than that normally used in reinforced concrete.

One variation is that the use of calcium chloride is forbidden in prestressed concrete members because of its effect on the tendons.

9-3. Buildings. ACI Standard Building Code Requirements for Reinforced Concrete (ACI 318-63), published in June, 1963, includes a chapter on prestressed concrete and ties it in with all other parts of the code with respect to load factors, material specifications, etc. ACI 318-63 is recommended as the basis of design for buildings using prestressed concrete. It is the basis of design for numerical examples in this book and portions of it are reproduced in Appendix A.

A large majority of the best and most economical prestressed concrete buildings are precast at a casting yard and trucked to the job site for erection. As these members are placed in the structure they must be attached to each other with connections that are both economical and structurally sound. Details of typical connections are presented in the literature of most of the casting yards and in several publications which are readily available.[1-3]* ACI-ASCE Committee 512 has prepared a report, Suggested Design of Joints and Connections in Precast Structural Concrete, which appears on pages 921 to 937 of the ACI journal for August, 1964. This report spells out which practices are good and which are bad, lists allowable stresses for specific conditions, and otherwise guides the engineer to a good joint design. It is used as a basis of examples of joint design presented in this book and parts of it are reproduced in Appendix B.

9-4. Bridges. As stated in Art. 9-1 the first specification on prestressed concrete published in the United States dealt with bridges and was published by the Bureau of Public Roads in 1954. From that beginning the point has been reached where requirements for prestressed concrete bridges are an integral part of Standard Specifications for Highway Bridges published by the American Association of State Highway Officials. The latest edition of these specifications is used as the basis for the bridge-design examples presented in this book.

Many of the states have complete sets of standard drawings and specifications for prestressed concrete bridges so that bridge designs can be picked directly from them and vary only in minor details to suit local conditions.

A joint committee composed of some members from the American Association of State Highway Officials and some members from the Prestressed Concrete Institute was set up to establish standards for prestressed concrete members for bridges. They have prepared standards for precast prestressed I beams plus poured-in-place deck bridges, for

*Superscript numbers indicate references listed in the Bibliography at the end of the chapter.

precast prestressed slab and hollow slab bridges, and for numerous sizes and shapes of piles. In addition to guiding engineers to economical designs, the standard cross-sections established enable the casting-bed operator to use the same forms for job after job. The joint AASHO-PCI Committee continues in existence to keep the standards up to date with latest developments.

9-5. Quality Control. Although prestressed concrete is superior to reinforced concrete in many ways and is usually cast in steel forms under plant-controlled conditions, it is still concrete and is subject to some of the control problems of that material. It is therefore the responsibility of the engineer, the architect, and the prestressed concrete fabricator to establish criteria which will ensure a product that is structurally and architecturally satisfactory but which can still be produced economically. This should be done by persons who are familiar with the design and erection of structures as well as with the fabrication of prestressed concrete members.

Most of the codes and specifications referred to in previous sections of this chapter include instructions concerning testing of materials, methods of fabrication, or requirements for the finished product. These should be understood and adhered to.

One of the most sensible and comprehensive documents covering quality control of prestressed concrete is a 35-page manual entitled Inspection of Prestressed Concrete, which is available from the Prestressed Concrete Institute for a nominal fee.

In prestressed concrete the finished member itself is to some degree a check on proper fabrication and adequate materials. The tension placed in the tendons when they are first stretched is appreciably greater than any tension that will exist in them under full design load after the member is completed. If the tendons are not in about the proper place at about the proper tension, the camber will not be correct. If the concrete is not strong enough, the camber will be excessive.

Another excellent guide to quality control of prestressed concrete has been prepared by a joint committee of AASHO and PCI. It is Tentative Standards for Prestressed Concrete Piles, Slabs, I-Beams and Box Beams for Bridges and an Interim Manual for Inspection of Such Construction. It is available from AASHO.

BIBLIOGRAPHY

1. Connection Details for Precast-Prestressed Concrete Buildings, Prestressed Concrete Institute, Chicago, Ill.
2. Cazaly, Lawrence and M. W. Huggins: "Prestressed Handbook," Canadian Prestressed Institute, Toronto, Ontario.
3. Preston, H. Kent: "Prestressed Concrete for Architects and Engineers," McGraw-Hill Book Company, New York, 1964.

CHAPTER 10

Other Design Considerations and General Information

10-1. Bond of Tendons. Bond, or lack of bond, between a prestressing tendon and the concrete member that it prestresses is one of the factors entering into the structural analysis of the member. Practically all tendons which are bonded are of uncoated steel—either wires, strands, or bars. Since concrete bonds readily to clean bright or slightly rusty steel our only need is to establish the conditions which will develop enough bond for the particular member under consideration.

In a pretensioned member the entire initial prestressing force in the strand is transferred by bond from the strand to the concrete in a relatively short length at each end of the concrete beam. This length is referred to as the transfer length. The length over which a strand must be bonded before its full force can be considered transferred to the concrete member is specified in the codes. Section 2611(a) of ACI 318-63 covers transfer length in building members, and Sec. 1.13.17 of AASHO Standards for Bridges covers it for bridges.

Within the range of strengths usually specified for prestressed concrete the strength of the concrete does not seem to have an appreciable effect on bond. Minimum strengths required at load transfer in pretensioned members are usually 3,500 to 4,000 psi, and minimum final ultimates are usually 5,000 psi or greater, although some building members require only 4,000 psi at final strength.

Specifications involving bond transfer are based on clean seven-wire strand made in accordance with ASTM A 416. The theory that it would be desirable to redesign or treat the tendon so that it would develop full bond in a shorter distance is not necessarily correct. In post-tensioned members the entire force from a tendon is applied at one point, and it is usually necessary to use reinforcing steel in the form of spirals, mats, or stirrups to keep the concrete from splitting in the area of stress concentration. Pretensioned strands transfer their load to the concrete gradually over their transfer length. Load transferred from strand to concrete at the end of the beam is being disbursed to other parts of the cross-section along the beam as additional load is being transferred from the strand. Load concentration at any one spot is thus reduced. For lightly prestressed members it may be desirable to develop bond in a short distance. Use of smaller diameter strands which have a correspondingly shorter transfer length is helpful here.

Stress concentrations at the ends of pretensioned members can be reduced by unbonding some of the strands for a short distance. This is accomplished by covering the strand to be unbonded with a length of plastic tube, greasing it, and wrapping it with paper or other means.

Bond between tendon and concrete influences the flexural strength of a prestressed concrete member. If the tendon is bonded, the additional tension it develops at each point along its length is in proportion to the increase in bending moment at that point. Flexural bond stresses are not high and are not a problem in pretensioned members which meet code requirements for transfer length or in post-tensioned members that are properly grouted.

A seven-wire strand with a light coat of rust will have a transfer length approximately two-thirds that of a clean strand of the same diameter. A heavy coat of rust can reduce bond and also be harmful to the strand. Rust should be permitted but kept within the limits set by ASTM A 416 which says: "The strand shall not be oiled or greased. Slight rusting, provided it is not sufficient to cause pits visible to the naked eye, shall not be cause for rejection."

10-2. Using Unbonded Tendons. This section deals with tendons which are unbonded for their full length. There is no association with the pretensioned tendons which are unbonded for a short distance to reduce stress concentrations as discussed in the preceding section.

It is usually less expensive to purchase, install, and tension unbonded post-tensioned tendons than grouted post-tensioned tendons. Unbonded tendons are shipped to the job site lubricated and encased in a permanent plastic tube or waterproof paper wrapping ready for placement in forms. After the concrete member has been poured and cured they are tensioned and anchored, and the job is complete. Lubrication between tendon and encasement keeps friction loss to a minimum during tensioning.

Lack of bond between concrete and tendon has a definite influence on the structural behavior of the member, and this influence should be considered in a design analysis.

In the previous section discussing bonded tendons we said that "the additional tension it [the tendon] develops at each point along its length is in proportion to the increase in bending moment at that point." If the tendon is not bonded, the increase in its length that takes place at points of increased moment is distributed over the full length of the tendon, thereby developing only a slight increase in tension. It is this action which results in a lower ultimate flexural strength for members with unbonded tendons. Section 2608(a)3 of ACI 318-63 and Sec. 1.13.10(c) of AASHO Standard Specifications each give two formulas for computing f_{su}, the stress in the prestressing steel at ultimate load. One equation applies to members with bonded tendons and the other to members with unbonded tendons. In typical structures the value of f_{su} for unbonded tendons will range from 60 per cent to 75 per cent of that for bonded tendons.

Other factors not covered in current codes should also be considered in the design of members with unbonded tendons. The entire subject of design of members with unbonded tendons is being studied by a committee associated with the engineering societies. When their report is published it should be referred to. In the interim the authors present their own suggestions.

One undesirable characteristic of members with unbonded tendons is the mode of flexural failure. When the tensile stress in the concrete at the point of maximum moment reaches its modulus of rupture, one crack will occur. Since there is no bonded steel to carry tension across the crack, it will widen rapidly and the member will fail at a load only slightly higher than that which caused the crack to occur. If the member being loaded is continuous rather than simple span, the opening of the crack will reduce the moment of inertia of the member and there may be a redistribution of moments. Failure at the cracked section will then occur when the moment has again been built up to something slightly greater than that which caused the crack. Failure due to the rapid opening of a single crack can be avoided by the addition of untensioned reinforcing steel bars near the tensile face of the concrete member. The total area of these bars should be large enough to distribute flexural cracks, and the area of the bars at their yield strength can be added to the allowable load in the tendons when computing ultimate flexural strength of the member.

Ultimate shear strength is reduced by lack of bond between tendon and concrete. Until data are available for more accurate analysis, it is suggested that the following modifications be adopted when computing ultimate shear capacity of building members that do not have enough bonded reinforcing bars in the tensile area to distribute cracks. Refer to ACI 318-63, Sec. 26-10. When computing M_{cr} for use in either Eq. (26-12 ACI) or (26-12A ACI) use

$$M_{cr} = \frac{I}{y}(f_{pe} - f_d)$$

When computing V_{cw}, drop the quantity $3.5\sqrt{f'_c}$ from Eq. (26-13 ACI) and drop the quantity $0.5F_{sp}\sqrt{f'_c}$ from Eq. (26-13A ACI).

Bridges constructed with unbonded tendons should have an adequate amount of bonded reinforcing bars in the tensile zone to avoid a "single-crack" failure from a severe overload.

When unbonded tendons are used, details of their anchor fittings are of extreme importance. The steel in all prestressing tendons is in the high-carbon range, and such steels are sensitive to stress concentrations. Many of the anchor fittings in common use cause stress concentrations in the tendons by nicking them or creating secondary stresses. In an unbonded tendon every increase in tension due to an increase in moment at any point is transmitted along the tendon to the anchor fittings, and repeated changes in tension become fatigue loadings at the points of stress concentration.

Anchor fittings which cause stress concentrations do not develop the full strength of the tendon. Failure always occurs at the fitting. Downgrading the tendon (i.e., assuming it has an ultimate strength equal to that developed by the anchor fitting but less than the actual strength of the tendon) is not necessarily a conservative approach. Specifications for all prestressing tendons require a large amount of elongation in the tendon before it fails. For a seven-wire strand, as an example, ASTM A 416 requires a minimum elongation of 3.5 per cent in a 24-in. length. Most of this occurs just before the tendon fails and is eliminated if the tendon fails at the anchor fitting due to stress concentrations.

Until a code or report covering the subject is published the following approach is suggested in the use of unbonded tendons:

1. Design for ultimate moment using formulas from the codes referred to near the beginning of this section.
2. Provide sufficient bonded reinforcement to meet the requirements of ACI 318-63, Sec. 2609(c). (See Appendix A.)
3. Modify the ultimate shear capacity of building members as suggested in this section. (This requirement can be ignored if, in the designer's judgement, there is enough bonded reinforcing steel in the tensile zone to distribute adequately any cracks which form.)
4. Consider the details of the anchor fitting to be sure it will give satisfactory performance in the structure being designed.

10-3. Cracking Load and Crack Control. Cracking load is the load which raises the tensile stress in the concrete to its modulus of rupture and causes a crack to occur. In ACI 318-63, Sec. 2600, M_{cr} is defined as "net flexural cracking moment." Section 2610(b) gives formulas for computing M_{cr} in both normal-weight and light-weight concrete. Both of these formulas are a bit on the conservative side for actual cracks. The actual modulus of rupture in normal-weight concrete is about $7.5\sqrt{f'_c}$ rather than $6\sqrt{f'_c}$.

In building design cracking load plays a part in the criteria for ultimate moment [ACI 318-63, Sec. 2609(c), and the criteria for ultimate shear, Sec. 2610(b)]. Except in these formulas or in very special cases, cracking load has no further significance. In the early days of prestressed concrete it was common practice to require the cracking load to be at least a certain amount greater than the design load. As it became apparent that this requirement often increased the cost of a member without accomplishing any necessary function the requirement was dropped. After all, reinforced concrete members are universally accepted, and many of them are cracked under their own dead weight.

In bridge design for more than a decade it was required that the structure carry full design load without tensile stress in the concrete. More recently AASHO Standards have been revised to permit up to 250 psi tensile stress in pretensioned bridges. This is slightly less than halfway from zero stress to cracking load for 5,000-psi concrete. This revision was made after a study of AASHO road tests[1]* in which prestressed beams were deliberately loaded until they cracked and then subjected to a million and a half cycles of load which created an equivalent of 800 psi tensile stress in the bridge member with no undesirable effects.

There are a few instances where it is desirable to prohibit cracks to protect the steel in a corrosive atmosphere, to prevent damage to a special finish on the underside of the member, etc. These are cases where prestressed concrete is performing a function beyond the reach of reinforced concrete. Cracks can be prevented by setting an adequate factor of safety on the cracking load and providing sufficient prestress to meet it.

It is not necessary that a prestressed concrete beam be free of cracks—even under just its own dead weight plus prestress. It is desirable that all cracks be controlled to some point beyond full design load.

Undesirable flexural cracks are seldom a problem when the tendons are bonded and are located near the tensile side of the member. If the center of gravity of the tendons is higher at the ends of a member than at mid-span, some of the tendons should be kept close to the bottom for the full length of the beam or reinforcing bars should be placed near the bottom when there are no tendons there.

Horizontal cracks are often found at the ends of pretensioned prestressed concrete beams. They occur at the neutral axis of the member or along the line where web meets bottom flange and are due to unequal distribution of the prestressing force across the end face of the member. Typical conditions resulting in such cracks are the location of strands at the end of the beam in one group or in one group of straight strands at the bottom and one group of deflected strands near the top. Horizontal cracks at beam ends cannot always be eliminated, but they can be completely controlled with a few properly placed stirrups.

The Research and Development Laboratories of Portland Cement Association have conducted a series of tests on control of horizontal cracking at ends of pretensioned beams and have presented a report on their findings.[2] The report suggests the use of the following formula:

* Superscript numbers indicate references listed in the Bibliography at the end of the chapter.

$$A_t = \frac{0.021F_I h}{f_s l_t} \tag{10-1}$$

in which A_t = total cross-sectional area of stirrups needed, sq in.
F_I = total initial prestressing force in tendons, kips
h = overall depth of girder, in.
f_s = maximum allowable stress in stirrups, ksi (use $f_s = 20$)
l_t = strand transfer length, in. (use 50 times strand diameter)

Stirrups equal to A_t should be placed within the distance $0.20h$ from the end of the member. They should be uniformly spaced with the first stirrup as near to the end of the girder as possible. Test results indicate that these stirrups will keep cracks fine and short. Additional stirrups should be provided in the remainder of the end zone of the girder to meet vertical shear requirements.

Equation 10-1 was developed by PCA from tests "on the end zones of girders in which the strand was divided into two groups located respectively in the top and bottom of the section." It seems logical that stirrup requirements would be reduced if strand distribution were more uniform across the end face of the girder.

10-4. Partial Prestress.[3] The term "partially prestressed" refers to a prestressed concrete member which has flexural tension in the precompressed tensile zone under the design load condition. Partial prestress is often used to reduce excessive camber, and in some cases it permits a reduction in the amount of steel required. Requirements for ultimate strength of a partially prestressed member are the same as for a fully prestressed member.

Since there is no established specification for partially prestressed members, their design can be based on ACI 318-63 with a few variations. In fact the allowable tensile stress of $6\sqrt{f'_c}$ in Sec. 2605(b) is already a big step into partial prestress. It represents the highest tensile stress that can be used if the designer wishes to maintain a relatively crack-free structure. The ultimate tensile strength of concrete is approximately equal to $7.5\sqrt{f'_c}$.

Just how will the use of partial prestress affect a typical prestressed concrete member? A tensile failure in the concrete will occur at a stress of about $7.5\sqrt{5{,}000} = 7.5 \times 70 = 525$ psi. If this were a reinforced concrete member in which $E_s/E_c = 7$, the tensile stress in the steel bar when the crack occurred would be $525 \times 7 = 3{,}675$ psi. But reinforcing bars are used at stresses up to 20,000 psi. Therefore much higher tensile stresses can be used before cracks occur which are equal to those that are standard in reinforced concrete.

Consider a prestressed member designed by the usual criteria and then found to have an excessive amount of camber. How should it be redesigned as a partially prestressed member with adequate structural properties?

1. Establish an amount of camber that is permissible, compute the prestressing force that will create this camber, and select the prestressing tendons required for this force.

2. Check ultimate strength of member using tendons selected in step 1. If ultimate is not sufficient, add enough tendons to bring it up to specification. Tension all tendons to a uniform stress such that the total prestressing force will be that determined in Step 1. This means that the tendons will not be tensioned to their full allowable stress, but they will have enough ultimate strength to give the member its specified ultimate strength. Another approach is to use enough tendons to give the member the prestressing force determined in step 1 and add high-yield-strength reinforcing bars to bring the ultimate strength of the member up to specification.

3. The amount of shear steel required in a prestressed concrete beam is a function of the magnitude and location of the prestressing force, and the amount of shear steel required increases as the prestressing force decreases. The formulas in Sec. 2610 of ACI 318-63 are still applicable to partially prestressed members because the magnitude and location of the prestressing force are included in the computation of V_c in Sec. 2610(b).

4. Step 1 established the prestressing force and Step 2, the amount of steel needed for ultimate. Some criterion should be established that will limit tensile stress under full design load. Since there is no formal specification covering this item at present, it is suggested that it be based on allowable stresses in reinforced concrete design. These computations can be carried out in two steps. First, apply load to the prestressed member until the stress in the bottom fiber is reduced to zero. Second, apply the remaining live load and compute the resulting stresses in the member as if it were ordinary reinforced concrete. If the stress developed in the steel by this second loading does not exceed the 18,000 or 20,000 psi allowed in reinforcing bars, the design is satisfactory.

5. Maximum deflection should be checked. Here again the computation can be carried out in two steps. First, compute deflection in the prestressed member for the load which brings the bottom-fiber stress to zero. Second, considering the member as reinforced concrete, compute deflection caused by application of the remaining live load.

Partial prestress is not covered in any of the formal specifications familiar to the authors. However, if reinforced concrete members are permitted in a specific structure, then partially prestressed members designed as suggested herein should also be acceptable.

10-5. Stress Losses. The causes of stress loss—shrinkage, elastic compression, creep, and relaxation—are discussed in Chap. 2. Section 2607 of ACI 318-63 also lists slip at anchorage and friction loss but these occur during the tensioning operation and can be compensated for at that time.

AASHO Standard Specifications says: "Losses of prestress due to all causes except friction may be assumed to be as follows:

Pretensioned members .... 35,000 psi
Post-tensioned members .... 25,000 psi"

It goes on to indicate that, when more exact data on losses are available, they may be used in place of these arbitrary values. ACI 318-63 Sec. 2607 lists the causes of losses to be considered but does not indicate any specific values. The AASHO values are the same as those listed in Sec. 208.3.2, Method 2, of Tentative Recommendations for Prestressed Concrete which was the guide for design until publication of ACI 318-63.

The values of 35,000 psi for pretensioned and 25,000 psi for post-tensioned tendons will be applied to the design of normal members; i.e., for members that fall in the following classification:

1. Members are made of 5,000 psi normal-weight concrete.
2. Concrete strength is 3,500 to 4,000 psi at release of pretensioned tendons or 4,000 psi or greater when post-tensioned tendons are tensioned.
3. Initial tension in tendons is approximately 70 per cent of their ultimate strength.
4. Stresses in concrete are 75 to 100 per cent of those permitted by codes.

Let us analyze the stress losses in a precast pretensioned single T beam which meets the four conditions just itemized for a normal member.

Relaxation of Tendons. Test data indicate that seven-wire strands made to ASTM A 416 and tensioned to 70 per cent of their ultimate strength will undergo a stress loss of approximately 4,000 psi between the time they are tensioned and the time their load is released from the casting bed anchors into the prestressed concrete member.

Elastic Compression of Concrete. Stress loss in the tendons due to elastic compression of concrete can be computed using Eq. (10-2) or, if the beam meets the four requirements for a "normal member," the value of 11,000 psi computed here can be used.

$$f_E = \frac{E_s}{E_c} f_{c.g.s.} \tag{10-2}$$

where f_E = stress loss in tendons due to elastic compression of concrete
E_s = modulus of elasticity of tendons
E_c = modulus of elasticity of concrete (Fig. 6-1)
$f_{c.g.s.}$ = average compressive stress in concrete along the c.g.s. of the tendons

From Art. 7-2, E_s = 28,000,000 psi

From Fig. 6-1, $E_c = 3{,}840{,}000$ psi for 4,000-psi concrete (the strength assumed at time of release of strands)

For this example we will assume $f_{c.g.s.} = 1{,}500$ psi. At first glance this may seem a rather low value but this is "average stress along the c.g.s." Allowable stresses listed in the codes are extreme fiber stresses and the c.g.s. of a group of strands is some distance from the extreme fiber even at mid-span. The distance is greater towards the ends of the beam. Also the dead weight of the beam is now acting to reduce compressive stress in the bottom fiber.

Substituting numerical values in Eq. (10-2) we get

$$f_E = \frac{28{,}000{,}000}{3{,}840{,}000} 1{,}500 = 11{,}000 \text{ psi}$$

In a pretensioned member, relaxation of tendons of approximately 4,000 psi takes place before the tendons are cut, and elastic compression of concrete takes place as soon as the tendons are cut so that *the stress loss in the tendons from initial tension to immediately after cutting the tendons* is the sum of these two values. In our numerical example this is

$$4{,}000 + 11{,}000 = 15{,}000 \text{ psi}$$

One of the conditions that must be investigated for critical stresses in the design of a precast pretensioned member is the condition immediately after cutting the strands. It is suggested that the value of 15,000 psi be used for stress loss at this point where the four requirements for a "normal member" are met unless the designer prefers to make a separate calculation for each individual member.

Shrinkage and Creep. Shrinkage of concrete takes place as the concrete dries out. Under normal conditions any shrinkage which takes place before the tendons are cut is negligible, and we will assume all shrinkage takes place after the tendons are cut.

Creep is the result of constant compression in the concrete and therefore does not begin until the tendons are cut.

Reports on tests designed to establish the stress losses due to shrinkage and creep[4,5] are not in close agreement on the amount of the loss that is due to creep and the amount due to shrinkage, but they are in reasonably good agreement on the total loss due to creep plus shrinkage. Since these losses occur simultaneously there is no need for us to try to separate them. Stress loss in tendons of a "normal member" due to creep plus shrinkage is approximately 18,000 psi.

Although the stress in the tendons is decreased about 15,000 psi when they are cut, stress loss due to relaxation of the tendons does not stop completely at this point. Until more complete data are available we will assume that they undergo a further relaxation of 2,000 psi while the concrete is shortening due to creep and shrinkage.

Approximate total stress loss in a normal member is therefore 4,000 + 11,000 + 18,000 + 2,000 = 35,000 psi.

Next let us consider a member which meets conditions 2, 3 and 4 for "normal members" but which is made of 5,000-psi *lightweight concrete.*

Relaxation of tendons immediately after they are cut will still be 4,000 psi.

Elastic Compression of Concrete. This can be computed using Eq. (10-2). The value of E_c can be computed from data in Fig. 6-1 and will be less than E_c in the foregoing computation because the unit weight of the concrete is less. Since E_c is less, the stress loss will be greater. For a concrete with a unit weight of 120 lb per cu ft and an ultimate strength of 4,000 psi (at time of strand release) Fig. 6-1 shows $E_c = 2{,}750{,}000$ psi. Using other values the same as in the previous computation and substituting in Eq. 10-2 we get

$$f_E = \frac{28{,}000{,}000}{2{,}750{,}000}\,1{,}500 = 15{,}250 \text{ psi.}$$

For this lightweight concrete and these conditions the total stress loss immediately after release of tendons would be 4,000 + 15,250 = 19,250 psi. A conservative value to use in design analysis would be 19,000 psi.

Shrinkage and Creep. The stress loss in the tendons due to creep and shrinkage in lightweight concrete is greater than in normal-weight concrete. It also varies from one aggregate to another. For approximations only it can be assumed that the creep and shrinkage in two different concretes varies in the inverse ratio of their modulus of elasticity. On this particular member, therefore, we would compute a stress loss of

$$\frac{3{,}840{,}000}{2{,}750{,}000}\,18{,}000 = 25{,}000 \text{ psi}$$

The approximate total stress loss in this lightweight concrete member would be 19,000 + 25,000 + 2,000 = 46,000 psi.

There is some variation in the creep and shrinkage properties of normal weight concretes and a greater variation in lightweight concretes (see Art. 6-4). Stress losses computed here are intended only to give the designer an idea of what to expect. Variations in intensity of stress, concrete strength, type of aggregate, proportion of cement paste in the mix, etc., will alter the magnitude of the stress loss.

10-6. Camber and Deflection. As compared with reinforced concrete, prestressed concrete has certain characteristics which makes it stiffer or less flexible. It is made of a much higher strength concrete which gives it a higher modulus of elasticity. All the concrete in the cross section contributes to the moment of inertia because it is crackless. In a reinforced concrete member only the compression concrete in the top and the tensile reinforcing steel in the bottom contribute to moment of inertia. The cracked concrete from the neutral axis down does not add to the stiffness of the member. On the other hand prestressed members usually have a large span-depth ratio which decreases their comparative stiffness.

Computations for camber and deflection of prestressed members are no more difficult than for steel structures of similar shape. The accuracy of these calculations is dependent upon the designer's knowledge of the properties of the concrete going into the member.

As pointed out in Chap. 1, one purpose of prestressing concrete is to create as much negative moment as possible in the member. What type of deflection is produced by a negative moment? It is an upward deflection or camber. As a result, practically all prestressed concrete beams have a camber whose magnitude is a function of moment of inertia of concrete section, strength of concrete, magnitude of prestress, eccentricity of prestress, and climatic conditions.

A characteristic of prestressed concrete members difficult for some designers to understand is that camber in such a member will *increase* with the passage of time. Concrete under constant compression undergoes a permanent inelastic shortening called creep. In a fully prestressed member not carrying its live load the bottom fibers are under a constant high compressive stress while the top fibers are under little or no stress. With the passage of time creep causes a shortening of the bottom surface of the member which, since there is no corresponding change in length of the top surface, results in additional camber.

As part of the design of any prestressed member, camber should be computed and camber growth estimated so that its influence on other details of the structure can be considered. Members with a small span-depth ratio—girders, for instance—very seldom have too much camber. As the span-depth ratio increases, as with single and double T's, we find more instances where camber is large—sometimes so large that it cannot be economically accommodated by other details of the structure.

There are several approaches to the control of excessive camber. Partial prestress is the only one that does not increase cost, but a cost analysis will show that the additional cost of any of the suggested methods is not great. Any one or a combination of two or more may be the best solution to the problem for a specific structure.

1. *Larger member.* A relatively small increase in depth with a correspondingly small increase in volume of concrete can show an appreciable increase in moment of inertia and decrease in camber.

2. *Larger prestressing force.* In unsymmetrical members, such as single and double T's where the neutral axis is near the top of the member, increasing the prestressing force and decreasing its eccentricity will reduce the camber without creating excessive stresses.

3. *Stronger concrete.* The modulus of elasticity of concrete increases as the strength increases. Thus curing the concrete to a higher strength before applying the prestressing force will decrease camber.

4. *Prompt loading.* Camber increases with time. This growth is greater for the member under its own dead weight than for the loaded member. Therefore camber growth is minimized by erecting the member and applying poured-in-place topping or other dead loads as soon as feasible after prestressing. This approach is sometimes impractical because of other scheduling requirements.

5. *Partial prestressing.* See discussion in Art. 10-4.

Deflections are computed in the same manner as camber. The shallow depth of some members with respect to the span engenders comparatively large deflections under live load. If the code being followed does not limit maximum deflections, some logical criteria should be established for the structure and the shallower members checked against it. One important item is proper roof drainage. If sufficient water collects on a flat long-span shallow roof to cause noticeable deflection, the deflection permits a greater depth of water which causes more deflection, etc.

10-7. Safety Precautions. In addition to the normal precautions required on any construction work it must be remembered that a prestressed concrete tendon under tension as high as 175,000 psi contains a tremendous amount of energy. The tendon and/or any fittings or equipment connected to it can whip across working areas with lethal results if its pent-up energy is suddenly released by some type of equipment failure.

The good safety record established by prestressed concrete producers during the first years of operation in the United States can be maintained if both old and new operators will remember the dangers involved and continue to observe the logical precautions even though they have not had any difficulties. There are several factors which merit special attention.

The wedge-type temporary grips used for holding seven-wire pretensioned strands under load should be handled in accordance with the instructions issued by their suppliers. One requirement, for example, is that they be kept clean. Dirt between a wedge and the steel case around it will retard the motion of the wedge into the case and can cause premature failure or slippage of the strand.

As discussed in Art. 7-4, excessive heat or an electric arc can destroy the high strength of a tendon. Such damage is not necessarily apparent to the naked eye and is discovered only when the tendon fails during the tensioning operation. This danger can be eliminated by proper storage and handling of tendons.

Operation and maintenance of hydraulic jacking equipment are not new problems. Proper procedures are important in prestressed concrete work because of the damage that can be done by the sudden release of load in a tendon due to an equipment failure.

If a flaw or weak spot exists in a tendon, fitting, or the equipment, it is most likely to show up during the tensioning operation. It is therefore a reasonable safety precaution to remove personnel from the area until tensioning is complete. In a post-tensioning operation the tendon is usually encased in the concrete member and cannot whip around. The chief danger is at the ends, and the normal tendency here is to fly away from the end of the member. The strands in a pretensioned bed are not confined and can whip in any direction.

The foregoing should not cause anyone to be afraid of prestressed concrete production. Properly operated plants have excellent safety records because the management is fully aware of the conditions discussed here and takes the necessary precautions.

Prestressed concrete is still comparatively new, and some contractors are not as familiar with it as with other structural materials. One important factor should be kept in mind in the erection of a structure composed of precast members. The structure probably is not stable until field welds have been made and poured-in-place concrete has been poured and cured. Therefore, as installation of precast members proceeds, there should be a definite schedule involving use of temporary bracing, making of field welds, and pouring concrete so that the partially erected structure is at all times stable and never subject to excessive stresses or danger of collapse. It is almost impossible to avoid temporary eccentric loading during the erection process, but the condition should be recognized, the schedule planned to minimize such conditions, and adequate temporary bracing placed where needed.

In some buildings the possibility of differential settlement must be faced and provided for. Where this possibility exists, members and joints should be designed to survive the motions involved without losing their ability to carry their loads. For simple spans, joints must be truly hinged so that rotation can take place without developing end moments in the beams, or else the joint, the beam, and the beam support must be made capable of taking the moment developed—which means it is no longer a true simple span. The best approach for designing rigid frames subject to possible differential settlement is to make sure that the overstressed sections will yield, without failure, and redistribute some of the excess load that has been placed on them.

10-8. Fire Resistance. By their very nature all prestressed concrete members are fire-resistant because the prestressing tendons are protected from the heat of the fire by a layer of concrete of some thickness. The fire rating of a member is a function of the thickness of this layer of concrete and some other factors.

A complete résumé of all fire tests on prestressed concrete in this country, including drawings, loading, and test results on 47 specific members, is presented on pages 14 to 43 of *Journal of the Prestressed Concrete Institute* for October, 1962. This article, An Interpretation of Results of

Fire Tests of Prestressed Concrete Building Components by A. H. Gustaferro and C. C. Carlson, both of Portland Cement Association, summarizes the information gained from the tests. It says

> . . . the most important factors affecting the fire resistance of prestressed concrete flexural members are:
>
> 1. Thickness of concrete cover between the prestressing steel and the surface exposed to fire (cover).
> 2. Degree of end restraint.
> 3. Size of cross section of the member.
> 4. Shape of the member.
> 5. Aggregate type.
> 6. Moisture content of the concrete.

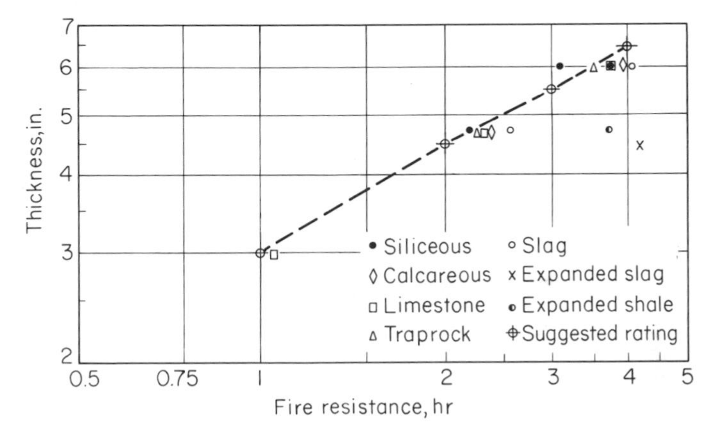

FIG. 10-1. Effect of slab thickness on fire resistance of concrete floors when failure is due to heat transfer.

The article points out that the fire rating of most prestressed members which have some degree of end restraint is determined by "heat transmission" rather than structural failure. Under the requirements of ASTM E 119 for standard fire tests a member is considered to have failed at the time it collapses or at the time the average temperature of its unexposed surface has risen 250° F. Figure 10-1, taken from the article, shows ratings as a function of heat transmission. Table 10-1, from the same article, shows recommended ratings based on type of cross section, total area, and concrete cover, and states that "on the basis of tests performed to date, the recommendations appear to be reasonably conservative."

Underwriters' Laboratories, Inc. have conducted a number of fire tests on various types of prestressed concrete members and copies of their reports on these tests are available from PCI at a nominal charge. Pre-

Table 10-1. Cover Required for Various Fire Ratings in Prestressed Concrete Members

Type of unit	Cross-sectional area,* sq in.	Recommended rating			
		1 hr	2 hr	3 hr	4 hr
Girders, beams, and joists	40–150	2 in.			
	150–300	1½ in.	2½ in.†		
	Over 300	1½ in.	2 in.	3 in.†	4 in.†
Slabs, solid or cored.............	....	1 in.	1½ in.	2 in.	

* In computing the cross-sectional area for joists, the area of the flange shall be added to the area of the stem, and the total width of the flange, as used, shall not exceed three times the average width of the stem.

† Adequate provisions against spalling shall be provided by means of wire mesh.

stressed concrete members bearing the Underwriter's label are available in most sections of the country.

10-9. The Prestressed Concrete Institute. The Prestressed Concrete Institute (PCI) has its principal office in Chicago. Its purpose is outlined in Article II of the PCI By-Laws as follows:

ARTICLE II

The purposes of this corporation are to stimulate and advance the common interests and general welfare of the Prestressed Concrete Industry and the Precast Concrete Industry;

To collect and disseminate knowledge, statistics, ideas and information relating to design, manufacture and use of prestressed concrete and precast concrete;

To advance prestressed concrete and precast concrete acceptance and use through investigations and research relative to new applications of prestressed concrete and precast concrete and engineering processes for improvement of the design, manufacture and use of prestressed concrete and precast concrete;

To establish industry-wide standards of design and production of prestressed concrete and precast concrete to improve quality and design of product;

To perform all lawful and desirable activities within the State of Illinois and elsewhere, to promote the efficient, constructive and beneficial operation of the Prestressed Concrete Industry and the Precast Concrete Industry.

The *PCI Journal,* published at regular intervals and mailed to all members, carries articles on prestressed concrete structures, plants, research, and structural analysis.

P C Items, also furnished to all members, brings news of current events in the industry.

At irregular intervals as the need develops, short courses are sponsored. These run for two or three days and are conducted by individuals who are proficient in the particular subjects to be covered in the course.

BIBLIOGRAPHY

1. Fisher, J. W.: Behavior of AASHO Road Test Prestressed Concrete Bridge Structures, *J. Prestressed Concrete Inst.*, February, 1963.
2. Marshall, W. T., and Alan H. Mattock: Control of Horizontal Cracking in the Ends of Pretensioned Prestressed Concrete Girders, *J. Prestressed Concrete Inst.*, October, 1962; or Research and Develop. Labs. Portland Cement Assoc., *Research Dept. Bull.* D58.
3. Abeles, P. W.: Partial Prestressing and Possibilities for Its Practical Application, *J. Prestressed Concrete Inst.*, June, 1959, pp. 35–51.
4. Hanson, J. A.: Prestress Loss as Affected by Type of Curing, Research and Develop. Labs. Portland Cement Assoc., *Research Dept. Bull.* D75.
5. Creep and Drying Shrinkage of Lightweight and Normal-Weight Concretes, *Natl. Bur. Std. Monograph* 74.

CHAPTER 11

Design of a Post-tensioned Girder

11-1. Choosing a Post-tensioned Member. The decision to use a post-tensioned member instead of a pretensioned one is influenced by so many different conditions that there are no rules which can be applied to determine the economical dividing line between the two. When the member is too large to be shipped from a casting yard to the job site, it is obvious that it must be cast at the job site and post-tensioned tendons will be required. When the size of the member is within shipping limitations, the following factors should be included in comparing the cost of the two methods:

1. Capacity of local casting beds. On long-span members the use of deflected tendons to offset dead weight is important. If facilities for deflecting enough strands for this purpose are not available, either a post-tensioned design or a pretensioned–post-tensioned combination is indicated.

2. Cross section of member. The cross section of a post-tensioned member is more efficient than that of a pretensioned member for the same loading if the web of the pretensioned member must be thickened appreciably to accommodate the deflected strands.

Post-tensioned members are used in bridges and buildings and for many special applications such as pile caps for piers. The girder designed in this chapter is a 100-ft-span roof girder such as might be used over a school gym.

11-2. Design Conditions. Design in accordance with ACI 318-63. Specifications for tendons are covered in Secs. 405(e) and (f) of ACI 318-63. (Portions of ACI 318-63 are reproduced in Appendix A.)

Span: 100 ft 0 in. center to center of bearings
Live load: 30 psf
Roofing: 10 psf
Double-T roof deck: 33 psf
Girder spacing: 30 ft center to center
Concrete: $f'_c = 5{,}000$ psi

Concrete: f'_{ci} = to be determined by calculations
Standard concrete at 150 lb per cu ft
Tendons to be grouted after tensioning operation is complete

11-3. Design Calculations. Calculations will follow the steps outlined in Chap. 3. The reader should thoroughly understand Chap. 4 before attempting to follow this analysis.

Step 1. Compute the properties of the cross section in Fig. 11-1.
Take moments about the bottom.

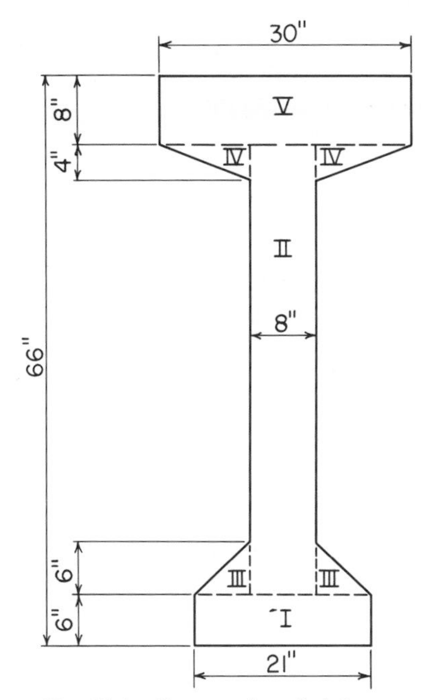

FIG. 11-1. Cross section of girder.

Section	*Area, A*	*y*	*Ay*	*Ay²*	*I₀*
I = 21 × 6	= 126	3	378	1,134	378
II = 8 × 52	= 416	32	13,300	426,000	93,750
III = 2 × 6.5 × 6 × ½	= 39	8	312	2,496	78
IV = 2 × 11 × 4 × ½	= 44	56.7	2,500	141,500	40
V = 30 × 8	= 240	62	14,900	923,000	1,280
	865		31,390	1,494,130	95,526
				95,526	
				1,589,656	

$$y_b = 31{,}390 \div 865 = 36.3$$
$$y_t = 66 - 36.3 = 29.7$$
$$I = 1{,}589{,}656 - 865(36.3^2) = 448{,}500$$

$$Z_t = 448{,}500 \div 29.7 = 15{,}100$$
$$Z_b = 448{,}500 \div 36.3 = 12{,}350$$
$$\text{Weight} = \frac{150 \times 865}{144} = 900 \text{ lb per ft}$$

Step 2. Compute stresses in the member due to its own dead weight.

$$M_G = \frac{900(100^2) \times 12}{8} = 13{,}500{,}000 \text{ in.-lb}$$
$$f^t{}_G = 13{,}500{,}000 \div 15{,}100 = +894 \text{ psi}$$
$$f^b{}_G = 13{,}500{,}000 \div 12{,}350 = -1{,}093 \text{ psi}$$

Step 3. Compute stresses in the member due to applied loads.

$$\begin{aligned} \text{Double T} &= 33 \text{ psf} \\ \text{Roofing} &= \underline{10} \text{ psf} \\ & \quad 43 \text{ psf} \end{aligned}$$

$$w_S = 43 \times 30 \text{ ft} = 1{,}290 \text{ lb per lin ft}$$
$$M_S = \frac{1{,}290(100^2) \times 12}{8} = 19{,}350{,}000 \text{ in.-lb}$$
$$f^t{}_S = 19{,}350{,}000 \div 15{,}100 = +1{,}281$$
$$f^b{}_S = 19{,}350{,}000 \div 12{,}350 = -1{,}567$$
$$\text{Live load} = 30 \times 30 = 900 \text{ lb per lin ft}$$
$$M_L = \frac{900(100^2) \times 12}{8} = 13{,}500{,}000 \text{ in.-lb}$$
$$f^t{}_L = 13{,}500{,}000 \div 15{,}100 = +894 \text{ psi}$$
$$f^b{}_L = 13{,}500{,}000 \div 12{,}350 = -1{,}093 \text{ psi}$$

Step 4. Determine the magnitude of the prestressing force and select the number of tendons.

This computation is based on conditions after all stress losses.

From ACI 318-63 Sec. 2605(b) the maximum allowable compressive stress after all losses is $0.45 f'_c$ or $0.45 \times 5000 = 2{,}250$ psi.

From the same section the maximum allowable tensile stress (when certain conditions are met) is $6\sqrt{f'_c}$ or $6\sqrt{5{,}000} = 425$ psi.

At this point we have sufficient information to check stresses in the bottom fiber of the girder and determine whether or not the section chosen is adequate under design-load conditions. Total tensile stresses in the bottom fiber are

$$f^b{}_G + f^b{}_S + f^b{}_L = 1{,}093 + 1{,}567 + 1{,}093 = 3{,}753 \text{ psi}$$

Since the specification permits a tensile stress of 425 psi under design load, the compressive stress that must be developed by the tendons is 3,753 − 425 = 3,328 psi. This is in excess of the 2,250 psi permitted by the specification, but we are working with a post-tensioned member and can place the tendons in a parabola so they are near the bottom of the girder at mid-span and near the neutral axis of the girder at the ends of the span. If the prestressing tendons are located and tensioned to provide a final bottom-fiber stress due to prestress of $f^b{}_F = 3{,}328$ psi, then the net bottom fiber stress including the girder's own dead weight will be

$$f^b{}_F - f^b{}_G = 3{,}328 - 1{,}093 = 2{,}235 \text{ psi}$$

which is just 15 psi under the allowable.

In a member of this type where the dead weight of the double-T roof members is always acting once the structure is completed, it might be argued that the temporary allowable stresses should apply until the girder is in place and supporting the roof members. If this approach were applied to this girder the net stress under final dead-load conditions would be

$$f^b{}_F - f^b{}_G - f^b{}_S = 3{,}328 - 1{,}093 - 1{,}567 = 668 \text{ psi}$$

Of course the most critical condition may be immediately after post-tensioning when the bottom-fiber stress due to prestress is greater than 3,328 psi because stress losses have not taken place and when the ultimate strength of the concrete may still be less than 5,000 psi. This condition will be checked after f'_{ci} has been determined.

In Chap. 3, Step 4 it is suggested that a first approximation of the prestressing force F be made using an approximate value for eccentricity of

$$e_{\text{approx}} = y_b - 0.15h$$

This approximate value of e is a good starting point for pretensioned members in which the individual seven-wire strands are spread out in a pattern on 2-in. centers both horizontally and vertically, but it is seldom applicable to post-tensioned tendons in which a large number of wires or strands are bunched together into a small cable. Another approach can be taken to establish an approximate e value for post-tensioned members.

The diameter of the flexible metal hose around a post-tensioned tendon seldom exceeds 2½ in. From ACI 318-63, Sec. 2616(a) the minimum cover over tendons in beams and girders is 1½ in. The distance from the bottom of a girder to the center of the bottom row of tendons is therefore 1¼ in. + 1½ in. = 2¾ in. If two rows of tendons are needed the vertical spacing between them would be the hose diameter = 2½ in. plus one and one-half times the aggregate size, or about 2½ in. + 1½ in. = 4 in.; and the distance from the bottom of the beam to the center of the entire group of tendons would be 2¾ in. + 2 in. = 4¾ in. (See ACI 318-63, Sec. 2617 for permissible arrangements of post-tensioned tendons.) With a little experience the designer will be able to make a reasonably accurate guess about the number of rows of tendons required in a particular member. For this girder we will estimate a full bottom row of tendons plus a second row that is not quite full so that the c.g.s. is a little below the midpoint between the two rows and we will use an approximate value of $e = y_b - 4.5 = 36.3 - 4.5 = 31.8$ in.

We can now substitute numerical values in Eq. (3-1) and get an approximate value for the final prestressing force F:

$$\frac{F}{A} + \frac{Fe}{Z_b} = f^b{}_G + f^b{}_S + f^b{}_L - f_{tp} \qquad (3\text{-}1)$$

$$\frac{F}{865} + \frac{31.8F}{12{,}350} = 1{,}093 + 1{,}567 + 1{,}093 - 425 = 3{,}328 \text{ psi}$$

Multiply both sides of the equation by 865:

$$F + 2.227F = 2{,}879{,}000$$
$$F = 892{,}000 \text{ lb}$$

In designing a pretensioned member it is standard procedure to select the size of seven-wire strand to be used and to work out a satisfactory pattern. This can be done because the properties of seven-wire strands have been standardized and they are available from a number of suppliers. Choosing the tendons for a post-tensioned member is another matter. There are several systems or methods of post-tensioning involving different types of tendons and anchor fittings, most of which are patented. The drawings and specifications must be so prepared that the proper magnitude and location of the prestressing force will be assured and also that each post-tensioning system can be used to the best advantage.

One common procedure is to specify the prestressing force and its location as shown in Fig. 11-2. (A would be given as a numerical value after calculations for Step 8 were completed.) The disadvantage to the method shown in Fig. 11-2 is that it is inflexible and therefore does not permit the most efficient use of the various systems. As an example let us assume that using the tendons of one system at full capacity, it takes 10.10 tendons to produce an F of 892,000 lb but that 10 tendons of this system can be arranged so that their c.g.s. is less than 4½ in. from the bottom, which gives a larger e than shown. The economical procedure would be to use 10 tendons and lower them enough to give the required compressive stress in the bottom fiber as long as this does not create excessive tensile stress in the top fiber. Working from the information given in Fig. 11-2 the bidder has no way of telling whether or not this method would work.

In order to obtain maximum efficiency in the choice of tendons the design drawings should give the properties of the concrete cross section and the permissible range of stresses due to the prestressing force.

The minimum value of $f^b{}_F$ must be large enough to keep the stress in the bottom fiber from exceeding -425 psi.

$$\text{Minimum } f^b{}_F = -f^b{}_G - f^b{}_S - f^b{}_L + f_{tp}$$
$$= 1{,}093 + 1{,}567 + 1{,}093 - 425 = +3{,}328$$

The maximum value of $f^b{}_F$ must keep the compressive stress under prestress plus dead load from exceeding $+2{,}250$ psi.

$$\text{Maximum } f^b{}_F = 2{,}250 - f^b{}_G$$
$$= 2{,}250 + 1{,}093 = +3{,}343 \text{ psi}$$

The minimum tensile stress $f^t{}_F$ must be large enough so that the compressive stress under full load does not exceed the allowable $+2{,}250$ psi.

$$\text{Minimum } f^t{}_F = 2{,}250 - f^t{}_G - f^t{}_S - f^t{}_L$$
$$= 2{,}250 - 894 - 1{,}281 - 894 = -819$$

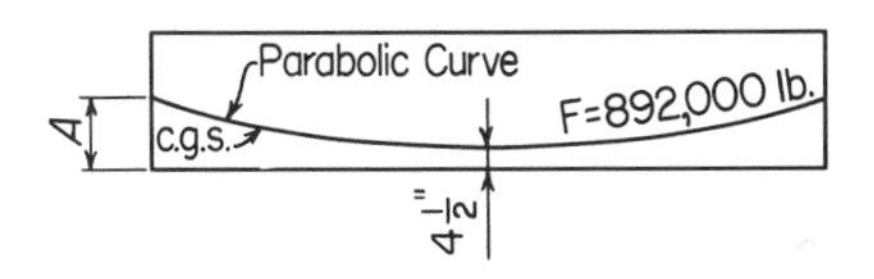

FIG. 11-2. Computed magnitude and location of prestressing force.

The maximum tensile stress in the top fiber must not exceed that permitted by ACI 318-63. Since the weight of the double T's will create a compressive stress in the top fiber and this weight will be acting all the time the structure is in service, the critical condition for tensile stress in the top fiber will be dead weight of girder only immediately after prestressing. ACI 318-63, Sec. 2605(a)2 gives an allowable value of $3\sqrt{f'_{ci}}$. Since the value of f'_{ci} has not yet been established we will be conservative and assume 4,000 psi. Then $3\sqrt{4{,}000} = 190$ psi. This value applies to the condition of initial tension and dead weight of girder or

$$f^t{}_{F_0} + f^t{}_G = -190$$
$$f^t{}_{F_0} = -190 - 894 = -1{,}084$$

All of the other values we have just established have been under the condition of final prestress. To convert this value to final prestress we will assume that the final prestress after all losses is 80 per cent of the initial prestress. Then

$$f^t{}_F = 80\% f^t{}_{F_0} = 80\%(-1{,}084) = -867 \text{ psi}$$

We have made two assumptions in computing $f^t{}_F$, but both are conservative and can be checked after we have established exact values for f'_{ci} and the ratio between F_o and F.

In summary, at the center of span

$$f^b{}_F = +3{,}328 \text{ to } +3{,}343 \text{ psi}$$
$$f^t{}_F = -819 \text{ to } -867 \text{ psi}$$

When the member is submitted to contractors for bids, these values as well as the properties of the cross section should be shown as illustrated in Fig. 11-3.

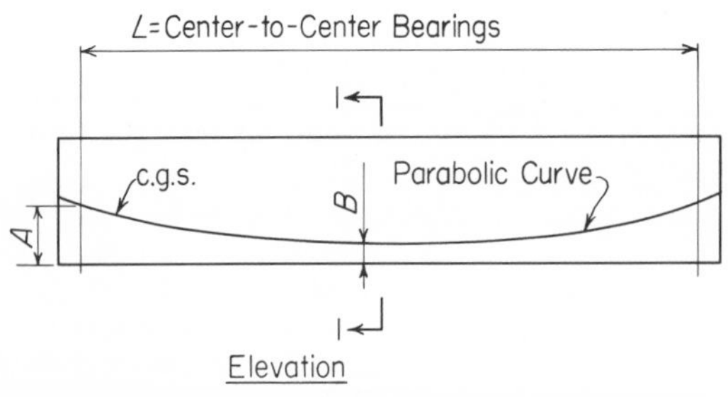

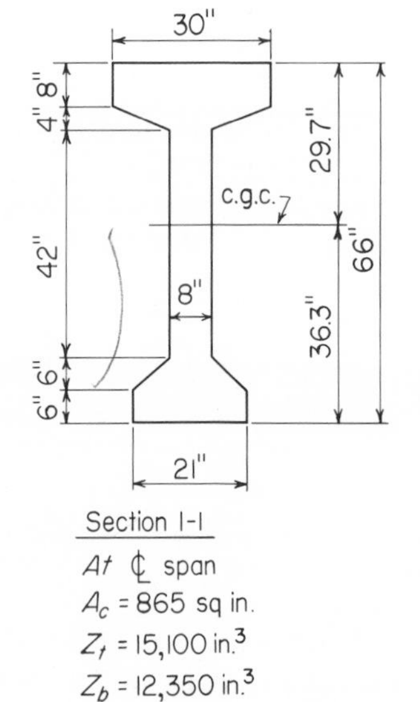

FIG. 11-3. Method recommended for showing prestressing requirements when submitting post-tensioned members for competitive bids.

Another method of presentation is to list several different values for B (Fig. 11-3) and to list the corresponding F required for each value of B. With this method, all calculations are done by the designer and the bidder need only select the most economical combination of B and F which the details of his tendons can suit.

Step 5. Locate the tendons in the cross section of the girder and compute the eccentricity provided.

Any of the post-tensioning systems illustrated in Chap. 8 could be considered. For this example we will choose the Prescon system and select tendons from Table 8-2.

As a general rule it is more economical to use a few large tendons than a number of small ones. The largest tendon shown in Table 8-2 has 30 wires and a final force of 212.0 kips at $0.6f'_s$. To provide 892 kips we would need 892 ÷ 212 = 4.21 cables; thus four would not be enough and five would be much more than needed. Apparently the smallest number of this type cable would be five, in which case each cable should have a final capacity of 892 ÷ 5 = 178.4 kips. A cable of 26 wires has a capacity of 183.8 kips at $0.6f'_s$. We will try five Prescon cables composed of twenty-six wires each, giving an F of 5 × 183.8 = 919 kips.

Note that Table 8-2 specifies that the force listed for each cable is at $0.6f'_s$. We must keep this in mind to be sure that conditions do not develop that would make the final tension less than this. The specification for the wire used in these tendons is ASTM Designation A 421 which calls for 0.250-in.-diameter wire to have a minimum ultimate strength of 240,000 psi. ACI 318-63, Sec. 2606(a)2 specifies a maximum initial tension of $0.70f'_s$ which would be 240,000 × 0.70 = 168,000 psi. In Art. 10-5 we suggest using 25,000 psi as the stress loss in a post-tensioned tendon. This would give a final stress of 168,000 − 25,000 = 143,000 psi, which is 143,000 ÷ 240,000 = 59.58% of f'_s and is close enough to 60 per cent to use the values given in Table 8-2. The tension in a post-tensioned cable at mid-span is less than at the jacks because of friction. Overtensioning is permitted to offset this problem. At a later point we will check to see if we can overtension these cables enough to completely offset the friction.

From Table 8-2 the outside diameter of the flexible metal conduit on a 26-wire cable is 2⅛ in. If we assume a maximum aggregate size of 1 in., the horizontal space between conduits must be 1½ in. and the minimum distance center to center of cables will be 2⅛ in. + 1½ in. or 3⅝ in. If all five cables are placed in one row the total width center to center of outside cables will be 4 × 3⅝ in. = 14.5 in. Since the bottom flange of the girder is 21 in. wide this leaves an edge distance of (21 − 14.5) ÷ 2 = 3.25 in. From ACI 318-63, Sec. 2616(a) the minimum required cover is 1½ in. One-half the conduit diameter is 1$\frac{1}{16}$ in. plus 1½ in. is 2$\frac{9}{16}$ in., the minimum distance from face of concrete to center of cable; thus 3¼ in. is more than enough. If necessary we could get maximum eccentricity by putting all the cables in one line 2$\frac{9}{16}$ in. above the bottom. At present we will keep the c.g.s. at 4½ in. above the bottom and use the pattern shown in Fig. 11-4. If friction or other factors make it necessary we can lower the c.g.s.

Compute c.g.s. for the cable pattern in Fig. 11-4:

$$\begin{array}{ll} 2 \times 3 & = 6 \\ \underline{3} \times 5.5 & = \underline{16.5} \\ 5 & \quad\;\; 22.5 \end{array}$$

$$22.5 \div 5 = 4.5 \text{ in.}$$

$$e = 36.3 - 4.5 = 31.8 \text{ in.}$$

Step 6. Establish the strength of the concrete and check top and bottom stresses at mid-span.

From Sec. 2605(a) of ACI 318-63 the temporary stresses immediately after transfer, before losses due to creep and shrinkage, shall not exceed $0.60f'_{ci}$ in compression or $3\sqrt{f'_{ci}}$ in tension. For the girder in this example the calculated stresses immediately after transfer include the weight of the girder.

The examples in Chaps. 4 and 5 dealt with pretensioned members in which the tendons were tensioned to $0.70f'_s$ while anchored to external

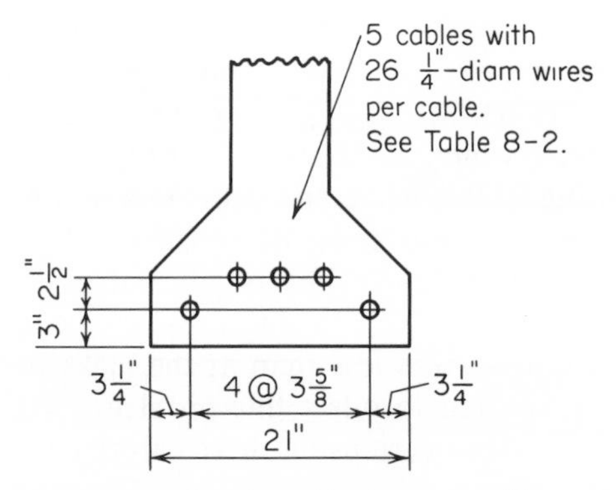

FIG. 11-4. Tentative cable pattern at mid-span.

abutments but lost approximately 15,000 psi at the transfer of load to the concrete because of elastic shortening of the concrete and relaxation of the tendons. In a pretensioned member, therefore, the prestressing force F_o immediately after transfer is less than that due to $0.70f'_s$. In a post-tensioned member elastic shortening takes place during the tensioning operation and is provided for by slight overtensioning of those tendons which are tensioned first. The full prestress force is effective as soon as the tendon is tensioned, so there is no loss at first due to tendon relaxation. Thus F_o in a post-tensioned member is the full initial tension at which the tendons are anchored.

It is necessary to establish the properties of the tendons and compute the initial prestress. From Step 5 there are five cables, each composed of twenty-six ¼-in.-diameter wires having a minimum ultimate strength of 240,000 psi. The desired final prestressing force is 892,000 lb.

$$A_s = 5 \times 26 \times 0.049088 = 6.38 \text{ sq in.}$$

The stress in the cables after all stress losses will be

$$f_{se} = 892{,}000 \div 6.38 = 139{,}800 \text{ psi}$$

Using the stress loss of 25,000 psi suggested in Art. 10-5 the initial stress in the tendons will be

$$139{,}800 + 25{,}000 = 164{,}800 \text{ psi}$$

from which

$$F_o = 164{,}800 \times 6.38 = 1{,}051{,}000 \text{ lb}$$

The initial stress is $164{,}800 \div 240{,}000 = 0.687f'_s$ which is less than $0.70f'_s$ and therefore within specification.

At mid-span stresses due to F_o are

$$f^t{}_{F_o} = \frac{F_o}{A_c} - \frac{eF_o}{Z_t} \quad \text{and} \quad f^b{}_{F_o} = \frac{F_o}{A_c} + \frac{eF_o}{Z_b}$$

$$f^t{}_{F_o} = \frac{1{,}051{,}000}{865} - \frac{31.8 \times 1{,}051{,}000}{15{,}100} = 1{,}215 - 2{,}214 = -999 \text{ psi}$$

$$f^b{}_{F_o} = \frac{1{,}051{,}000}{865} + \frac{31.8 \times 1{,}051{,}000}{12{,}350} = 1{,}215 + 2{,}706 = +3{,}921 \text{ psi}$$

Net stresses at mid-span from F_o plus dead weight of girder are

$$f^t{}_{F_o+G} = -999 + 894 = -105 \text{ psi}$$
$$f^b{}_{F_o+G} = 3{,}921 - 1{,}093 = +2{,}828 \text{ psi}$$

Inspection of these two stresses suggests that the bottom-fiber stress is most critical. In order to meet the specification this must not exceed $0.60f'_{ci}$; therefore the minimum value of f'_{ci} will be

$$f'_{ci} = 2{,}828 \div 0.60 = 4{,}710 \text{ psi}$$

For this value of f'_{ci} the allowable tensile stress is $3\sqrt{4{,}710} = 3 \times 68.6 = 206$ psi, which is greater than the existing 105 psi and therefore satisfactory.

In a post-tensioned member the pressure under the bearing plates must also be considered in establishing f'_{ci}. Section 2605(c) of ACI 318-63 says bearing stress shall not exceed

$$f_{cp} = 0.60f'_{ci}\sqrt[3]{\frac{A'_b}{A_b}} \qquad \text{(26-1 ACI)}$$

but not greater than f'_{ci}.

From the end block details in Fig. 11-7 we get

$A_b = 7 \times 11\frac{1}{2} = 80.5$ sq in.
$A'_b = 10 \times 14\frac{1}{2} = 145$ sq in.
F_o for one cable $= 1{,}051{,}000 \div 5 = 210{,}200$ lb
$f_{cp} = 210{,}200 \div 80.5 = 2{,}610$ psi actual

Using the f'_{ci} of 4,710 psi computed on the basis of flexural stress requirements and applying Eq. (26-1 ACI) we get an allowable bearing pressure of

$$f_{cp} = 0.6(4{,}710)\sqrt[3]{\frac{145}{80.5}}$$
$$= 0.6(4{,}710)1.215 = 3{,}430 \text{ psi allowable}$$

which is greater than the actual pressure of 2,610.

Thus the bottom-fiber stress at mid-span is the governing factor and $f'_{ci} = 4{,}710$ psi.

The only loading condition that is more critical under F_o than under F is F_o plus dead weight of girder only, and the top-fiber and bottom-fiber stresses for that condition have just been checked.

Check top-fiber and bottom-fiber stresses under final prestress plus girder only and final prestress plus all applied loads.

$$f^t{}_F = \frac{892{,}000}{865} - \frac{892{,}000 \times 31.8}{15{,}100} = 1{,}031 - 1{,}879 = -848 \text{ psi}$$
$$f^b{}_F = \frac{892{,}000}{865} + \frac{892{,}000 \times 31.8}{12{,}350} = 1{,}031 + 2{,}297 = +3{,}328 \text{ psi}$$
$$f^t{}_F + f^t{}_G = -848 + 894 = +46 \text{ psi}$$
$$f^b{}_F + f^b{}_G = 3{,}328 - 1{,}093 = +2{,}235 \text{ psi}$$
$$f^t{}_{F+G+S+L} = -848 + 894 + 1{,}281 + 894 = +2{,}221 \text{ psi}$$
$$f^b{}_{F+G+S+L} = 3{,}328 - 1{,}093 - 1{,}567 - 1{,}093 = -425 \text{ psi}$$

All of these stresses are within the allowables of +2,250 to −425 psi.

Step 7. Check ultimate flexural strength to make sure it meets the requirements of ACI 318-63, Secs. 2608 and 2609. Check percentage of prestressing steel.

Section 2608 gives two equations for computing the ultimate flexural capacity M_u of a prestressed concrete member. Equation (26-4 ACI) is applicable to rectangular sections or flanged sections in which the neutral axis lies within the flange. Equation (26-5 ACI) is applicable to flanged sections in which the neutral axis lies outside the flange. The location of the neutral axis is determined by evaluating the term

$$1.4dp\frac{f_{su}}{f'_c} \tag{4-2}$$

If the flange thickness is more than Eq. (4-2), Eq. (26-4 ACI) is applicable. If the flange thickness is less than Eq. (4-2), Eq. (26-5 ACI) is applicable.

Establish numerical values for the symbols in Eq. (4-2) and evaluate the term.

$$d = y_t + e = 29.7 + 31.8 = 61.5 \text{ in.}$$
$$p = \frac{A_s}{bd} = 6.38 \div (61.5 \times 30) = 0.00346$$

From ACI 318-63, Sec. 2608(a)3,

$$f_{su} = f'_s\left(1 - 0.5\frac{pf'_s}{f'_c}\right) \qquad \text{(26-6 ACI)}$$

Substituting numerical values for Eq. (26-6 ACI),

$$f_{su} = 240{,}000\left(1 - \frac{0.5 \times 0.00346 \times 240{,}000}{5{,}000}\right)$$

$$= 240{,}000(1 - 0.083) = 220{,}000 \text{ psi}$$

Substituting numerical values in Eq. (4-2)

$$\frac{1.4 \times 61.5 \times 0.00346 \times 220{,}000}{5{,}000} = 13.11$$

Since the flange thickness is less than this, Eq. (26-5 ACI) is applicable.
Solve Eq. (26-5 ACI):

$$M_u = \phi\left[A_{sr}f_{su}d\left(1 - \frac{0.59A_{sr}f_{su}}{b'df'_c}\right) + 0.85f'_c(b - b')t(d - 0.5t)\right] \qquad \text{(26-5 ACI)}$$

From ACI 318-63, Sec. 1504(b), $\phi = 0.90$.
From ACI 318-63 Sec. 2608(a)2,

$$A_{sr} = A_s - A_{sf} \qquad \text{and} \qquad A_{sf} = 0.85f'_c(b - b')\frac{t}{f_{su}}$$

From Fig. 11-1, $b = 30$ in. and $b' = 8$ in.
For t use average flange thickness which is equal to the area of the flange divided by b.

$$\text{Area of flange} = (30 \times 8) + (2 \times 11 \times 4 \times ½) = 284$$

$$t = 284 \div 30 = 9.47$$

$$A_{sf} = \frac{0.85(5{,}000)(30 - 8)(9.47)}{220{,}000} = 4.02$$

$$A_{sr} = 6.38 - 4.02 = 2.36$$

Substituting numerical values in Eq. (26-5 ACI),

$$M_u = 0.90[2.36 \times 220{,}000 \times 61.5\left(1 - \frac{0.59 \times 2.36 \times 220{,}000}{8 \times 61.5 \times 5{,}000}\right) + 0.85 \times 5{,}000(30 - 8)9.47(61.5 - 0.5 \times 9.47)]$$

$$= 0.90[27{,}955{,}000 + 50{,}265{,}000] = 70{,}398{,}000 \text{ in.-lb}$$

From ACI 318-63 the required ultimate moment is

$$U = 1.5D + 1.8L \qquad \text{(15-1 ACI)}$$

For this example,

$$M_D = M_G + M_S = 13{,}500{,}000 + 19{,}350{,}000 = 32{,}850{,}000 \text{ in.-lb}$$
$$M_L = 13{,}500{,}000 \text{ in.-lb}$$

Therefore the required capacity is

$$\begin{aligned} M_u &= 1.5(32{,}850{,}000) + 1.8(13{,}500{,}000) \\ &= 49{,}275{,}000 + 24{,}300{,}000 = 73{,}575{,}000 \text{ in.-lb} \end{aligned}$$

Since this is slightly larger than the actual capacity of the girder which was computed as 70,398,000 in.-lb using Eq. (26-5 ACI), it will be necessary to revise some of the details. We can provide additional ultimate flexural capacity by adding some reinforcing steel in the critical region or by increasing the size of the tendons. If the size of the tendons is increased, a new initial tension should be computed so that the new tendons still give the desired F of 892,000 lb. This means that the tendons will be working at less than full capacity under design load but that their additional strength will be available to give the required ultimate.

For this example we will add some reinforcing steel bars with a yield strength f_y of 40,000 psi. [See ACI 318-63, Sec. 2608(a)4.]

The total tension in the cables at ultimate moment is

$$A_s f_{su} = 6.38 \times 220{,}000 = 1{,}403{,}600 \text{ lb}$$

This developed an ultimate moment capacity of 70,398,000 in.-lb.

The moment capacity to be developed by the reinforcing bars is 73,575,000 – 70,398,000 = 3,177,000 in.-lb. The approximate tensile force which the reinforcing bars must develop is therefore

$$\frac{3{,}177{,}000 \times 1{,}403{,}600}{70{,}398{,}000} = 63{,}350 \text{ lb}$$

At 40,000 psi the required area is

$$A'_s = 63{,}350 \div 40{,}000 = 1.58 \text{ sq in.}$$

Three #7 bars give $A'_s = 3 \times 0.60 = 1.80$ sq in.

The cable pattern at mid-span can be rearranged as shown in Fig. 11-5 to keep the same eccentricity as shown in Fig. 11-4 and used in all the calculations and still provide room for the reinforcing steel. Compute c.g.s. of cables in Fig. 11-5.

$$\begin{aligned} 2 \times 2\tfrac{5}{8} &= 5.25 \\ 3 \times 5\tfrac{3}{4} &= 17.25 \\ \overline{5} \qquad & \overline{22.50} \div 5 = 4.50 \text{ in.} \end{aligned}$$

The reinforcing bars should not be needed for the full length of the girder. After the path of the cables has been established, in a later step the ultimate flexural capacity of the girder with cables only should be computed at some point along the girder, perhaps the one-quarter point, and compared with the required ultimate moment at that point to see if all the bars are still needed. Be sure to compute the new value of d for the point being considered. It will be less than at mid-span.

Check percentage of steel in accordance with ACI 318-63, Sec. 2609(a). Since this is a flanged section the applicable requirement is: "For flanged sections, p shall be taken as the steel ratio of only that portion of the total tension steel area which is required to develop the compressive strength of the web alone." In other words the ratio $A_{sr}f_{su}/b'df'_c$ shall not exceed 0.30.

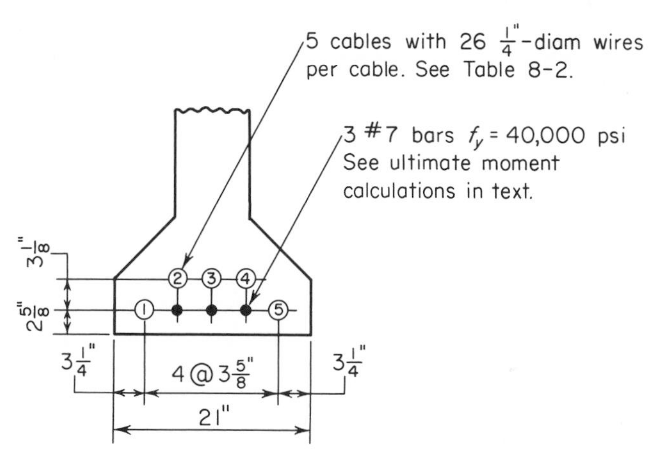

FIG. 11-5. Final cable pattern plus reinforcing steel at mid-span.

Substituting numerical values in this ratio gives

$$(2.36 \times 220{,}000) \div (8 \times 61.5 \times 5000) = 0.211$$

which is less than 0.30.

ACI 318-63, Sec. 2608(a)4 must also be satisfied for a beam with non-prestressed reinforcement. It says the following ratio must not exceed 0.30.

$$\frac{pf_{su}}{f'_c} + \frac{p'f_y}{f'_c}$$

From previous calculations,

$$p = 0.00346 \qquad f_{su} = 220{,}000 \text{ psi} \qquad f'_c = 5{,}000 \text{ psi}$$
$$f_y = 40{,}000 \text{ psi} \qquad A'_s = 1.80 \text{ sq in.}$$
$$p' = 1.80/(30 \times 61.5) = 0.00098$$

Substituting in the ratio,

$$\frac{0.00346 \times 220{,}000}{5{,}000} + \frac{0.00098 \times 40{,}000}{5{,}000} = 0.152 + 0.008 = 0.160$$

which is less than 0.30.

ACI 318-63, Sec. 2609(c) says: "The total amount of prestressed and unprestressed reinforcement shall be adequate to develop an ultimate load in flexure of at least 1.2 times the cracking load calculated on the basis of a modulus of rupture of $7.5\sqrt{f'_c}$." See Step 7, Chap. 4 for further discussion.

$$\begin{aligned} \text{Cracking moment} &= (7.5\sqrt{f'_c} + f^b{}_F)Z_b \\ &= (7.5\sqrt{5{,}000} + 3{,}328)\,12{,}350 = 47{,}646{,}000 \text{ in.-lb} \\ 1.2 \times 47{,}646{,}000 &= 57{,}175{,}000 \text{ in.-lb} \end{aligned}$$

Since the ultimate moment capacity of the member as computed earlier in this step is greater than 57,175,000 in.-lb, the member meets the requirement of Sec. 2609(c).

Step 8. Establish path of tendons and check critical points along the girder both for initial and final prestressing forces.

Since all the loads on this member are uniform loads and the bending-moment curve of a uniform load is a parabola, the logical curve for the tendons is also a parabola. The lowest point of the parabola at the center of span is already established. If we can establish the highest points, at the ends of the member, and know the equation of the curve, the entire curve will be established. Within certain limits the location of the elevation of the ends of the curve is usually a matter of choice rather than design. It must be high enough so that the stresses it creates at the end of the member are within the allowable, and it must permit suitable details for the tendon anchors. Unless some condition indicates otherwise, a most convenient location is the c.g.c. For this example, we shall locate the end of the curve on the c.g.c.

The parabolic bending-moment curve due to a uniform load is shown in Fig. 11-6*a*. The parabolic curve of the cables is shown in Fig. 11-6*b*. These curves have the same equation except that in the bending-moment curve B and y are measured in foot-pounds, whereas in the cable curve B and y are measured in inches.

Now that the curve of the tendons is established, we can plot the stresses at points along the span. Table 11-1 shows the critical condition under initial prestress plus the dead load of the girder only. Since the section modulus of the girder is constant for its full length, the stresses due to dead load will vary in accordance with the equation in Fig. 11-6, and we already know the stresses at the center of span.

We can use Eq. (1-1) to compute the stresses due to F_o

$$f_{F_o} = \frac{F_o}{A_c} \pm \frac{F_o e}{Z} \qquad (1\text{-}1)$$

Substituting in Eq. (1-1) for stresses due to F_o at the ends of the span (where e equals zero because the cables are on the c.g.c.), we get

$$f_{F_o} = \frac{F_o}{A_c} \pm \frac{F_o(\text{zero})}{Z} = \frac{F_o}{A_c}$$

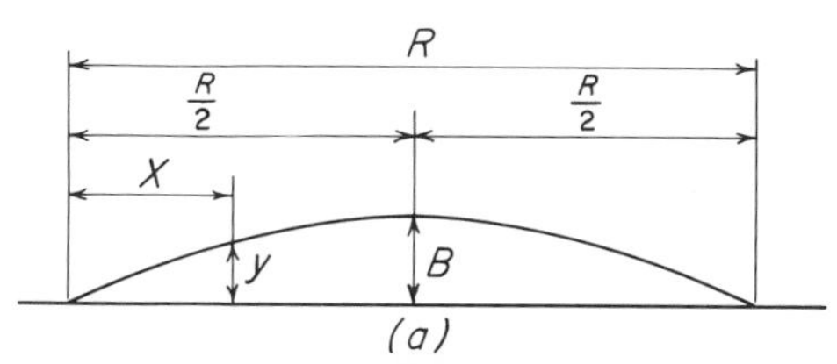

Curve of Bending Moment Due to Uniform Load

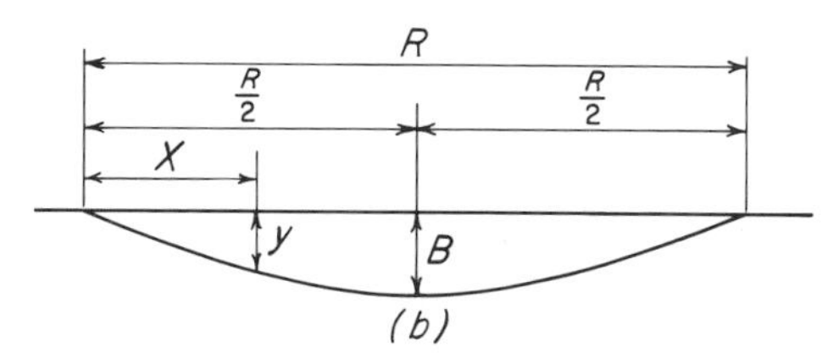

Curve of Parabolic Path of Tendons

Equation for curves *(a)* and *(b)*

$$y = 4B\,\frac{X}{R}\left(1 - \frac{X}{R}\right)$$

Values of y for X at Tenth Points of Span as Computed from Above Equation

X	y
0	0
0.1R or 0.9R	0.36B
0.2R or 0.8R	0.64B
0.3R or 0.7R	0.84B
0.4R or 0.6R	0.96B
0.5R	B

$L = R\,(1 + 2.67\,n^2)$

L = length along parabolic curve

$n = \frac{B}{R}$

For tangent to curve see Eq. 11-1

For unsymmetrical parabola see Fig. 14-3

FIG. 11-6. Parabolic curves for uniform bending moment and path of tendons.

which means that the stress is uniform over the full depth of the member and equal to

$$\frac{F_o}{A_c} = \frac{1{,}051{,}000}{865} = +1{,}215 \text{ psi}$$

In the span the stress in the top fiber due to F_o is

$$f^t{}_{F_o} = \frac{F_o}{A_c} - \frac{F_o e}{Z_t} \qquad (1\text{-}1a)$$

In this equation F_o/A_c is a constant value of $+1{,}215$ psi and is so listed in Table 11-1. $-F_o e/Z_t$ is made up of two constants F_o and Z_t and the variable e. In Fig. 11-6*a*, e is equal to Y, and at mid-span $e = Y = B$. Thus at mid-span,

$$-\frac{F_o e}{Z_t} = -\frac{1{,}051{,}000 \times 31.8}{15{,}100} = -2{,}214 = B$$

$$\frac{F_o e}{Z_b} = \frac{1{,}051{,}000 \times 31.8}{12{,}350} = +2{,}706 = B$$

The values of $-F_o e/Z_t$ and $F_o e/Z_b$ at the tenth points of the span can be computed using the formula or table of coefficients in Fig. 11-6.

We now have sufficient data to complete Table 11-1.

The figures in column 10 of Table 11-1 show that the compression stress at points along the span is always less than that at mid-span, which is 2,828 psi. This is within allowable stresses because it is the value used in Step 6 to establish f'_{ci} as 4,710 psi. All of the top-fiber stresses in column 6 are well within the allowable.

Table 11-2 shows the critical condition under final prestress plus all applied loads. Stresses at the tenth points of the span are based on the parabolic curves in Fig. 11-6 in the same manner as those in Table 11-1.

$$\frac{F}{A_c} = \frac{892{,}000}{865} = +1{,}031$$

At the center of span,

$$-\frac{Fe}{Z_t} = \frac{892{,}000 \times 31.8}{15{,}100} = -1{,}878$$

$$+\frac{Fe}{Z_b} = \frac{892{,}000 \times 31.8}{12{,}350} = +2{,}297$$

Using these values we shall complete Table 11-2.

The figures in columns 6 and 10 of Table 11-2 show that the stresses are within the allowable for the full length of the member.

If the magnitude of the net stresses is not clear to the reader from Tables 11-1 and 11-2, he should plot stress diagrams similar to those in Chaps. 4 and 5 using the stresses listed in these tables as a basis for the diagrams.

The elevation of the c.g.s. at the tenth points of the span is developed using the equation in Fig. 11-6.

Step 9. Design of Shear Steel. We will use the simplified method which is demonstrated in Step 9B of Chap. 4. As shown in Chap. 4 three calculations are generally all that are required.

Calculation 1. Solve Eq. (26-11 ACI) from ACI 318-63 for A_v. This is the minimum shear steel permitted and will be used for the full

Table 11-1. Critical Conditions under Initial Prestress

X (1)	$\frac{F_o}{A_c}$ (2)	$-\frac{F_oe}{Z_t}$ (3)	$f^t{}_{F_o} = 2 + 3$ (4)*	$f^t{}_G$ (5)	$f^t{}_{F_o+G}$ (6)	$\frac{F_oe}{Z_b}$ (7)	$f^b{}_{F_o} = 2 + 7$ (8)†	$f^b{}_G$ (9)	$f^b{}_{F_o+G}$ (10)
0	+1,215	0	+1,215	0	+1,215	0	+1,215	0	+1,215
$0.1L$ or $0.9L$	+1,215	−797	+418	+322	+740	+974	+2,189	−393	+1,796
$0.2L$ or $0.8L$	+1,215	−1,417	−202	+572	+370	+1,732	+2,947	−700	+2,247
$0.3L$ or $0.7L$	+1,215	−1,860	−645	+750	+105	+2,273	+3,488	−918	+2,570
$0.4L$ or $0.6L$	+1,215	−2,125	−910	+858	−52	+2,598	+3,813	−1,050	+2,763
$0.5L$	+1,215	−2,214	−999	+894	−105	+2,706	+3,921	−1,093	+2,828

* $f^t{}_{F_o}$ = column 2 plus column 3.
† $f^b{}_{F_o}$ = column 2 plus column 7.

Table 11-2. Critical Conditions under Final Prestress

X (1)	$\frac{F}{A_c}$ (2)	$-\frac{Fe}{Z_t}$ (3)	$f^t{}_F = 2 + 3$ (4)*	$f^t{}_{G+S+L}$ (5)	$f^t{}_{F+G+S+L}$ (6)	$\frac{Fe}{Z_b}$ (7)	$f^b{}_F = 2 + 7$ (8)†	$f^b{}_{G+S+L}$ (9)	$f^b{}_{F+G+S+L}$ (10)
0	+1,031	0	+1,031	0	+1,031	0	+1,031	0	+1,031
$0.1L$ or $0.9L$	+1,031	−676	+355	+1,105	+1,460	+827	+1,858	−1,351	+507
$0.2L$ or $0.8L$	+1,031	−1,202	−171	+1,964	+1,793	+1,470	+2,501	−2,402	+99
$0.3L$ or $0.7L$	+1,031	−1,578	−547	+2,578	+2,031	+1,930	+2,961	−3,153	−192
$0.4L$ or $0.6L$	+1,031	−1,803	−772	+2,946	+2,174	+2,205	+3,236	−3,603	−367
$0.5L$	+1,031	−1,878	−847	+3,069	+2,222	+2,297	+3,328	−3,753	−425

* $f^t{}_F$ = column 2 plus column 3.
† $f^b{}_F$ = column 2 plus column 7.

length of the beam except at points where subsequent calculations indicate that more is required.

From ACI 318-63,

$$A_v = \frac{A_s}{80} \times \frac{f'_s}{f_y} \times \frac{s}{d}\sqrt{\frac{d}{b'}} \qquad \text{(26-11 ACI)}$$

For this example the terms in Eq. (26-11 ACI) have the following values

$A_s = 6.38 \qquad f'_s = 240{,}000$ psi
f_y will be set $= 40{,}000$ psi
$b' = 8$ in.

From ACI 318-63, Sec. 2610(c) the maximum spacing of shear reinforcement is three-fourths the depth of the member but not more than 24 in.; thus for this example the maximum allowable is $s = 24$ in. From ACI 318-63, Sec. 2610(a)(1) the value of d is the effective depth at the section of maximum moment, thus $d = 61.5$ in.

Substituting numerical values in Eq. (26-11 ACI)

$$A_v = \frac{6.38 \times 240{,}000 \times 24}{80 \times 40{,}000 \times 61.5}\sqrt{\frac{61.5}{8}} = 0.52 \text{ sq in.}$$

Two #4 bars $= 2 \times 0.20 = 0.40$ sq in.

$$\frac{0.40 \times 24}{0.52} = 18.46 \text{ in.}$$

Minimum stirrups at any point along the girder will be two #4 bars at 18.46 in. center to center.

Calculation 2. Solve Eq. (26-13 ACI) of ACI 318-63 for V_{cw} at station $d/2$. Substitute this value of V_{cw} for V_c in Eq. (26-10 ACI) of ACI 318-63 and compute A_v. If this value of A_v is less than that computed from Eq. (26-11 ACI), Calculation 1, Eq. (26-11 ACI) governs and V_{cw} should not be critical at any station.

From ACI 318-63,

$$V_{cw} = b'd(3.5\sqrt{f'_c} + 0.3f_{pc}) + V_p \qquad \text{(26-13 ACI)}$$

The terms in Eq. (26-13 ACI) have the following values:

$b' = 8 \qquad f'_c = 5{,}000$

The value of d used in Eq. (26-13 ACI) is not the same as that used in Eq. (26-11 ACI). From ACI 318-63, Sec. 2610(b): "When applying Eq. (26-13) and (26-13A) the effective depth, d, shall be taken as the distance from the extreme compression fiber to the centroid of the prestressing tendons, or as 80 percent of the over-all depth of the member, whichever is the greater." At station $d/2$ the value of d will be slightly larger than at the support where it is 29.7 in. Eighty per cent of the overall depth is $0.80 \times 66 = 52.8$ in. which is obviously greater; thus $d = 52.8$.

See ACI 318-63, Sec. 2600 for the definition of f_{pc}. For a noncomposite member such as the one in this example, $f_{pc} = F/A_c = 1{,}215$ psi. (In the example in Chap. 4 the value of F was decreased because pretensioned tendons were being used and their full tension had not yet been transferred to the concrete by bond at the point $d/2$ being considered. In this post-tensioned member the full prestress force is applied through the bearing plate at the end of the girder and is effective from that point on.

V_p = vertical component of the effective prestress force at the section considered.

Using the symbols in Fig. 11–6*b*, the tangent of the angle of the curve with the horizontal at a point X from the end of the curve is

$$\tan \phi = \frac{4B}{R}\left(1 - \frac{2X}{R}\right) \tag{11-1}$$

Substituting numerical values in Eq. (11-1) for the point under consideration,

$$X = 66 \div 2 = 33 \text{ in.} \div 12 = 2.75 \text{ ft}$$
$$B = 31.8 \div 12 = 2.65 \text{ ft}$$
$$\tan \phi = \frac{4 \times 2.65}{100}\left(1 - \frac{2 \times 2.75}{100}\right) = 0.1002$$

The shear carried by the tendons at this point is equal to the vertical component of the final tension in the tendons, which is

$$V_p = 0.1002 \times 892{,}000 = 89{,}400 \text{ lb}$$

Substituting numerical values in Eq. (26-13 ACI),

$$V_{cw} = 8 \times 52.8(3.5\sqrt{5{,}000} + 0.3 \times 1{,}215) + 89{,}400$$
$$= 258{,}500 + 89{,}400 = 347{,}900 \text{ lb}$$

This value for V_{cw} is now substituted for V_c in ACI 318-63, Eq. (26-10 ACI) which is

$$A_v = \frac{(V_u - \phi V_c)s}{\phi d f_y} \tag{26-10 ACI}$$

From ACI 318-63, Sec. 2610(a) the value for d to be used in Eq. (26-10 ACI) is (for this example) $d = 61.5$.

From Sec. 1504(b) for diagonal tension $\phi = 0.85$.

We have already set $f_y = 40{,}000$ psi.

The maximum allowable spacing, used in applying Eq. (26-11 ACI) is $s = 24$ in.

From ACI 318-63, Sec. 1506(a) the required ultimate load capacity (where wind and earthquake are not applied) is $U = 1.5D + 1.8L$.

For this example

$$w_{G+S} = 900 + 1{,}290 = 2{,}190 \text{ lb per ft}$$

and $$w_L = 900 \text{ lb per ft}$$

Thus the ultimate applied load is

$$1.5(2{,}190) + 1.8(900) = 4{,}905 \text{ lb per ft}$$

At $X = 2.75$,

$$V_u = 4{,}905(50 - 2.75) = 231{,}800 \text{ lb}$$

Substituting numerical values in Eq. (26-10 ACI),

$$A_v = \frac{(231{,}800 - 0.85 \times 347{,}900)24}{0.85 \times 61.5 \times 40{,}000}$$

Since $0.85 \times 347{,}900$ is greater than 231,800 the result is a negative number which means that the shear carrying capacity of the tendons plus the concrete without shear steel is greater than the applied ultimate shear and therefore no stirrups are needed on the basis of diagonal tension. The shear steel computed from Eq. (26-11 ACI) will be used in this area.

Calculation 3. Compute V_{ci} at station $L/4$ using Elstner's formula which says that at station $L/4$

$$V_{ci} = 0.6b'd\sqrt{f'_c} + (0.33w_{S+L} + 0.250w_G)L$$

Substitute this value of V_{ci} in Eq. (26-10 ACI) and solve for A_v.

The simplified method for computing V_{ci} which was developed by Elstner is discussed in Appendix C. Formulas for other stations along the beam and examples of their application are given in Chap. 4, Step 9B and Appendix C. In these formulas

$$w_G = \text{weight of beam in kips per linear foot}$$
$$w_{S+L} = \text{all applied load in kips per linear foot}$$

The units to be used in these equations for w_G and w_{S+L} are design load values, they are not ultimate load values.

$$w_{S+L} = 1.29 + 0.90 = 2.19 \text{ kips per ft}$$
$$w_G = 0.90 \text{ kips per ft}$$

Substituting in Elstner's formula for $L/4$,

$$V_{ci} = 0.6 \times 8 \times 61.5\sqrt{5{,}000} + (0.333 \times 2.19 + 0.25 \times 0.90)100$$
$$= 20.87 + 95.43 = 116.30 \text{ kips}$$

At station $L/4$

$$V_u = 4{,}905(50 - 25) = 122{,}600 \text{ lb}$$

Substituting in Eq. (26-10 ACI) from ACI 318-63

$$A_v = \frac{(122{,}600 - 0.85 \times 116{,}300)24}{0.85 \times 61.5 \times 40{,}000}$$
$$= 0.27 \text{ sq in.}$$

This is less than the 0.52 sq in. computed in Calculation 1 using Eq. (26-11 ACI) and the same spacing of 24 in. Equation (26-11 ACI) is therefore the governing factor at $L/4$, $d/2$, and thus for the full length of the girder.

Step 10. Compute camber and deflection.

In Step 8 we deliberately set the curve of the c.g.s. in a parabolic path with e equal to zero at the ends of the member and 31.8 in. at mid-span. The bending moment due to F_o is F_oe. Since e varies in a parabolic curve, the bending moment due to F_o will also vary in a parabolic curve from zero at the ends of the member to F_oe at mid-span. Since a uniform load produces a parabolic bending-moment curve, it is apparent that the tendons exert a uniform vertical load w_T against the member. Find w_T.

$$F_oe = 1{,}051{,}000 \times 31.8 = 33{,}422{,}000 \text{ in.-lb} = 2{,}785{,}000 \text{ ft-lb}$$

The formula for bending moment due to a uniform load is

$$M = \frac{wl^2}{8} \qquad \text{or} \qquad w = \frac{8M}{l^2}$$

Substituting,

$$w_T = \frac{8 \times 2{,}785{,}000}{100^2} = 2{,}288 \text{ lb per ft}$$

The vertical uplift of 2,228 lb per ft exerted by the tendons is offset by the 900 lb per ft dead weight of the girder so that the net uplift causing camber is 2,228 − 900 = 1,328 lb per ft.

From Fig. 6-1 and also from ACI 318–63, Sec. 1102(a),

$$E_c = w^{1.5}\, 33\sqrt{f'_c}$$

We are using concrete at 150 lb per cu ft and have set its strength at time of prestressing as 4,710 psi. Then

$$E_c = 150^{1.5}(33)\sqrt{4{,}710} = 4.16(10^6)$$

The standard formula for deflection (or camber) due to a uniform load is

$$\Delta_{G+F} = \frac{5wl^4}{384EI} = \frac{5 \times 1{,}328(100^4)12^3}{384 \times 4.16(10^6)448{,}500} = 1.60 \text{ in.}$$

Thus the camber at the completion of the prestressing operation will be approximately 1.60 in. The erection schedule will have a definite influence on the total camber growth of the girder. If the dead load of the roof is applied shortly after the girder is prestressed the camber growth in that short interval will be small. If the girder is allowed to stand unloaded for a relatively long period after prestressing it will probably experience a considerable increase in the 1.60-in. camber.

We will assume that the roof members are placed within a few days after the prestressing operation.

For $f'_c = 5{,}000$ psi, $E_c = 4.29(10^6)$

The roof weight is $w_S =$ 1,290 lb per ft

$$\Delta_S = \frac{5 \times 1{,}290(100^4)12^3}{384 \times 4.29(10^6)448{,}500} = 1.51 \text{ in.}$$

The net camber under prestress plus dead load of member plus dead load of roof (assuming no camber growth before placement of roof) would be

$$\Delta_{G+F} - \Delta_S = 1.60 - 1.51 = 0.09 \text{ in.}$$

If camber growth is 150 per cent, the final camber will be

$$0.09 + 1.5(0.09) = 0.22 \text{ in.}$$

Deflection under live load of 900 lb per ft will be

$$\Delta_L = \frac{5 \times 900(100^4)12^3}{384 \times 4.29(10^6)448{,}500} = 1.05 \text{ in.}$$

11-4. End-block Details. Section 2614 of ACI 318-63 indicates that end blocks shall be provided when they are needed and that they shall be reinforced to resist the stresses to which they are subjected. It does not include formulas for computing stresses or required steel dimensions.

Where a complete analysis of the stresses in the end block of a post-tensioned member is desired, the theory developed by Y. Guyon is recognized as being one of the most accurate available. It is presented in his book "Prestressed Concrete" published by John Wiley & Sons, Inc.

Formulas, charts and examples covering design of reinforcing in the end blocks of post-tensioned members on the basis of Guyon's theory are presented in Appendix E of the "Prestressed Handbook" published by the Canadian Prestressed Concrete Institute in Toronto, Ontario, Canada. Charts showing computed spalling and bursting stresses in the end of a post-tensioned beam are included in the article Practical Analysis of the Anchorage Zone Problem in Prestressed Beams by Rolf J. Lenschow and Mete A. Sozen, published in the *Journal of the American Concrete Institute,* November, 1965, pages 1421 to 1439.

In the Tentative Recommendations for Prestressed Concrete prepared by the ACI-ASCE Committee 323 and published in 1958, Sec. 214.3 begins with the sentence: "End blocks are usually proportioned by experience." This is still true in most cases. A few empirical values are suggested herewith.

The different post-tensioning systems with different anchorage details will naturally develop different stress patterns in the concrete under the anchors. The supplier of the anchor fitting should always be asked for his recommendation of reinforcing steel to be used in connection with his anchor.

In lieu of a recommendation from the supplier of the anchor, a grid of #2 bars at 3-in. centers in each direction may be placed about 1½ in. from the inside face of the anchorage bearing plate.

Vertical stirrups, in addition to those required for shear, should be placed in the end block. If the anchorages are spread with reasonable uniformity over the face of the end block, the stirrups should have a capacity equal to 3 per cent of the initial prestress force when the stress in the stirrups is 20,000 psi. If the anchorages are grouped in a space that is two-thirds or less of the depth of the beam, the capacity of the stirrups should be 4 per cent of the initial prestress force when the stress in the stirrups is 20,000 psi. These stirrups should extend the full depth of the beam to tie together the portions beneath the bearing plates and the portions not beneath bearing plates.

Horizontal bars having a total area approximately equal to that of the vertical stirrups are placed near the faces of the end blocks for the length of the end block.

The length of the end block is generally equal to three-fourths the beam depth for deep beams and equal to the beam depth for shallow beams. The length of the end block is the distance from the beginning of the anchorage area to the point where the end block intersects the narrowest width of the member. The width of the end block is usually equal to the width of the narrow flange unless greater width is needed to accommodate the anchorages of the tendons.

For this example details of the end block are shown in Fig. 11-7. The bearing plates are 7 by 1¼ by 11½ in. and made of steel having a minimum yield of 59,000 psi and a minimum ultimate of 90,000 psi as recommended by the tendon manufacturer in Table 8-2. Since the anchorages are spread over more than two-thirds of the beam depth the requirement for vertical stirrups is

$$3\% \times 1{,}051{,}000 = 31{,}530 \text{ lb}$$

$$\text{Area} = \frac{31{,}530}{20{,}000} = 1.58 \text{ sq in.}$$

Since we are using #4 bars for shear we will try #4 bars here also.

$$1.58 \div .20 = 7.9 = 8 \ \#4 \text{ bars or 4 pairs}$$

The 39-in.-long end block should include two pairs of #4 bars for shear steel (maximum spacing 18½ in. from Step 9, Calculation 1) so we need a total of six pairs of #4 bars in the end block as shown in Fig. 11-7.

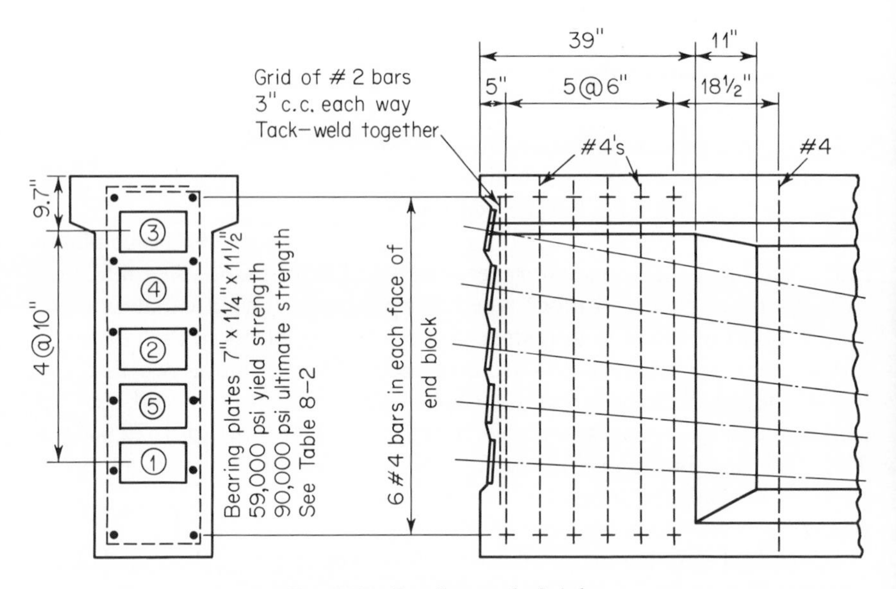

FIG. 11-7. Details at end of girder.

11-5. Friction in Tendons. When a tendon is post-tensioned, it elongates and therefore moves with respect to the tube or cored hole in which it is encased. The sliding of the tendon along the encasement creates friction which reduces the tension in the tendon. The drop in tension due to friction increases with the distance from the jack and is a function of several factors as indicated in Sec. 2607 of ACI 318–63.

Previous calculations in this chapter show that each cable in the girder should have an initial tension of 210,200 lb. This is the tension required at mid-span. We will make calculations for cable 3 which has the deepest sag and therefore the most friction loss. (See Figs. 11-5 and 11-7.)

It is first necessary to evaluate $KL + \mu\alpha$, Sec. 2607(b), to see whether to apply Eq. (26-2 ACI) or (26-3 ACI). From the Table in Sec. 2607(c),

$$K = 0.0015 \qquad \text{and} \qquad \mu = 0.25$$

L is defined as "length of prestressing steel element from jacking end to any point x." In this case it is from the jack at the end to mid-span so

$$L = 50 \text{ ft}$$

α is defined as "total angular change of prestressing steel profile in radians from jacking end to any point x."

Since the cable is horizontal at mid-span, the angular change is the angle between the cable and the horizontal at the end of the girder. This angle can be computed using Eq. (11-1)

$$\tan \phi = \frac{4B}{R}\left(1 - \frac{2X}{R}\right) \qquad (11\text{-}1)$$

From Figs. 11-5 and 11-7,

$$B = 66 - (5.75 + 9.7) = 50.55 \text{ in.}$$
$$X = 0 \qquad R = 100 \times 12 = 1{,}200$$

Substituting numerical values in Eq. (11-1),

$$\tan \phi = \frac{4 \times 50.55}{1{,}200}\left(1 - \frac{0}{1{,}200}\right) = 0.1685$$

The angle whose tangent is 0.1685 is 9°34′ or 0.167 radian. Thus

$$\alpha = 0.167$$

Substituting numerical values,

$$KL + \mu\alpha = 0.0015 \times 50 + 0.25 \times 0.167 = 0.0750 + 0.0418 = 0.1168$$

Since this is less than 0.3, Eq. (26-3 ACI) is applicable:

$$T_o = T_x(1 + KL + \mu\alpha) \qquad (26\text{-}3 \text{ ACI})$$

T_x, the prestress force required at mid-span, is 210,200 lb. Substituting in Eq. (26-3 ACI),

$$T_o = 210{,}200(1 + 0.0750 + 0.0418) = 234{,}750 \text{ lb}$$

For cable 3 the force applied at the jack will be 234,750 lb. The force applied to the other cables will be less because their angular change is less.

The initial stress in cable 3 at the jack will be

$$234{,}750 \div 1.276 = 184{,}000 \text{ psi}$$

This is $184{,}000 \div 240{,}000 = 76.67\%$ of its ultimate. Section 2606(a) permits 80 per cent if approved by the manufacturer of the tendon.

11-6. Transformed Section. Throughout the calculations in this chapter we have used the properties computed for the section in Fig. 11-1 as a solid concrete member. Actually this member has five holes cored in it as shown in Fig. 11-5. Tendons are placed in these holes and tensioned, after which the holes are filled with grout. Obviously the actual stresses are not the same as those we have computed. We shall check the actual stresses at center of span.

Compute the properties of the section shown in Fig. 11-1 except with holes out for cables as shown in Fig. 11-5. The hole diameter is 2⅛ in. and the area is 3.55 sq in. Take moments about the bottom. Begin with totals from calculations in Step 1.

	A	y	Ay	Ay^2	I_o
	865		31,390	1,494,130	95,526
$-2 \times 3.55 =$	$-$ 7.1	2⅝	-19	-49	
$-3 \times 3.55 =$	-10.7	5¾	-62	-354	
Total	847.2		31,309	1,493,727	95,526
				95,526	
				1,589,253	

$$\begin{aligned} y_b &= 31{,}309 \div 847.2 = 36.96 \\ y_t &= 66 - 36.96 = 29.04 \\ I &= 1{,}589{,}253 - 847.2(36.96)^2 = 432{,}300 \\ Z_t &= 432{,}300 \div 29.04 = 14{,}876 \\ Z_b &= 432{,}300 \div 36.96 = 11{,}688 \end{aligned}$$

The tendons will be grouted before the double T's are erected, but the girder will carry its own dead weight with the holes out after the tendons are tensioned and before they are grouted. Check stresses due to dead weight of girder with holes out using the moment from Step 2.

$$\begin{aligned} f^t{}_G &= 13{,}500{,}000 \div 14{,}876 = +908 \text{ psi} \\ f^b{}_G &= 13{,}500{,}000 \div 11{,}688 = -1{,}155 \text{ psi} \end{aligned}$$

When the tendons have been grouted, they will work with the concrete section, and their transformed area should be added to its properties. Since the grout is not prestressed, its area will not be included. The modulus of elasticity for steel wire is $E_s = 29{,}000{,}000$ from ACI 318-63, Sec. 1103. (If seven-wire strands were used the value of E_s would be approximately 28,250,000). From previous calculations $E_c = 4{,}290{,}000$,

$$n = \frac{E_s}{E_c} = \frac{29{,}000{,}000}{4{,}290{,}000} = 6.76$$

The area of each cable is 1.276 sq in. and it is equivalent to a concrete area of $1.276 \times 6.76 = 8.63$ sq in. Compute the properties of the section with holes out and transformed area of tendons added. Begin with properties of the section with holes out.

	A	y	Ay	Ay^2	I_o
	847.2		31,309	1,493,727	95,526
$2 \times 8.63 =$	17.3	2⅝	45	119	
$3 \times 8.63 =$	25.9	5¾	149	856	
Total	890.4		31,503	1,494,702	95,526
				95,526	
				1,590,228	

$$y_b = 31{,}503 \div 890.4 = 35.38$$
$$y_t = 66 - 35.38 = 30.62$$
$$I = 1{,}590{,}228 - 890.4(35.38)^2 = 475{,}675$$
$$Z_t = 475{,}675 \div 30.62 = 15{,}535$$
$$Z_b = 475{,}675 \div 35.38 = 13{,}445$$

Check stresses from applied loads using moments from Step 3.

$$f^t{}_S = 19{,}350{,}000 \div 15{,}535 = +1{,}245$$
$$f^b{}_S = 19{,}350{,}000 \div 13{,}445 = -1{,}439$$
$$f^t{}_L = 13{,}500{,}000 \div 15{,}535 = +869$$
$$f^b{}_L = 13{,}500{,}000 \div 13{,}445 = -1{,}004$$

Adding these stresses we get

$$f^t{}_{G+S+L} = +908 + 1{,}245 + 869 = +3{,}022$$
$$f^b{}_{G+S+L} = -1{,}155 - 1{,}439 - 1{,}004 = -3{,}598$$

From Table 11-2 the values obtained for these stresses based on a solid cross-section were

$$f^t{}_{G+S+L} = +3{,}069 \qquad \text{and} \qquad f^b{}_{G+S+L} = -3{,}753$$

Check stresses due to F in the section with holes out.

$$e = 36.96 - 4.5 = 32.46$$

$$f^t{}_F = \frac{892{,}000}{847.2} - \frac{892{,}000 \times 32.46}{14{,}876} = -893$$

$$f^b{}_F = \frac{892{,}000}{847.2} + \frac{892{,}000 \times 32.46}{11{,}688} = +3{,}530$$

Then

$$f^t{}_{F+G+S+L} = -893 + 3{,}022 = +2{,}129$$
$$f^b{}_{F+G+S+L} = 3{,}530 - 3{,}598 = -68$$

From Table 11-2 the values obtained for these stresses using a solid concrete cross section were

$$f^t{}_{F+G+S+L} = +2{,}222 \qquad \text{and} \qquad f^b{}_{F+G+S+L} = -425$$

The foregoing calculations show that, in this case as in most cases, design based on a solid concrete cross section results in lower actual stresses than those found by computation. It is common practice to base prestressed concrete computations on a solid cross section, but the effect of the holes should be given consideration in each design and should be checked if there is any doubt about their influence on actual stresses.

CHAPTER 12

Compression Members

12-1. Piles. The most widely used prestressed concrete compression members are piles. Standards for these piles have been prepared by the Joint AASHO-PCI Committee and are presented in Art. 15-2.

Prestressed concrete piles have several properties which make them preferable to other types for many applications.

They withstand severe driving conditions without cracking or spalling. A nonprestressed concrete pile will crack as it recoils from a heavy hammer blow. When the hammer strikes, its force creates compression in the pile which causes an elastic shortening. As soon as the energy of the hammer is expended and there is no force to maintain the elastic shortening, the pile springs back to its original length. By the time it reaches its original length, it has gathered momentum, causing further elongation which in turn creates tensile stresses in the pile. Under a heavy hammer blow these tensile stresses cause cracks, and under repeated blows spalling develops. Prestressed concrete piles are fabricated with sufficient compressive stress to offset the tensile stresses caused by recoil. Experience indicates that a final compressive stress of 700 psi is normally sufficient for this purpose.[1*] Where especially severe driving conditions exist, the magnitude of the prestress should be increased. Final determination of the amount required is a matter of trial and experience with the existing driving conditions.

Prestressed concrete piles are easy to lift and transport. Their compressive stress enables them to resist large bending moments, so that they can be lifted with simple one- or two-point picks where reinforced concrete piles of the same dimensions require complicated rigging for multipoint picks.

Prestressed concrete piles are durable because they are crackless. They are seldom cracked by driving or handling, and even if they should be, the compressive stress will keep the cracks tightly closed once the pile is in place.

A prestressed concrete pile has a much higher moment of inertia than a reinforced pile of the same dimensions. In a prestressed pile the entire

* Superscript numbers indicate references listed in the Bibliography at the end of the chapter.

cross-sectional area of the concrete contributes to the moment of inertia, while in a reinforced pile the concrete in the tensile side has no function and only the reinforcing contributes to the moment of inertia. This additional moment of inertia is important in piles whose capacity is determined by their slenderness ratio.[2]

AASHO-PCI criteria for the design of piles are presented in Art. 15-2 under the heading Pile Capacities and Loading. They cover conditions

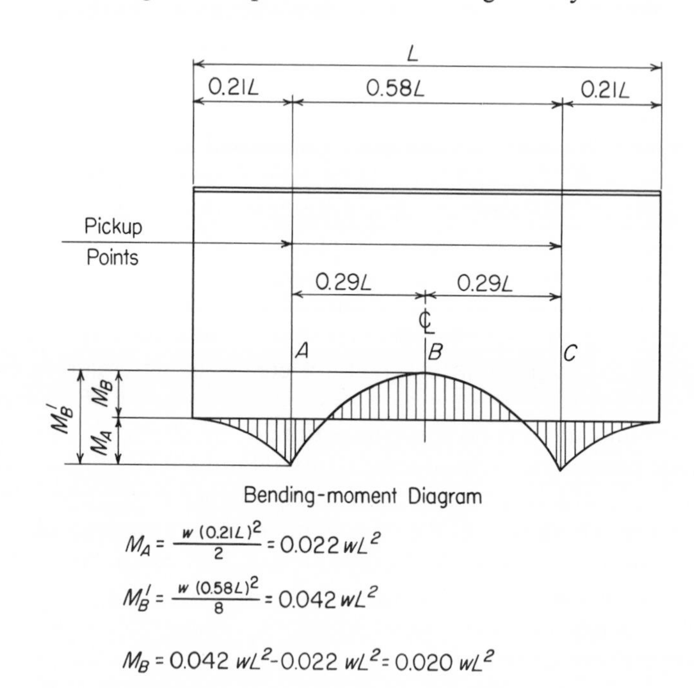

FIG. 12-1. Bending moments for lifting a pile with a two-point pickup.

up to the point where L/D equals 25. Where L/D is greater than 25 it is suggested that the design criteria shown in Fig. 12-4 be followed.

12-2. Handling Piles.[3] As an example we will use an 18-in.-square pile with details as recommended in Figs. 15-1*a* and 15-1*c* and compute the maximum length that can be handled by two-point and one-point picks.

From Fig. 15-1*a*

$A_c = 324$ sq in.
$Z = 972$ in.^{3}
$F = 227{,}000$ lb
Weight $= 335$ lb per ft
Use $f'_c = 5000$ psi
$f_F = 227{,}000 \div 324 = 701$ psi

Refer to Art. 15-2, General Notes—Prestressed Concrete Piles; Pick-Up and Handling. To allow for impact in lifting and transporting the applied load is specified as 1.5 times full dead load. Allowable tensile stress is $6\sqrt{f'_c}$. From this

$$w = 1.5 \times 335 = 502$$
$$f_t = 6\sqrt{5{,}000} = 424$$

From the foregoing data the maximum allowable stress change in the pile will be 701 + 424 = 1,125 psi. This represents a bending moment of (1,125 × 972) ÷ 12 = 91,100 ft-lb.

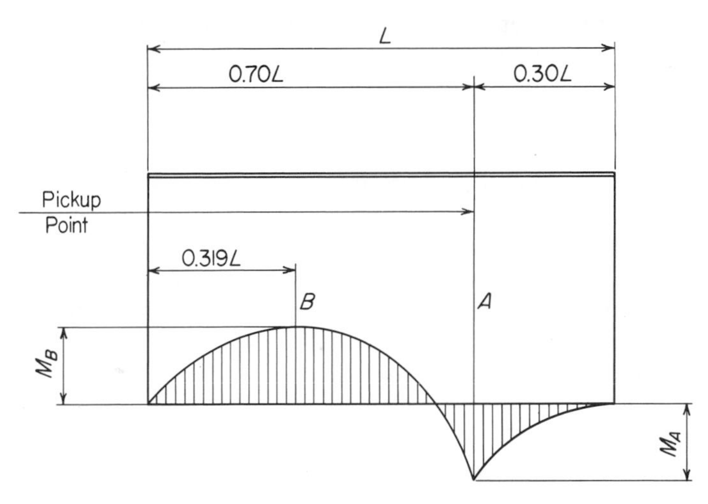

FIG. 12-2. Bending moments for lifting a pile with a one-point pickup.

Recommended pickup points are shown in Fig. 15-1*d* and corresponding bending moments in Figs. 12-1 and 12-2.

Compute maximum length for two-point pickup using data from Fig. 12-1.

$$M_A = 0.022wL^2$$

Substituting numerical values,

$$91{,}100 = 0.022 \times 507L^2$$
$$L^2 = 8{,}160 \qquad L = 90.4 \text{ ft}$$

Compute maximum length for one-point pickup using data from Fig. 12-2.

$$M_A = 0.045wL^2$$

Substituting numerical values,

$$91{,}100 = 0.045 \times 507L^2$$
$$L^2 = 3{,}990 \qquad L = 63.2 \text{ ft}$$

12-3. Sheet Piles. Although no universal standards have been developed yet, prestressed concrete sheet piles have found wide application.[4-7]

The most common detail is a rectangular cross section with a tongue on one side and a groove on the opposite side. At the bottom, one side is beveled so that it will press against the adjacent pile during driving.

Except where bending moments to be resisted are high, the prestressing strands are placed symmetrically about each axis so there is no tendency to camber. Spiral ties similar to those in AASHO-PCI Standards, Fig. 15-1*c*, are used. Magnitude of prestress is determined by handling and driving stresses or by stresses due to loads that will be applied in service.

When the wall being made from the piles is straight the tongue and groove are directly opposite each other; that is, the angle between them is 180°. When the wall turns, special piles are cast in which the angle between the tongue face and the groove face is 45°, 60°, 90°, or any other necessary to furnish the desired contour.

Conditions which require bearing piles as well as sheet piles can be dealt with by a combination pile of the type shown in Fig. 12-3.

12-4. Prestressed Concrete Columns. The subject of details and allowable stresses for prestressed concrete columns is being studied by one of the technical committees. When their recommendations are published they should be followed. In the interim the authors present herewith their suggestions on the subject.

In many of the instances where precast prestressed concrete columns are used in preference to precast reinforced concrete columns they are selected because the casting yard finds them more economical to fabricate, ship, and erect rather than because they have superior structural qualities. Spiral ties are similar to, but not necessarily at the same spacing as, those shown in Fig. 15-1*c*. For a casting yard equipped to make pretensioned members the placing and tensioning of the strands and the placement of the machine-made spiral ties can be accomplished with much less labor than the preparation of a reinforcing cage for a column. In addition, a prestressed column will remain crackless during handling if it is designed in accordance with Art. 12-2, whereas it requires appreciable care in handling to keep a reinforced column crackless.

It is our aim to establish conservative criteria for the design of prestressed columns using the data presently available to us. We will assume an ultimate concrete strength of 5,000 psi since this is the strength used for the vast majority of prestressed concrete building members. If the designer wishes to use a different strength of concrete he can apply the same reasoning using different values. The pattern of the prestressing strands will be symmetrical about each axis and the prestress will seldom exceed 1,000 psi. Since a bonded tendon can never cause buckling (Art. 1-5), the only effect from the prestressing force is the compressive stress. If we subtract

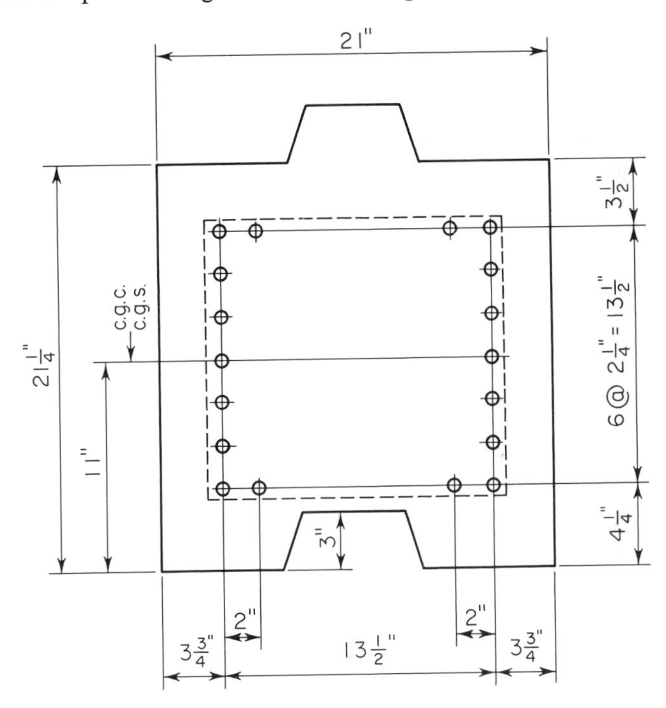

FIG. 12-3. Combination sheet pile and bearing pile. Note that strand pattern is arranged to provide a group of strands near the tensile side of the pile no matter on which side the load is applied.

this from the ultimate strength of the concrete we have the remaining strength to work with, and the column is a homogeneous elastic member because it is prestressed. The available working strength of the concrete is therefore 5,000 − 1,000 = 4,000 psi.

Section 1402 of ACI 318-63 gives the following formula for use in the design of columns:

$$P = A_g(0.25f'_c + f_s p_g)$$

in which P = allowable load on column

A_g = gross cross-sectional area of concrete

$f_s p_g$ is a term for the load carried by the reinforcing steel

Since there is seldom any reinforcing steel in a prestressed column the term $f_s p_g$ is zero and the equation becomes

$$P = 0.25f'_c A_g$$

We have already stated that the available working strength of the column is 4,000 psi so we get

$$P = 0.25(4000)A_g = 1{,}000A_g$$

This says, in effect, that the allowable load in a short prestressed column where buckling is not a consideration is 1,000 psi on the gross area of concrete.

It has always been standard procedure to design reinforced concrete columns and piles on the basis of L/D (ratio of unsupported length to diameter or width of column cross section). A prestressed concrete column however is an elastic member and like other elastic compression members should be designed on the basis of L/r (ratio of unsupported length to radius of gyration of column cross section).

For values of L/r up to 40, buckling has a neglegible influence and the allowable applied stress f_{pc} can be taken as 1,000 psi.

For values of L/r equal to 120 or more the ultimate strength of a column is a function of its elastic properties and the stiffness of its cross section. In this range Euler's formula with a factor of safety of two represents standard design procedure. Euler's formula is

$$P' = \frac{\pi^2 EI}{L^2}$$

in which P' = load at which column will buckle. Since $I = Ar^2$, the formula can be written

$$P' = \frac{\pi^2 EAr^2}{L^2}$$

or

$$\frac{P'}{A} = \frac{\pi^2 E}{(L/r)^2}$$

In this equation P'/A is the unit compressive stress at which the member will buckle. Using P as the allowable load on the column and a factor of safety of two, the equation becomes

$$\frac{P}{A} = \frac{0.5\pi^2 E}{(L/r)^2}$$

The value of E for use in this equation is the long time modulus of elasticity of the concrete. Section 916(d) of ACI 318-63 suggests that this value be not more than one-third the value of E_c given by Sec. 1102 (Fig. 6-1). When this factor is applied the equation becomes

$$\frac{P}{A} = \frac{0.5\pi^2 E_c/3}{(L/r)^2}$$

Substituting f_{pc} for P/A and combining the numerical values we get the following equation for columns with L/r of 120 or more:

$$f_{pc} = \frac{1.65E_c}{(L/r)^2} \tag{12-1}$$

We have now established the allowable stress for L/r from zero to 40, line AB in Fig. 12-4, and for L/r from 120 and up, line CD in Fig. 12-4.

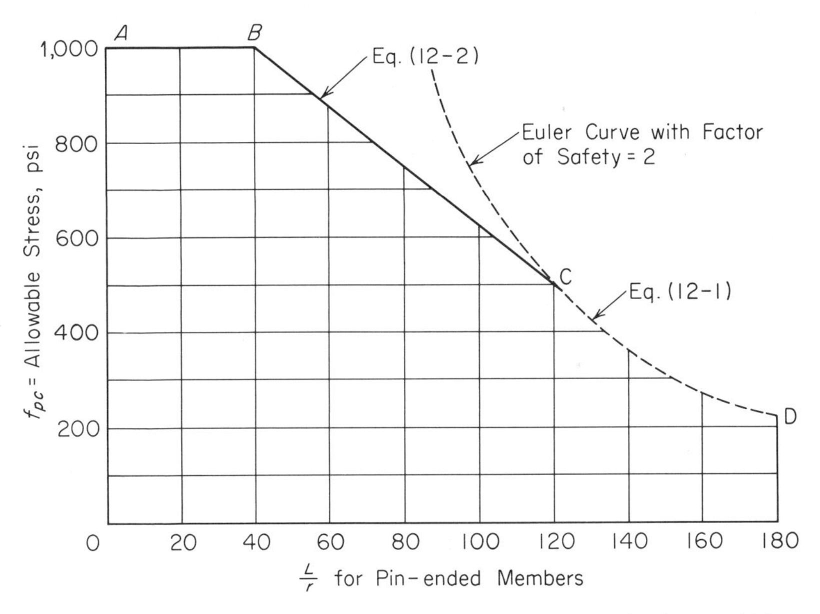

FIG. 12-4. Suggested allowable unit compressive stress in prestressed columns which meet following criteria: $f'_c = 5{,}000$ psi; compressive stress from prestressing force does not exceed 1,000 psi; and spiral ties are provided similar to those in Fig. 15-1c.

The empirical column formulas for other materials are of the "straight-line" type and a straight line from B to C will complete our curve and give us values of allowable stress in line with the criteria we have established. For 5,000 psi concrete and L/r of 120, Eq. (12-1) gives

$$f_{pc} = \frac{1.65 \times 4{,}300{,}000}{120^2} = 495 \text{ psi}$$

To simplify our straight-line equation we will call this 500 psi. Then when $L/r = 40$, $f_{pc} = 1{,}000$, and when $L/r = 120$, $f_{pc} = 500$. From this the equation for L/r from 40 to 120 can be written

$$f_{pc} = 500 + \frac{120 - L/r}{80} 500$$
$$= 500(1 + 1.5 - 0.0125L/r)$$
$$= 1{,}250 - 6.25L/r \qquad (12\text{-}2)$$

The values given in Eqs. (12-1) and (12-2) and in Fig. 12-4 are for pin-ended members. The length or effective length to be used for L can be determined in accordance with Sec. 915 of ACI 318-63.

12-5. Combined Bending and Direct Load. For members subjected to a combination of bending plus direct compression the condition governing maximum allowable loads is usually the compressive stress in the concrete. The limiting condition is expressed by Eq. (12-3).

$$\frac{f_L}{f_{pc}} + \frac{f_M}{f_{MA}} \leqq 1 \qquad (12\text{-}3)$$

in which

f_L = compressive stress due to direct load
f_M = compressive stress due to bending moment
f_{pc} = allowable compressive stress under direct load (see Fig. 12-4)
f_{MA} = allowable compressive stress due to bending moment

When the bending moment is high and the direct load is low, the stress on the tensile side of the member may be the critical factor. In this case the ultimate strength of the member in bending should be computed and should show a factor of safety of at least 2.

BIBLIOGRAPHY

1. Dean, W. E.: Prestressed Concrete—Difficulties Overcome in Florida Bridge Practice, *Civil Eng.*, June, 1957, pp. 60–63. See p. 61.
2. Pretensioned Piles Are 132 Feet Long, *Eng. News-Record*, July 19, 1956, pp. 33–36.
3. Gerwick, Ben C., Jr.: Torsion in Concrete Piles during Driving, *J. Prestressed Concrete Inst.*, June, 1959, pp. 58–63.
4. Dean, W. E.: Prestressed Concrete Sheetpiles for Bulkheads and Retaining Walls, *Civil Eng.*, April, 1960, pp. 68–70.
5. Perez, Henry T.: Miami Causeway Sparks Innovations, *Construc. Methods and Equipment*, March, 1959.
6. Remington, William F.: Prestressed Sheetpiles Enclose Jetties, *Eng. News-Record*, May 14, 1959, pp. 59–63.
7. McGregor, T. H.: T-Shaped Concrete Sheet Piles Renew Deteriorated Bulkhead, *Eng. News-Record*. Jan. 3, 1963, p. 33.

CHAPTER 13

Design of Connections between Precast Members in Buildings

13-1. Scope. Precast concrete members either prestressed or reinforced can be economically assembled to form complete structures.[1]* Each joint can be rigid or pin type to suit the requirements of the designer. Some of the many possible types of connections are illustrated and discussed in Art. 15-7.

ACI-ASCE Committee 512 has made a study of the behavior of connections between precast concrete members and has issued a report entitled Suggested Design of Joints and Connections in Precast Structural Concrete. Excerpts from this report applicable to examples in this chapter are reprinted in Appendix B. The complete report appeared in the August, 1964 issue of the *Journal of the American Concrete Institute* and is available from ACI as Title No. 61-51. It is recommended that engineers contemplating the design of precast structures obtain and study the complete report.

A good basic philosophy for the location, detail, and design of joints between precast members is presented in Sec. 103 of the report by Committee 512 which says:

103—General considerations

It is recommended that joints and connections occur at logical locations in the structure, and when practical, at points which may be most readily analyzed and easily reinforced. Precautions should be made to avoid connection and joint details which would result in stress concentrations and the resulting spalling or splitting of members at contact surfaces. Liberal chamfers, steel edged corners, adequate reinforcement and cushioning materials are a few of the means by which stress concentrations may be avoided or provided for.

The strength of a partially completed or completed structure should be governed by the strength of the structural members rather than by the strength of the connections; the connection should not be the weak link in the structure."

*Superscript numbers indicate references listed in the Bibliography at the end of the chapter.

13-2. Design Considerations. Joints can be subjected to various types of loading and stress conditions. Each should be considered and, unless obviously not critical, analyzed.

In designing joints it is important to use details that will permit the joint to function in the manner that was assumed in the analysis of the frame. If it is considered pin connected, it should be detailed so that it will not take moment. If it is assumed rigid, it should be detailed to carry the full moment that will be applied to it. As discussed in Art. 15-7, the use of reinforced joints gives full rigid frame behavior just as well as the use of prestressed joints.

(Note: References in this chapter preceded by (512) are to sections of "Suggested Design of Joints and Connections in Precast Structural Concrete," and the items referenced will be found in Appendix B.)

13-2A. Transfer of Shear. This subject is covered in (512) Sec. 301. Methods discussed are the extension of reinforcing steel from a precast member into cast-in-place concrete, bearing of the supported member on the supporting member, embedded steel plates or shapes, brackets or corbels on columns, and prestressing across the connecting faces. Other methods not specifically discussed are permitted provided that "the method satisfies the principles of statics and the unit stress requirements of codes for the materials involved."

13-2B. Transfer of Moment. This subject is covered in (512) Sec. 302. Methods discussed are the extension of reinforcing bars, additional reinforcing bars, embedded structural steel shapes or plates, prestressing tendons, and "other methods that satisfy the principles of statics and the unit stress requirements of codes for the materials involved."

13-2C. Transfer of Torsion, Axial Tension and Axial Compression. These subjects are covered in (512) Secs. 303 to 305. See Appendix B and also the complete Committee 512 report.

13-3. The Complete Structure. Although this chapter deals chiefly with design and details of joints and connections to transfer shear, moment, etc., from one member to another, it is of primary importance that the performance of the structure as a whole be considered and the function of each joint be established before details of the joints are considered. In developing his design the engineer has almost complete freedom to decide which members shall be continuous, which simple span, where expansion joints shall be located, etc. Joint details will be simplified and cost minimized if the recommendations presented at the end of Art. 13-1 are kept in mind.

The details of rigid or continuous joints automatically tie together the members involved. For nonrigid joints and for expansion joints, steps should be taken to limit the motion of one member with respect to the other so there will be no danger of the members becoming separated under whatever extreme conditions might develop. One typical detail of this type uses one or more reinforcing bars projecting upward a short distance from the supporting member. A hole in the supported member fits over the dowel and is filled with mastic. The members can move with respect to each other, but this motion is halted when the dowel comes in contact with the edge of the hole (see Fig. 13-1).

Expansion joints are required in buildings of precast prestressed concrete members just as in buildings of other materials. They can be of the type where one member moves with respect to the other on elastomeric pads or Teflon coated pads, or they can be of the type where two sets of columns and beams are installed adjacent to each other permitting complete freedom of motion.

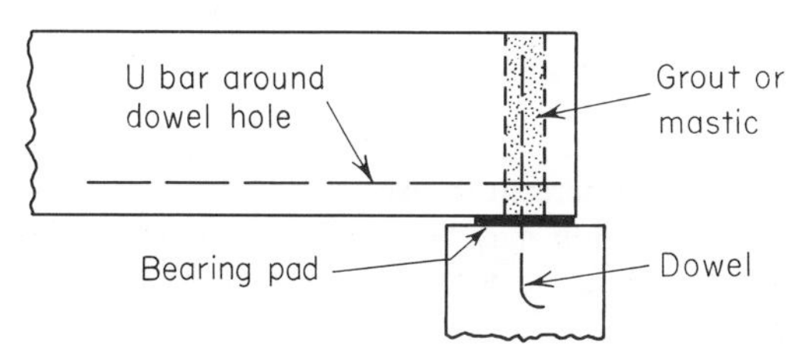

FIG. 13-1. Typical dowel connection. For fixed end, dowel hole is filled with grout. For expansion end, dowel hole is filled with mastic.

Section 301.3.5 of (512) says:

301.3.5—Other tensile stresses: Positive connections at the bearing of both ends of simply supported members by welding, bolts or any other means which prevent movements arising from creep, shrinkage or temperature change are not recommended. One end should be free to accommodate such movement. However, if such connections are made at both ends, reinforcement should be provided to resist the tensile forces which may develop in both the supporting and supported member from horizontal forces, and the bearing plates or other such devices should be designed to resist these forces.

When applying this recommendation to a structure which he is designing, the engineer should investigate local experience on similar structures. Continuous floor systems and roof systems of single T's or double T's have often been built in lengths in excess of 200 ft without the development of any problems. On the other hand, double T's 50 ft long welded at each end to rigid members have developed serious cracks in the bottom at the end due to shrinkage and temperature changes. Local prestressed concrete fabricators should be consulted for their experience and suggestions.

In many cases the shortening of a prestressed member due to creep and shrinkage is not a problem because the column or other member to which it is framed can bend sufficiently to compensate for the shortening without developing an appreciable horizontal force.

13-4. Design Requirements. Recommended criteria for design of joints are clearly spelled out in (512).

Stresses under design load conditions are covered in (512) Sec. 202.1 which says:

> **202.1—Design stress:** Design stresses, except as noted hereafter, should not exceed those provided in ACI (318-63) "Building Code Requirements for Reinforced Concrete," ACI-ASCE "Tentative Recommendations for Prestressed Concrete," AISC "Specifications for the Design, Fabrication and Erection of Structural Steel for Buildings" or AWS "Standard Code for Arc and Gas Welding in Building Construction," whichever is applicable. No flexural tension stresses should be permitted in the concrete of prestressed joints and connections.

Ultimate load conditions are covered in (512) Sec. 202.2 which says:

> **202.2—Ultimate load factors:** The ultimate strength capacity of joints and connections should be at least 10 percent in excess of that required of the members connected. This recommendation may be satisfied by proportioning the joint or connection to provide strength in the connection or joint 1.1 times the ultimate strength capacity required by ACI 318-63.

13-5. Bearing Pads. Bearing pads between precast concrete slabs, beams or girders, and the structural members which support them are designed to perform one or more of the following functions:

1. Provide a cushion that eliminates the slight irregularities on the surfaces of the precast concrete and distribute pressure evenly over the bearing surface. This conforms to (512) Sec. 301.3.3 which says: "Direct concrete to concrete bearing is not recommended between units unless the bearing is accomplished by cast-in-place concrete or grout."
2. Permit rotation at the ends of a beam as the beam deflects under applied load. As the beam deflects the pad can compress to the new slope of the end of the beam without causing excessive edge stresses and spalling of the supporting member.
3. Permit horizontal motion of the end of the beam with respect to the supporting member when temperature changes or other conditions induce such motion.

Elastomeric bearing pads can perform all three functions. Criteria for design of such pads for bridges are published in American Association of State Highway Officials Standard Specifications for Highway Bridges, Eighth Edition, as revised by AASHO Interim Specifications, 1963 and subsequent editions of AASHO Standards. A brochure, Design of Neoprene Bridge Bearing Pads published by E. I. du Pont de Nemours & Co., Inc. of Wilmington, Delaware, discusses neoprene pads and includes a number of examples of the design of pads for varying conditions.

As this is written there is no specification covering the use of elastomeric pads in buildings even though they are used extensively. Where the pads are designed to take horizontal motion in combination with bearing, many engineers apply the AASHO specification to building members.

The AASHO Specification permits a maximum allowable pressure of 500 psi under dead load only and 800 psi under dead plus live load. This is probably quite conservative when applied to buildings. In a pad providing rotation and/or horizontal motion the stresses due to these functions must be allowed for.

For pads thick enough to provide horizontal motion another property that must be considered is "shape factor." When an elastic pad is loaded its edges bulge. The magnitude of the bulging increases with the unit pressure on the pad, with the increase in thickness of the pad and with the increase in length of the pad. Bulging causes shearing stresses within the pad which can cause failure unless they are kept within certain limits. If the combination of loading and shape factor is critical, the situation can be remedied by laminating the pad. The designer determines how thin the pad must be so that bulging stresses are not critical and then calls for a pad composed of neoprene sheets of that thickness sandwiched between or laminated with thin steel sheets using enough alternate layers of pad and steel to give the total thickness of neoprene in the pad needed for the horizontal motion.

Elastomeric pads are available in different degrees of hardness (modulus of elasticity) expressed as 50, 60, and 70 hardness. Under the same unit pressure the 50 hardness will compress more than the 60 hardness. AASHO says that total elastic compression under dead plus live load shall not exceed 15 per cent of the unloaded thickness of the pad. Conformance with this requirement eliminates excessive creep during the life of the structure. The du Pont brochure presents a chart for each degree of hardness showing the percentage of compressive strain that will take place for a given bearing pressure and shape factor.

Additional data on design are presented in Design of Elastomer Bearings by Charles Rejcha, published in the October, 1964 issue of the *Journal of the Prestressed Concrete Institute.*

For determining approximate dimensions of elastomeric pads, two rules of thumb can be applied:

1. The pad thickness should be at least double the anticipated horizontal motion.

2. Compression due to vertical load plus rotation should not exceed 20 per cent. (The Rejcha article suggests using a minimum rotation of 0.01 radian around the axis perpendicular to the girder to allow for the effect of surfaces not exactly parallel plus the rotation due to deflection of the girder.)

In building work the only function of the elastomeric pad is often that of providing a cushion between the two precast surfaces. There is no horizontal motion or rotation to be provided for. For this purpose it is common practice to use a ¼-in.-thick pad of 60 hardness. Such a pad has an excellent shape factor and can be subjected to higher stresses than the thicker pads. In his article Rejcha says: "We recommend the use of a value of 1000 psi for average vertical stress or P/A. This stress suits the commonly used ½-in. ± elastomer layers. It is quite possible to exceed the 1,000 psi using a thinner elastomer layer." He also says that "some designers are using a value of 1,700 psi." Of course the bearing stress on the concrete specified by ACI 318-63 (512) should not be exceeded.

Another approach to joints where members must move with respect to each other involves the use of elastomeric pads coated with Teflon. This method, which was used on the A & P project described in Art. 15-8, has been quite successful. Teflon is another du Pont product. To provide an expansion joint two pieces of ¼-in. elastomeric material are used. One side of each is coated with Teflon and the pads are attached, one to the beam and one to the supporting member, with the uncoated sides against the concrete. When the structure is erected the Teflon-coated surfaces bear against each other. The coefficient of friction of Teflon on Teflon is higher at low pressures but at pressures of 100 psi and up it is in the range of 0.06.

Other materials used as cushions between precast members include roofing felt and fiberglass pads. The fiberglass pads are ½ in. thick and relatively soft so that they compress to a thickness of 1⁄16 to ⅛ in.

13-6. Steel Bearing Surfaces. Steel plates or shapes are frequently cast into the ends of beams, etc., and into the surface of the supporting member at the point where the beam rests. The bearing stress of the concrete on the steel should not exceed that permitted in the applicable specification as discussed in Art. 13-7. Bearing stresses of the one steel member on the other steel member can be as high as permitted by the applicable structural steel code as long as the embedded steel members are stiff enough to distribute the load evenly to the concrete in which they are embedded.

Prestressed concrete members seldom have a completely flat soffit; they usually have a camber. This means that there is a tendency for a beam to make a line contact at the very end of the beam rather than to rest flat for the length of the bearing surface. Since the slope of the bottom of the beam with respect to the horizontal is extremely small, there is a slight rotation, and the beam does rest on the bearing surface. However, the condition does exist and causes higher pressures at the very end of the beam. To keep the embedded steel firmly in place to distribute pressure, the reinforcing bar welded to the embedded steel should have more than a token tack weld and should extend both along the beam and up into the beam. One detail that has proven quite efficient in eliminating a tendency to spall the end of a double T is illustrated in Fig. 13-2.

13-7. Other Connection Materials. There are so many variations of possible joint details and choices of materials that the subject of joint detail and design could fill a book by itself. In fact there are two books, in addition to the Committee 512 report, which are almost mandatory for the engineer designing buildings of precast and/or prestressed concrete members.

Connection Details for Precast-Prestressed Concrete Buildings was prepared by the PCI Committee on Connection Details and is available from the Prestressed Concrete Institute in Chicago.

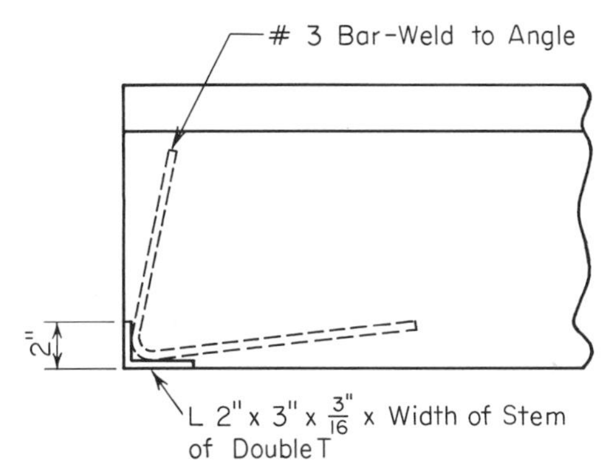

FIG. 13-2. Steel bearing angle at end of double T members. (*Courtesy Concrete Structures Inc., Richmond, Virginia.*)

The "Prestressed Handbook" was written by Laurence Cazaly and M. W. Huggins. It is available from the Canadian Prestressed Concrete Institute in Toronto, Ontario, Canada. This handbook, which covers all phases of prestressed concrete design, devotes more than 125 pages to joints and their location, details, stress analysis, and tables and charts to guide the selection of details that will assure structurally sound joints.

In previous articles we have discussed the conditions where one member bears on another through a pad and where each member has a steel plate or cast-in shape so that the steel surfaces bear on each other. Two other details are used.

In some areas the fabricators or contractors who are erecting the structure prefer to use a thin bed of cement mortar between the two members. This is entirely satisfactory as long as the beam can be placed before the mortar sets. If the bottom of the beam is on a slight slope due to camber, the mortar bed will be squeezed to conform and there will be uniform pressure across the bearing surface. Some erectors consider it difficult to get beams erected and leveled before the mortar sets and prefer to use a pad between the members.

Although (512) Sec. 301.3.3 says that "direct concrete to concrete bearing is not recommended between units unless the bearing is accomplished by cast-in-place concrete or by grout," the details shown in Figs. 15-22 and 15-23 are used regularly by many designers without problems. When one precast concrete member bears on another there are probably small high spots which cause stress concentrations and possibly even some crushing of the high spot, but this should not be critical unless the high

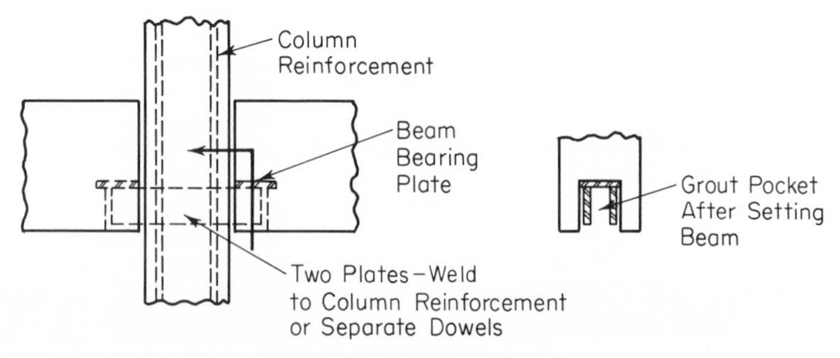

BC(*c*) Vertical Plate Haunch

This haunch type is preferred where the beams have heavy end reactions. The section modulus of the plates is much larger in the vertical position, and the long fillet welds are dependable.

1. The cantilever moment for the plates should be computed at the column bars.
2. The bearing plate in the beam should be wide enough to provide stability against overturning of the beam during erection. The area of the plate should be adequate to keep bearing stresses within allowable limits.
3. For fireproof construction, the cantilevered plates can be set up into the beam to allow for concrete cover below.

NOTE: This haunch type may be considered where concrete haunches are shown on the various beam-to-column connections.

FIG. 13-3. Precast beam supported on steel plates projecting from column. (Reprinted by permission of the Prestressed Concrete Institute.)

spot is near an edge and causes a spall or split. In details like Figs. 15-22 and 15-23 it is assumed that the poured-in-place concrete transfers much of the shear to the beam.

Details using steel plates, bars, or shapes projecting from a precast member to support or bear upon another member provide small neat connections and rapid erection. Examples are given in Figs. 13-3 and 13-4. As indicated in (512) Sec. 202.1 (Art 13-4) the stresses in the steel members should conform to the requirements of AISC in Specifications for the Design, Fabrication and Erection of Structural Steel for Buildings. Of course the steel members must be detailed to distribute their load without causing excessive stresses in the concrete member of which they are a part.

The bar type of hanger illustrated in Fig. 13-4 is called a Cazaly Hanger. The Canadian Prestressed Handbook, mentioned previously, includes an easily applied set of rules for designing the component parts of Cazaly Hangers.

When concrete is confined on both the horizontal surface and the vertical surface by a steel angle and the angle is anchored in the main body of the concrete by steel straps as illustrated in Fig. 13-4, the concrete will withstand extremely high bearing pressures. Of course the angle must be stiff enough to distribute the load to the concrete. Here again the

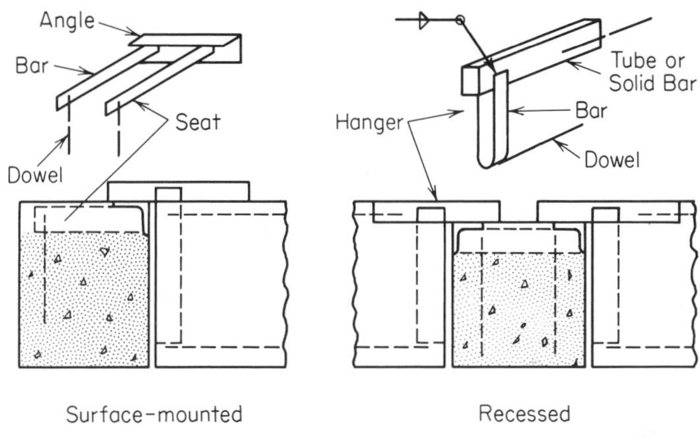

BG-7 Steel Hanger, Simple Spans

The steel hanger provides a means for fast erection in all weather. It is commonly used in conjunction with a steel bearing frame.

1. No other bearing pad is required.
2. To allow for abuse and *miscellaneous iron* workmanship, it is advisable to design the steel for a factor of safety of 4.
3. The *bearing stress* inside the projected area of the strap should not exceed approximately 2,500 psi.
4. Dowels welded to the hanger must be sufficient to develop longitudinal restraints. They should never be omitted entirely.

Fig. 13-4. Steel hanger projecting from beam. See discussion in text. (Reprinted by permission of the Prestressed Concrete Institute.)

Canadian Prestressed Handbook has tables of recommended dimensions. For an angle with proper stiffness it is safe to design to a bearing pressure of f'_c under ultimate load conditions. Some designers even use a bearing stress of 1.5 f'_c under ultimate load conditions. Data on load tests of concrete confined in such a manner are included in an article, Composite Designs in Precast and Cast-in-Place Concrete, by Arthur R. Anderson, published in *Progressive Architecture* September, 1960. When the concrete is not confined, as between the two legs of an angle, bearing pressure between concrete and steel should be figured on the basis of design load and limited to that permitted by ACI 318-63, Table 1002(a).

13-8. Design Example—Double T's Continuous over Supporting Beam. This example will illustrate double T's like those in Fig. 13-5 supported on a soffit beam and made continuous across the beam by the use of reinforcing bars in the cast-in-place slab on top of the double T's as shown in Fig. 13-6.

Precast pretensioned soffit beams 6 in. deep and 24 in. wide are erected on false work, double T's are set in place on the soffit beams with a 4½ by ¼ by 4½-in. neoprene pad under each stem, reinforcing steel is placed, and

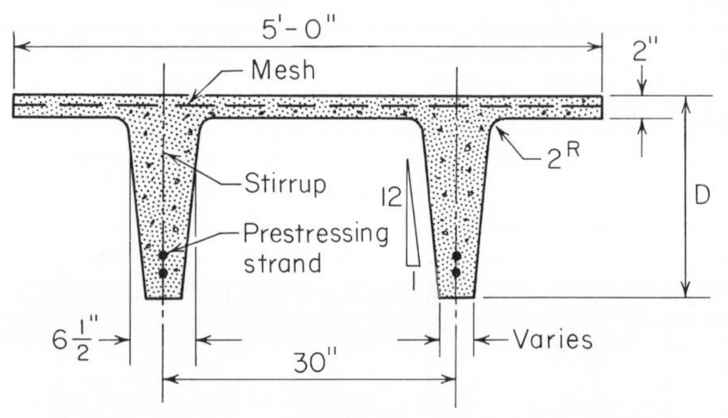

FIG. 13-5. Details of a typical 5-ft-wide double-T section.

concrete is cast. When the concrete has cured and the false work is removed, the resulting structure is a composite beam in one direction carrying the double T's which are perpendicular to it. The double T's carry their own dead weight and the dead weight of the cast-in-place concrete as simple span members. The composite system of double T's and cast-in-place concrete carries the live load as members continuous over the supporting beam.

Dead weight of the 5-ft double T is 240 lb per lin ft, and dead weight of the 3-in. slab is 37.5 lb per sq ft or 188 lb per lin ft. Total dead weight is $240 + 188 = 428$ lb per lin ft. Live load is 80 lb per sq ft or $80 \times 5 = 400$ lb per lin ft.

In the design of a complete structure the exact moment and shear would be determined as part of the frame analysis. For the purpose of this example we will assume that the moment is

$$M = \frac{w(l')^2}{12}$$

and the shear is

$$V = \frac{wl}{2}$$

in which l = distance center to center of beams and l' = clear span.

$$V = \frac{(428 + 400)(35)}{2} = 14{,}490 \text{ lb}$$

Each stem is 4½ in. wide at the bottom and bears on the neoprene pad for a distance of 4½ in. The total bearing area is therefore 4½ × 4½ × 2 = 40.5 sq. in., and the bearing pressure is 14,490 ÷ 40.5 = 358 psi. This is less than the $0.25f'_c = 0.25 \times 5{,}000 = 1{,}250$ psi permitted by ACI 318-63 Table 1002(a).

The neoprene pad is used to compensate for minor surface irregularities in accordance with (512) Sec. 301.3.3 which says that "direct concrete to concrete bearing is not recommended between units unless the bearing is accomplished by cast-in-place concrete or grout." Note that the edge

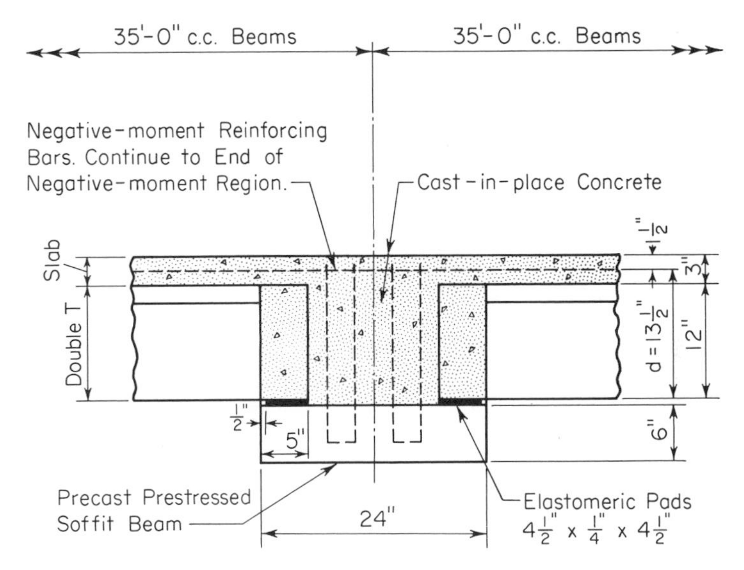

FIG. 13-6. Composite soffit beam and continuous double T's. See Art. 13-8 for analysis of joint.

of the neoprene pad is set back ½ in. from the edge of the soffit beam to minimize spalling.

Under working-load conditions only the live load is carried by the continuous member so the negative moment is

$$M_L = \frac{400(33^2)12}{12} = 435{,}600 \text{ in.-lb}$$

The double T and slab must be designed to carry a working-load negative moment of 435,600 in.-lb as a reinforced concrete member. The bottom side of the T is the compression side, so the width of the compression side is the width of the two stems or $b = 2 \times 4½ = 9$ in.

Use one layer of reinforcing bars in the slab as shown in Fig. 13-6. Use steel with $f_y = 60{,}000$ psi. Then from ACI 318-63 Sec. 1003(a) the allowable working stress is $f_s = 24{,}000$ psi.

From Fig. 6-1 the modulus of elasticity of 5,000 psi concrete is $E_c = 4{,}290{,}000$ psi. Then

$$n = \frac{E_s}{E_c} = \frac{29{,}000{,}000}{4{,}290{,}000} = 6.75$$

As shown in Fig. 13-6, $d = 13\frac{1}{2}$ in. At working load the moment carrying capacity of a reinforced concrete beam is

$$M = A_s f_s jd$$

Try #5 bars at 12 in. center to center. Area of one #5 bar = 0.31 sq in.

$$A_s = \frac{60 \times 0.31}{12} = 1.55 \text{ sq in.}$$

$$p = \frac{A_s}{bd} = 1.55 \div (9 \times 13.5) = 0.0128$$

$$pn = 0.0128 \times 6.75 = 0.0864$$

$$k = \sqrt{2pn + pn^2} - pn$$

$$= \sqrt{0.1728 + 0.0075} - 0.0864$$

$$= 0.4246 - 0.0864 = 0.3382$$

$$j = 1 - \frac{0.3382}{3} = 0.887$$

$$M = 1.55 \times 24{,}000 \times 0.887 \times 13.5 = 445{,}500 \text{ in.-lb}$$

This is slightly more than the computed value of M_L so the joint meets working-load requirements with #5 bars at 12 in. center to center.

Check ultimate-strength capacity. (512) Sec. 202.2 says that joints should have an ultimate capacity of 1.1 times the ultimate capacity required by ACI 318-63. From ACI 318-63, Sec. 1506(a) the ultimate capacity should be

$$U = 1.5D + 1.8L \qquad \text{(15-1 ACI)}$$

In this example the double T alone carries its own dead weight plus the dead weight of the cast-in-place slab as a simple span member. The only loads which create a negative moment over the beam are those which are applied after the cast-in-place concrete has cured. At ultimate loading therefore the loads which cause moment over the beam are $0.5D + 1.8L$ because $1.0D$ is already being carried by the double T as a simple span member. The numerical value of this load is

$$(0.5 \times 428) + (1.8 \times 400) = 934 \text{ lb per lin ft}$$

$$M_u = \frac{934(33^2)12}{12} = 1{,}017{,}000 \text{ in.-lb}$$

From ACI 318-63, Sec. 1601(a) the ultimate capacity of a reinforced concrete beam is

$$M_u = \phi\left[A_s f_y\left(d - \frac{a}{2}\right)\right] \qquad \text{(16-1 ACI)}$$

in which $a = A_s f_y / 0.85 f'_c b$.

In this example

$\phi = 0.90 \qquad A_s = 1.55$ sq ft.
$f_y = 60{,}000$ psi $\qquad d = 13.5$ in.
$b = 9$ in. $\qquad f'_c = 5{,}000$ psi
$a = (1.55 \times 60{,}000) \div (0.85 \times 5{,}000 \times 9) = 2.43$ in.

$$M_u = 0.90\left[1.55 \times 60{,}000\left(13.5 - \frac{2.43}{2}\right)\right] = 1{,}028{,}000 \text{ in.-lb}$$

This is slightly larger than the M_u required by ACI 318-63. However (512) Sec. 202.2 says:

> The ultimate strength capacity of joints and connections should be at least 10 per cent in excess of that required of the members connected. This recommendation may be satisfied by proportioning the joint or connection to provide strength in the connection or joint 1.1 times the ultimate strength capacity required by ACI 318-63.

Note that the 1.1 factor is applied to the ultimate-load condition but not to the working-load condition.

There is considerable question about whether or not the 1.1 factor should be applied to this particular example. Many "joints and connections" are simply devices for transfer of load from one member to another that would not have been needed if the structure had been cast as a unit instead of precast in separate pieces. The 1.1 factor should certainly be applied to such joints. The finished structure in this example is actually a continuous member with reinforcing bars carrying the negative moment and prestressing strands carrying the positive moment. There is no point in designing the negative-moment area to a greater factor of safety than the positive-moment area so we will not apply the 1.1 factor.

If the reinforcing bars chosen to satisfy the working-load requirements are not enough to satisfy ultimate-moment requirements, two courses are open.

1. Enough reinforcing steel can be added to satisfy the ultimate-moment requirement.
2. The effect of using the reinforcing steel designed on the working-load basis can be considered.

In the case of prestressed concrete double and single T's the second procedure is often the most desirable. As additional loads are applied to the system, the negative moment over the beam increases until the yield strength of the reinforcing steel is reached. From this point on the negative moment remains constant and moment due to additional loads is taken only as increased positive moment in the composite double T and cast-in-place deck. There are two advantages to this procedure. The capacity of the composite double-T section to carry positive moment is much greater than its capacity to carry negative moment. In the system being considered, for a fully elastic structure, the negative bending moment over the supporting beam is approximately double the positive moment at mid-span. Allowing the system to yield in the negative-moment area will minimize the negative moment at ultimate load and increase the positive moment which the structure is better able to carry.

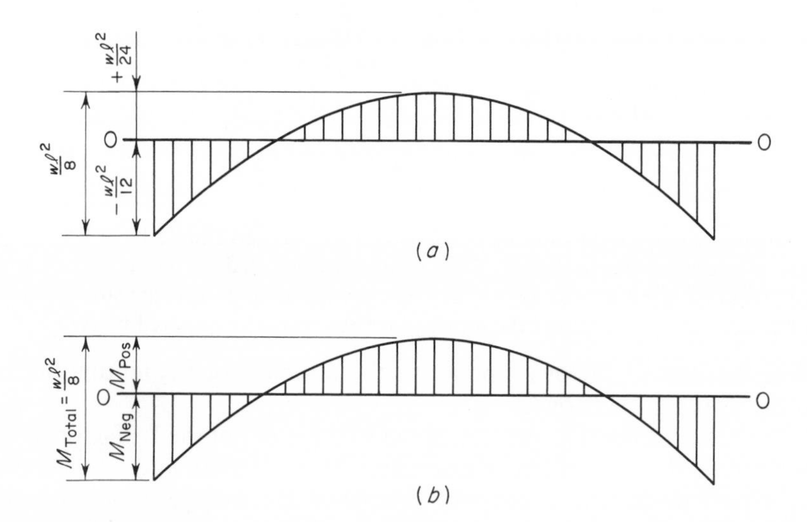

FIG. 13-7. Bending moment diagram for uniformly loaded beam. (*a*) Diagram for beam with complete fixity at each end; (*b*) diagram of beam with negative moment at each end but less than complete fixity.

Computation of ultimate moments in the system is a simple matter. If the double T's were completely fixed at each end the moment diagram would be as shown in Fig. 13-7*a*. The weight per linear foot w is the total of $1.5D + 1.8L$. Note that the total depth of the moment diagram is $wl^2/12 + wl^2/24 = wl^2/8$, which is the bending moment for a simple span beam.

Regardless of the magnitude of the negative moment the total depth of the moment diagram, M_{tot}, will be $wl^2/8$. Therefore if the magnitude of the negative moment is known, the magnitude of the positive moment can be computed from the following equation:

$$wl^2/8 = M_{tot} = M_{neg} + M_{pos} \tag{13-1}$$

or

$$M_{pos} = M_{tot} - M_{neg} \tag{13-1a}$$

The moment at which the reinforcing steel will yield, M_{neg}, is that computed by ACI 318-63, Eq. (16-1 ACI) as illustrated earlier in this section.

The foregoing computations have been based on the criteria for reinforced concrete beams. They have not considered the fact that there is compressive stress in the stem of the double T from the prestressing tendons. It would seem that the compressive stress in the stem of the double T should be subtracted from $0.85f'_c$ and only the remaining stress used in computing ultimate-moment capacity by ACI 318-63, Eq. (16-1 ACI). [$0.85f'_c$ is the stress assumed in the concrete at ultimate moment in Eq.

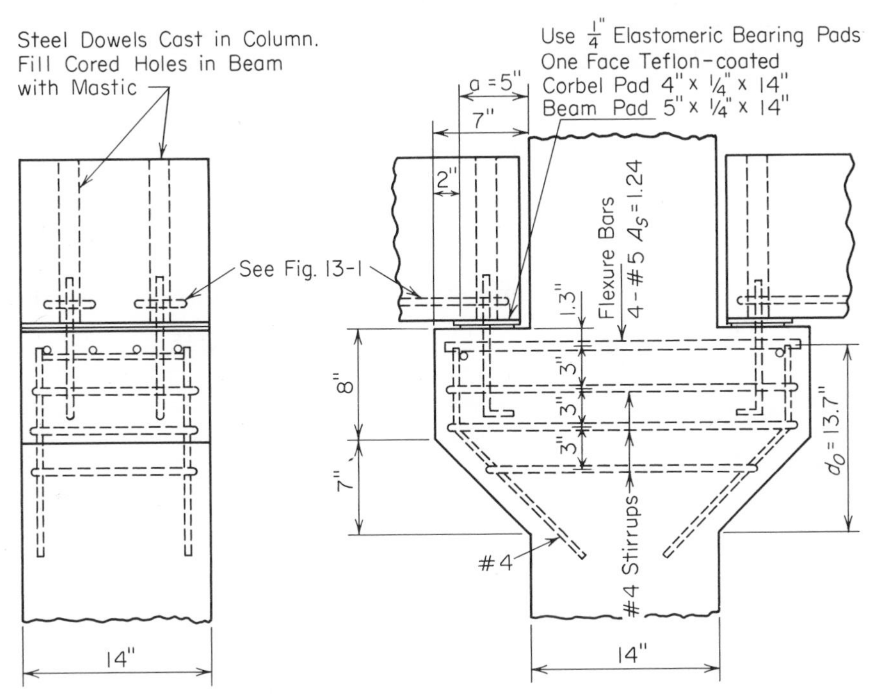

FIG. 13-8. Details of corbel designed in Art. 13-9 of text. See Fig. 301.3*a* in Appendix B for limiting dimensions, symbols, etc.

(16-1 ACI).] Test data have shown that "ultimate negative moment is governed by tension within the practical range of deck steel reinforcement.... This is so even for 0.6 per cent of straight prestressing reinforcement providing an effective concrete prestress of 2,000 psi." The foregoing quotes are from Portland Cement Association, Research and Development Laboratories Bulletin D34 entitled "Precast-Prestressed Concrete Bridges—Pilot Tests of Continuous Girders." The tests were made on I-beam and slab bridges. Bulletin D34 gives equations for checking ultimate-moment conditions, but this seems unnecessary in most cases unless there is a high percentage of negative-moment reinforcing steel in the system.

13-9. Design Example—Beam on Corbel Subject to Vertical Load Only. Design in accordance with requirements of (512) Sec. 301.3 and use same symbols as Fig. 301.3a in Appendix B.

Assume details as shown in Fig. 13-8 and the following conditions:

Dead load $= V_D = 24{,}000$ lb
Live load $= V_L = 37{,}500$ lb
Width of bracket = width of beam $= b = 14$ in.
Beam and corbel are separated by two ¼-in. elastomeric bearing pads with Teflon coated faces in contact with each other. The pad on the corbel is 4×14 in.

Compute ultimate-vertical-load capacity of corbel, V, using following equation from (512) Sec. 301.3.1.

$$V = \phi[6.5bd_o\sqrt{f'_c}(1 - 0.5^{d_o/a})(1{,}000p)^{1/3}]$$

From Fig. 13-8 and (512) numerical values for use in this equation for this example are:

$\phi = 0.85$ $\qquad b = 14$
$d_o = 13.7$ in. $\qquad f'_c = 5{,}000$ psi
$f_y = 40{,}000$ psi $\qquad a = 5$ in.
$A_s = 4$ #5 bars $= 4 \times 0.31 = 1.24$ sq in.
$p = A_s/bd_o = 1.24/(14 \times 13.7) = 0.0065$

$$d_o/a = \frac{13.7}{5} = 2.74$$

Substituting numerical values in the equation,

$$V = 0.85[6.5 \times 14 \times 13.7\sqrt{5{,}000}(1 - 0.5^{2.74})(1{,}000 \times 0.0065)^{1/3}]$$

$$\begin{array}{ll} \text{Log of } 0.5 = & 9.69897 - 10 \\ & \times 2.74 \\ \hline & 26.57518 - 27.4 \quad \text{or} \quad 9.17518 - 10 \end{array}$$

which is the log of 0.15.

Thus

$0.5^{2.74} = 0.15$
$(1{,}000 \times 0.0065)^{1/3} = 6.5^{1/3}$
$\text{Log } 6.5 = 0.81291 \div 3 = 0.27097 = 1.87$
$\sqrt{5{,}000} = 70.7$

Substituting in the equation,

$$V = 0.85[6.5 \times 14 \times 13.7 \times 70.7(1 - 0.15)1.87]$$
$$= 119{,}085 \text{ lb} = \text{corbel capacity at ultimate}$$

ACI 318-63, Sec. 1506 requires an ultimate capacity of $1.5D + 1.8L$, and Sec. 202.2 of (512) requires that joints have an ultimate capacity 10 per cent in excess of that required for the other parts of the structure. Thus the required ultimate capacity is

$$V_u = 1.10(1.5D + 1.8L) \qquad (13\text{-}2)$$

Substituting numerical values,

$$V_u = 1.10(1.5 \times 24{,}000 + 1.8 \times 37{,}500)$$
$$= 113{,}850 \text{ lb}$$

Since $V = 119{,}085$ lb the design of the corbel for flexure is adequate.

Section 301.3.1 of (512) says: "The reinforcement index 'q' for all corbels should have a maximum of 0.15 and a minimum of 0.04 where $q = pf_y/f'_c$." Substituting numerical values for this example

$$q = \frac{0.0065 \times 40{,}000}{5{,}000} = 0.052$$

which is within the specified limits.

Section 301.3.1 of (512) also says: "The depth of the bracket outer face should be at least from 0.4 to 0.5, the total bracket depth."

From Fig. 13-8,

$$\frac{8}{7 + 8} = 0.53, \text{ which is satisfactory.}$$

Section 301.3.2 of (512) says: "For a/d_o ratios less than 1.0 stirrups in the amount of $0.005bd_o$ should be used."

For our example,

$$\frac{a}{d_o} = \frac{5}{13.7} = 0.36 < 1.0$$

$$0.005bd_o = 0.005 \times 14 \times 13.7 = 0.96 \text{ sq in.}$$

Section 301.3.2 of (512) requires "at least three pairs" of bars, which means at least six bars.

$$\frac{0.96}{6} = 0.16 \text{ sq in. per bar}$$

Use #4 bars for stirrups.

There is no specified requirement for the vertical bars in the outside corners of the corbel, but they are essentially anchors for the stirrups and can be made the same size as the stirrups.

Total design reaction of the beam on the corbel is

$$24{,}000 + 37{,}500 = 61{,}500 \text{ lb}$$

Area of the bearing pad is

$$4 \times 14 = 56 \text{ sq in.}$$

Bearing pressure is

$$61{,}500 \div 56 = 1{,}098 \text{ psi}$$

Allowable bearing pressure from ACI 318-63, Table 1002(a) is

$$0.25f'_c = 0.25 \times 5{,}000 = 1{,}250 \text{ psi}$$

From Art. 13-5, the recommended bearing pressure in ½-in. elastomeric pads is 1,000 psi but "it is quite possible to exceed 1,000 using a thinner elastomer layer" and our pads are only ¼-in. thick, so this pressure is permissible.

13-10. Post-tensioned Joints. With reference to prestressed joints Sec. 301.7 of (512) says: "Such application should conform to the provisions of ACI-ASCE Committee 423(323), with adherence to stress limitations suggested for segmental elements." Evidently this refers to Tentative Recommendations for Prestressed Concrete presented by ACI-ASCE Committee 323 in 1958. A prestressed joint would necessarily carry moment as well as shear and would therefore be classed as continuous. Section 213 of the Tentative Recommendations deals with "continuity" but makes no specific requirements other than that normal allowable stresses be applied.

It is the recommendation of the writers that a prestressed joint be treated as a prestressed beam with the same shear and moment applied and that prestressing tendons, stirrups, etc., be supplied to keep stresses within the limits set by ACI 318-63. If shear is to be carried through the prestressed joint rather than by the bearing of one member upon another, Eq. (26-10 ACI) of ACI 318-63, Sec. 2610 must be satisfied.

Figure 15-25 illustrates a post-tensioned joint. For a detail such as this the corbel can be designed to carry all of the shear as in Art. 13-9 or the corbel can be considered as an erection platform and the prestressed joint designed to carry all the shear. Section 301.1 of (512) says: "The entire shear should be considered as transferred through one type of device." In either case the end of the beam must be designed in accordance with ACI 318-63 as a beam carrying both shear and moment.

There are several points to remember in the design of a post-tensioned joint such as that shown in Fig. 15-25.

1. The prestressing tendons apply a large negative moment to the end of the beam. The beam must contain reinforcing steel or prestressing ten-

dons which lap the post-tensioned tendons sufficiently to transfer the negative moment to the beam.

2. In computing ultimate negative moment capacity of the joint the compressive stress, due to prestress in the beam, on the bottom or compressive side of the beam can probably be ignored. See discussion near the end of Art. 13-8.

3. The column must be reinforced and/or prestressed so that it has adequate capacity to carry the moment and to withstand concentrated stress under bearing plates, etc.

BIBLIOGRAPHY

1. Birkeland, Philip W., and Halvard W. Birkeland: Connections in Precast Concrete Construction, *J. Am. Concrete Inst.,* March, 1966.

CHAPTER 14

Continuous Structures

STRUCTURES WITH CONTINUOUS TENDONS

14-1. Characteristics. For structures in which tendons are continuous over a pier and in which the tendons are curved so that they carry shear, the familiar equation (1-1) for stress in the concrete due to the prestressing force is not applicable. In a continuous member with continuous tendons the vertical loads applied to the member by the curvature of the tendons are not distributed to the supports in the same proportions as in the simple beam which is the basis for Eq. (1-1).

Mathematical procedures leading to a direct solution for the prestressing force and its eccentricity have been developed, but they are involved and seem difficult for some designers to follow. A procedure based on relatively simple mathematical formulas is suggested in the following section.

14-2. Design Procedure. Step 1. Establish a trial path for the center of gravity of the tendons (see Fig. 14-2).

At points of maximum negative moment, over the intermediate piers, place the tendons as close to the top of the member as details permit, and at points of maximum positive moment, near the centers of the spans, place the tendons as close to the bottom of the member as details permit. At the ends of the member the tendons can be placed at any elevation the details permit and which does not create excessive stresses. For simplicity in the first computation we shall place the tendons at the center of gravity of concrete at the ends of the member.

The tendons are usually placed in a parabolic curve between the control points which have been established. If, at some point in the calculations, it is found that the path of tendons chosen is not satisfactory, a new path should be established to correct the difficulty found with the first one.

Step 2. Assume an arbitrary final tension F_A in the tendons. Any value will do; it need not be close to the actual final value.

Step 3. Compute the vertical loads applied to the member by the tendons due to F_A.

Step 4. Compute the stresses in the member due to the vertical loads from the tendons found in Step 3.

Step 5. Compute the direct compressive stress F_A/A_c in the member.

Step 6. Compute bending moments and unit stresses due to eccentricity of tendons at the ends of the member. If tendons are on the c.g.c. at the ends of the member, eccentricity is zero and there are no moments or stresses from this factor.

Step 7. Add algebraically the stresses from Steps 4, 5, and 6 to get the net stresses f_{F_A} due to F_A.

Step 8. Compute moments and stresses due to dead load.

Step 9. Compute maximum positive moments and stresses and maximum negative moments and stresses due to live load.

Step 10. Use the stresses from Steps 8 and 9 to compute maximum dead- plus live-load stresses. Choose what appears to be the point on the structure which will require the maximum F, and establish the amount of compressive stress $f^b{}_F$ required.

Step 11. Compute the required final prestress F from Eq. (14-1).

$$F = \left(\frac{f^b{}_F}{f^b{}_{FA}}\right)F_A \tag{14-1}$$

Check the entire structure, combining this value with dead load only and also with dead load plus critical live loads. Consider the economy of the section. Would a smaller or differently shaped cross section be more economical? Check initial prestress plus dead load.

Step 12. Compute reactions on piers. Since the tendons carry some of the shear, the pier reactions will not be the same as for an unprestressed structure with the same elastic properties. The pier reactions due to prestress plus dead load can be computed in four steps.

a. Compute the reactions due to dead load, considering the structure as elastic.

b. Compute the pier reactions due to the upward forces of the tendons against the structure. These will be negative reactions.

c. Compute the reactions from the downward forces of the tendons over the piers.

d. Add algebraically the values computed in *a, b, and c* to get the pier reactions due to prestress plus dead load.

Since the structure is fully elastic under design live load, the formulas used for elastic structures are used in computing live-load reactions.

Step 13. Check ultimate strength. Some authorities feel that moment redistribution should be used in the computation of ultimate strength. Others feel that there are not enough available data and point out that moment redistribution has not yet been included in reinforced concrete specifications. Moment redistribution will not be used in this example.

Step 14. Compute shear. The shear is the same as for any elastic structure less the shear carried by the tendons.

Step 15. Check camber.

Step 16. Work out details of tendons, anchorages, shear steel, etc.

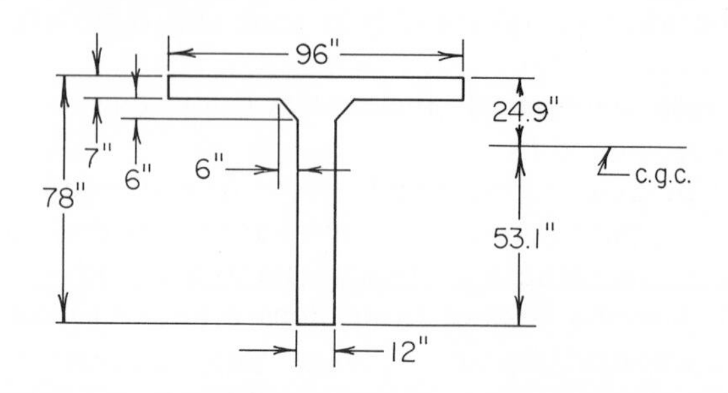

Section Properties – Taking Moments About Top

	A, sq in	y, in	Ay	Ay^2	I_o
96 x 7 =	672	3.5	2,350	8,220	2,745
12 x 71 =	852	42.5	36,200	1,538,000	358,000
6 x 6 x 2/2 =	36	9	324	2,915	72
	1,560		38,874	1,549,135	360,817
				360,817	
				1,909,952	

$Y_t = 38{,}874 \div 1{,}560 = 24.9''$

$Y_b = 78 - 24.9 = 53.1''$

$I = 1{,}909{,}952 - 1{,}560(24.9)^2 = 943{,}000$

$Z_t = 943{,}000 \div 24.9 = 37{,}880$

$Z_b = 943{,}000 \div 53.1 = 17{,}760$

$\text{Weight} = 1{,}560 \frac{150}{144} = 1{,}625 \text{ lb per ft}$

FIG. 14-1. Cross section and section properties of concrete section to be used in design example.

14-3. Numerical Example. For the purpose of this example we shall consider a bridge of two 125-ft spans with both the concrete and the tendons continuous over the center pier. The cross section is a poured-in-place beam and slab type with the beams 8 ft 0 in. on centers. Figure 14-1 shows one beam and the slab which functions with it. The design live load is 1,000 lb per lin ft. (a uniform live load has been used to simplify calculations. The only change for the design of an actual bridge would be computation of live-load moments and shears on the basis of truck loads or standard uniform plus concentrated loads.) Design is based on allowable stresses, etc., in the AASHO Specification

$$f'_c = 5{,}000 \text{ psi}$$

Follow the steps outlined in the foregoing procedure. An elevation of the structure is shown in Fig. 14-2. Slide-rule calculations are normally of sufficient accuracy for prestressed concrete design and have been used in this example.

Step 1. In order to locate the tendons as near to the concrete surfaces as possible, we should have some idea of the magnitude of F. The uniform compressive stress F/A_c in an efficient continuous bridge design is usually around 800 to 1,000 psi. For the section shown in Fig. 14-1 try an F of 1,200,000 lb. Select a group of tendons to give this force and plot typical patterns at top and bottom of the concrete cross section. The c.g.s. can easily be located within 8 to 10 in. of the concrete surface. We shall place the c.g.s. 10 in. from the top over the center pier, 10 in. from the bottom near the centers of span, and on the c.g.c. at the end piers.

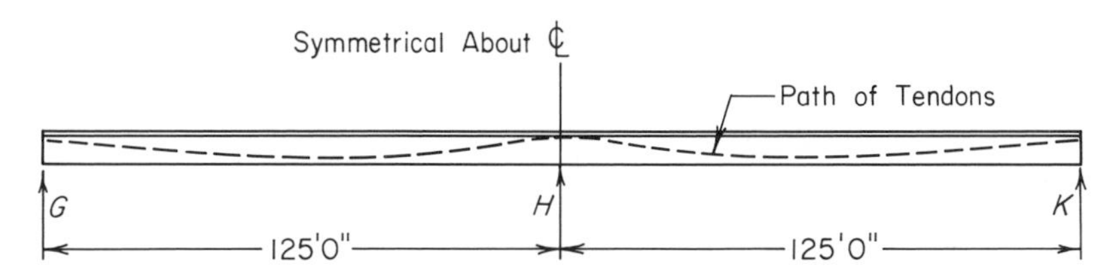

FIG. 14-2. Elevation of two-span bridge in design example.

A diagram for the path of the c.g.s. is shown in Fig. 14-3. The known dimensions on this diagram are

$$R = 125 \text{ ft } 0 \text{ in.} = 1{,}500 \text{ in.}$$
$$D = 10 \text{ in.}$$

Since the c.g.s. has been placed on the c.g.c. at the ends,

$$M = y_t - D = 24.9 - 10 = 14.9 \text{ in.}$$

When M is small with respect to R, as it is in this example, the difference in elevation between the lowest point on the curve and the curve at the center of span is negligible. For our first trial we shall assume these elevations are the same, which gives

$$B = E - 0.5M = 58 - 7.45 = 50.55 \text{ in.}$$

From Fig. 14-3 the lowest point on the curve occurs at

$$X = \frac{R}{2} + \frac{RM}{8B} = \frac{1{,}500}{2} + \frac{1{,}500 \times 14.9}{8 \times 50.55}$$

$$= 750 + 55 = 805 \text{ in.}$$

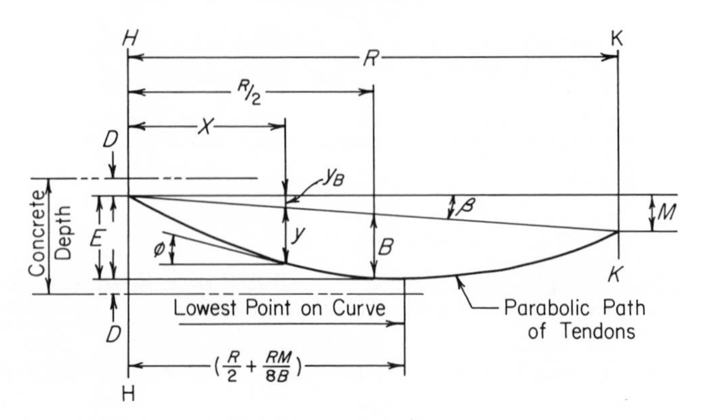

$y = 4B\frac{X}{R}(1 - \frac{X}{R})$ See tabulation Fig. 11-5.

$y_B = M\frac{X}{R}$

$\text{Tan } \phi = \frac{4B}{R}(1 - \frac{2X}{R}) + \frac{M}{R}$

Length of tendon along parabolic curve = L

$L = R(\text{Sec. } \beta + \frac{2.67n^2}{\text{Sec.}^3 \phi})$

$n = \frac{B}{R}$

w_T = uniform upward load applied to concrete by tendons

$w_T = \frac{8BF}{R^2}$

V_H and V_K = concentrated downward loads applied at supports by tendons = $F \tan \phi$

At H, $X = 0$ and $\tan \phi = \frac{4B + M}{R}$

At K, $X = R$ and $\tan \phi = \frac{M - 4B}{R}$

FIG. 14-3. Geometry of parabolic path of tendons where ends of parabola are at different elevations.

At this point from Fig. 14-3

$$y_B = 14.9\frac{805}{1{,}500} = 8.0 \text{ in.}$$

$$y = 4 \times 50.55\frac{805}{1{,}500}\left(1 - \frac{805}{1{,}500}\right) = 50.2 \text{ in.}$$

$$y_{B+y} = 58.2 \text{ in.}$$

The distance from the bottom of the concrete to the c.g.s. at $X = 805$ in. (the lowest point on the curve) is

$$78 - (10 + 58.2) = 9.8 \text{ in.}$$

This is so close to the 10 in. originally chosen that there should be no problem in arranging a satisfactory tendon pattern to suit this dimension. (For those problems in which this value proves to be too small, a new value of B should be chosen and the calculations repeated. The new value of B would be the old value of B minus the distance the c.g.s. needs to be raised. If we needed to raise the c.g.s. to 10 in. in this example, the amount to be raised would be $10 - 9.80 = 0.20$ in. and the value of B for the second trial would be $50.55 - 0.20 = 50.35$ in. We shall continue with the values in our first trial.

Step 2. Let $F_A = 1{,}000{,}000$ lb.

Step 3. From Fig. 14-3

$$w_T = \frac{8BF_A}{R^2} = \frac{8 \times 50.55 \times 1{,}000{,}000}{1{,}500^2} = 179.6 \text{ lb per in.}$$

Step 4. From Fig. 14-4 the bending moment M_T due to the uniform load w_T is the coefficient in the table for uniform load full length $\times$ $w_T(2R)^2$. In this case $w_T(2R)^2 = 179.6(2 \times 1{,}500)^2$. Since w_T is an upward load, the sign of the moment will be opposite to that given in the table in Fig. 14-4. Using the coefficients in the table and the section properties from Fig. 14-1 we get

	Station on girder					
	G or K	1	2	3	4	H
Moment, in.-kips	0	−22,620	−29,100	−19,400	+8,080	+50,100
f^t_T	0	−598	−768	−512	+213	+1,323
f^b_T	0	+1,275	+1,639	+1,092	−455	−2,820

Step 5. The direct compressive stress is $1{,}000{,}000 \div 1{,}560 = 641$ psi.

Step 6. Since the c.g.s. is on the c.g.c. at the ends of the member, there is no bending moment due to the eccentricity of the tendons.

If the c.g.s. were above the c.g.c. at the ends, this eccentricity would create a positive moment of F_Ae and the moment diagram would be that shown in Fig. 14-5. The stresses due to this moment should be computed for summation in Step 7. If the c.g.s. were below the c.g.c. at the ends, the moments at G and K would be $-F_Ae$ and at H would be $+0.5F_Ae$.

Figure 14-6 shows the moment diagram resulting from eccentricity of tendons at the ends of a three-span continuous structure such as that shown in Fig. 15-11.

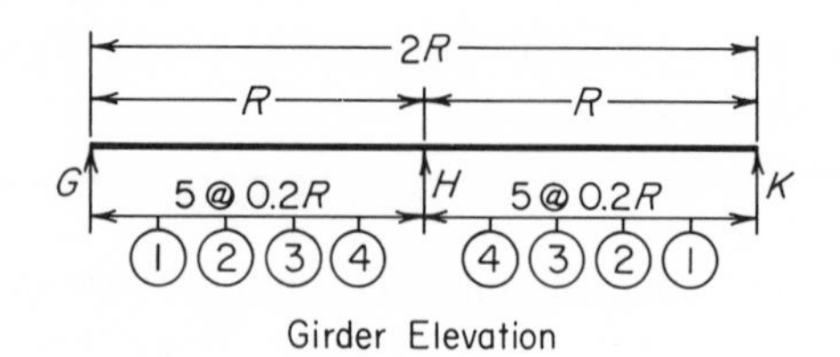

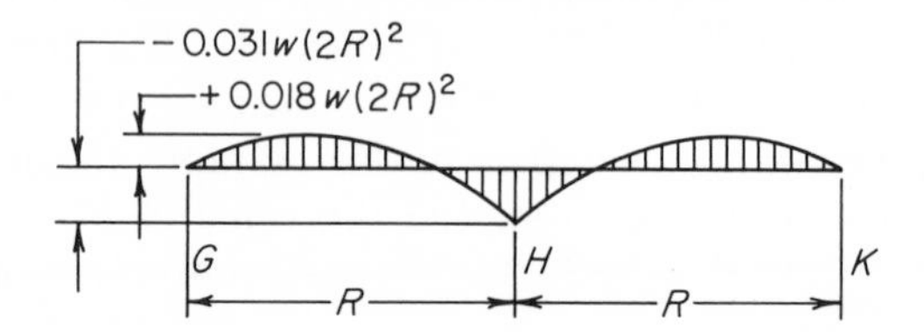

Table of Moment Coefficients

Station on girder	*G* or *K*	1	2	3	4	*H*
Uniform load full length	0	+0.014	+0.018	+0.012	−0.005	−0.031
Max positive moment	0	+0.017	+0.024	+0.021	+0.008	
Max negative moment	0	−0.003	−0.006	−0.009	−0.013	−0.031

Moment is coefficient from table × $w(2R)^2$.

FIG. 14-4. Moments in a two-span continuous beam having a constant moment of inertia.

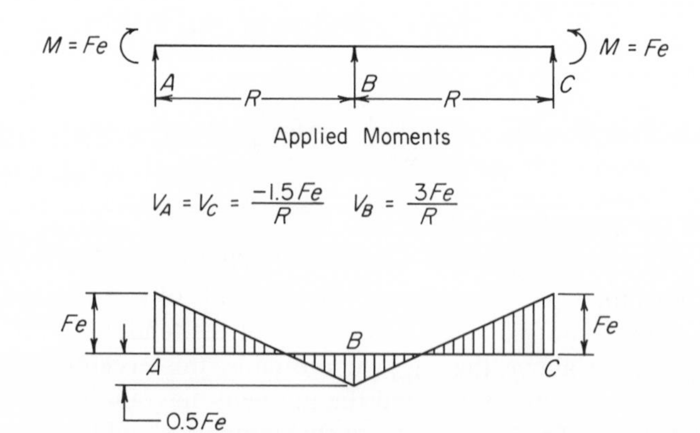

FIG. 14-5. Moments in a two-span continuous beam due to eccentric application of prestressing force at ends.

Step 7. The net stresses due to F_A are the sum of those computed in Steps 4, 5, and 6, which give

	Station on girder					
	G or K	1	2	3	4	H
$f^t{}_{F_A}$	0	+43	−127	+129	+854	+1,964
$f^b{}_{F_A}$	0	+1,916	+2,280	+1,733	+186	−2,179

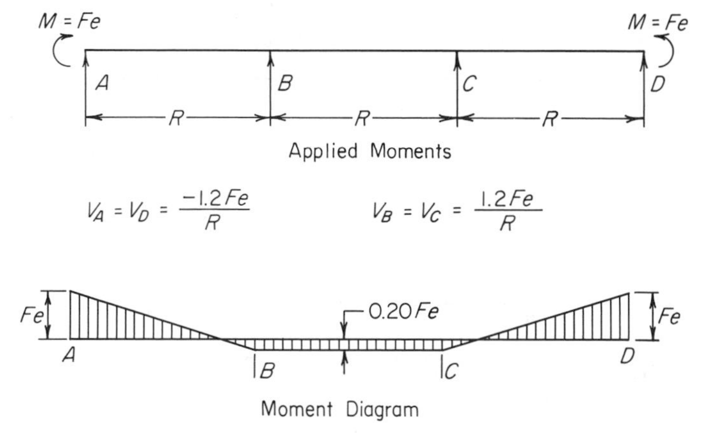

FIG. 14-6. Moments in a three-span continuous beam due to eccentric application of prestressing force at ends.

Step 8. The dead-load moments are found by multiplying the coefficients in the table in Fig. 14-4 for uniform load full length by $w(2R)^2$. In this case $w(2R)^2 = 1{,}625/12(2 \times 1{,}500)^2$. The moments and stresses are

	Station on girder					
	G or K	1	2	3	4	H
Dead-load moment, in.-kips	0	+17,060	+21,930	+14,620	−6,090	−37,800
$f^t{}_D$	0	+450	+599	+386	−161	−998
$f^b{}_D$	0	−961	−1,236	−823	+343	+2,130

Step 9. Maximum live-load moments are found using the coefficients in the table in Fig. 14-4. For this computation $w(2R)^2 = 1{,}000/12(2 \times 1{,}500)^2$. Maximum positive moments and stresses are

	Station on girder					
	G or K	1	2	3	4	H
Live-load moment, in.-kips.	0	+12,760	+18,000	+15,760	+6,000	Live load never causes positive moment at this point
$f^t{}_L$.	0	+337	+475	+416	+158	
$f^b{}_L$.	0	−718	−1,014	−887	−338	

Maximum negative moments and stresses are

	Station on girder					
	G or K	1	2	3	4	H
Live-load moment, in.-kips.	0	−2,250	−4,500	−6,750	−9,750	−23,250
$f^t{}_L$.	0	−59	−119	−178	−258	−614
$f^b{}_L$.	0	+127	+254	+380	+549	+1,310

Step 10. Adding dead-load stresses from Step 8 to stresses for maximum positive moment from Step 9 gives

	Station on girder					
	G or K	1	2	3	4	H
$f^t{}_{D+L}$	0	+787	+1,074	+802	−3	−998
$f^b{}_{D+L}$	0	−1,679	−2,250	−1,710	+5	+2,130

Adding dead-load stresses from Step 8 to stresses for maximum negative moment from Step 9 gives

	Station on girder					
	G or K	1	2	3	4	H
$f^t{}_{D+L}$	0	+391	+480	+208	−419	−1,612
$f^b{}_{D+L}$	0	−834	−982	−443	+892	+3,440

Comparison of the stresses computed in this step with the stresses produced by the prestressing force as summarized in Step 7 suggests that the governing condition is found at station 2 on the girder under dead load plus positive moment from live load. This stress is −2,250 psi.

Step 11. Using Eq. (14-1), find the required F.

$$F = \left(\frac{f^b{}_F}{f^b{}_{FA}}\right)F_A \tag{14-1}$$

$$F_A = 1{,}000{,}000 \qquad f^b{}_F = 2{,}250 \qquad f^b{}_{FA} = 2{,}280$$

$$F = \frac{1{,}000{,}000 \times 2{,}250}{2{,}280} = 987{,}000 \text{ lb}$$

Stresses due to this value of F will be

$$\frac{987{,}000}{1{,}000{,}000} = 98.7\% \text{ of those due to } F_A$$

Using this ratio, the stresses due to $F = 987{,}000$ lb are

	Station on girder					
	G or *K*	1	2	3	4	*H*
$f^t{}_F$	0	+42	−125	+127	+843	+1,938
$f^b{}_F$	0	+1,891	+2,250	+1,710	+184	−2,150

One of the critical conditions is F plus dead load only. The sum of the stresses due to F and the stresses due to dead load is

	Station on girder					
	G or *K*	1	2	3	4	*H*
$f^t{}_{F+D}$	0	+492	+474	+513	+682	+940
$f^b{}_{F+D}$	0	+930	+1,014	+887	+527	−20

A second critical condition is F plus dead load plus maximum positive live-load moments. The sum of the stresses under this loading is

	Station on girder					
	G or *K*	1	2	3	4	*H*
$f^t{}_{F+D+L}$	0	+829	+949	+929	+840	+940
$f^b{}_{F+D+L}$	0	+212	0	0	+189	−20

A third critical condition is F plus dead load plus maximum negative live-load moments. The sum of the stresses under this loading is

	Station on girder					
	G or K	1	2	3	4	H
$f^t{}_{F+D+L}$	0	+433	+355	+335	+424	+326
$f^b{}_{F+D+L}$	0	+1,057	+1,268	+1,267	+1,076	+1,290

Under all three critical conditions the compressive stresses are well below the allowable of 2,000 psi. The only tensile stress is 20 psi in the bottom fiber at H. Since this is negligible, we can say that the member is satisfactory under the design load condition.

At this point we should consider the economy of the section being designed. The maximum compressive stress under any of the critical conditions is 1,268 psi in the bottom fiber at station 2 under F plus dead load plus maximum negative moment. This is only 63.4 per cent of the allowable 2,000 psi.

What would happen if we decreased the depth and/or width of the 12-in. stem? We would save some concrete and reduce dead weight. For the same prestressing force the compressive stress F/A_c would increase because A_c would be smaller. Even so it would probably be necessary to provide a larger F to balance completely the increased stress in the smaller member. The savings resulting from less concrete should be compared with the cost of additional prestressing force.

Several factors would effect the 20-psi tensile stress at H. If the net result is an increase in tensile stress it can be offset by adjusting the path of tendons. Decreasing the value of B (Fig. 14-3) by 1 in. will decrease the tensile stress $f^b{}_T$ computed in Step 4 by about 56 psi. B can be decreased 1 in. by raising the c.g.s. at center of each span by 1 in. or by lowering the c.g.s. at H by 2 in. Raising or lowering the c.g.s. at each end of the structure has no effect on stresses at H because the stresses due to eccentricity at the ends (Fig. 14-5) are exactly balanced by the stresses due to the corresponding change in B. We shall continue with the section shown in Fig. 14-1 because previous experience with continuous structures of this type shows that the dimensions of the required section are determined by ultimate-moment requirements more often than by design-load requirements.

The 1963 Interim AASHO Specification, Sec. 1.13.7(B)(2) says that under design loads post-tensioned bridge members shall have zero flexural tension "in the precompressed tensile zone." As the tensile stress under consideration is not in this zone but in a zone which is subjected to compressive stress by any live or dead load applied to the structure, it is the authors' feeling that a reasonable amount of tensile stress ($6\sqrt{f'_c}$) can be permitted. Reinforcing bars should be provided to carry the tensile stress.

Now that the final value of F has been established, we must check the fourth critical condition, which is initial prestress plus dead load. Details of the tendons will not be worked out in this example, so we shall assume that they are of 0.196-in.-diameter wire with an initial stress of 175,000 psi and a final stress of 150,000 psi. This means that F_I will be $(175{,}000/150{,}000)F = 1.166 \times 987{,}000 = 1{,}151{,}000$ lb. The stresses due to F_I will be 1,151,000/1,000,000, or 1.151 times those due to F_A, which gives

	Station on girder					
	G or K	1	2	3	4	H
$f^t_{F_I}$	0	+49	−146	+148	+982	+2,260
$f^b_{F_I}$	0	+2,205	+2,623	+1,995	+214	−2,508

The stresses due to F_I plus dead load are

	Station on girder					
	G or K	1	2	3	4	H
$f^t_{F_I+D}$	0	+499	+453	+534	+821	+1,262
$f^b_{F_I+D}$	0	+1,244	+1,387	+1,172	+557	−378

All the compressive stresses are well within the allowable, and the only tensile stress is the 378 psi in the bottom fiber at H. AASHO, Sec. 1.13.7 (B)(1) permits a tensile stress of $3\sqrt{f'_{ci}}$ for members without reinforcing bars. For this problem this is $3\sqrt{5{,}000} = 212$ psi. We must provide reinforcing bars to carry the tension in accordance with the following computation.

The diagram in Fig. 14-7*a* shows the stress distribution which would exist if the tension were carried by the concrete. Under this condition the center of the compressive force is 20 in. below the top of the girder. For a trial computation we shall assume that the center of the group of reinforcing bars to be provided can be located within 4½ in. of the bottom, so that $y = 53½$ in.

The bending moment to be resisted by the reinforcing bars is the section modulus of the section for its bottom fiber times its tensile stress, or $17{,}760 \times 378$. Thus the tension the bars must carry is $(17{,}760 \times 378)/53.5 = 125{,}500$ lb. At 20,000 psi we need 6.275 sq in. of reinforcing steel. Use six #9 bars with a total area of 6 sq in. These can be placed in a satisfactory pattern as shown in Fig. 14-7*b*. Since the bottom fiber stress goes from −378 psi at *H* to +557 psi at station 4, the reinforcing bars will all be relatively short. Points at which each layer of bars can be stopped should be computed by plotting the curve for $f^b{}_{F_I+D}$ from *H* to 4.

As a check on the most economical design, computations should be made with the tendons at a lower elevation at *H* to determine the total force required when all tensile stress in the concrete is eliminated at *H*. The cost of these tendons for the full length of the structure should then be compared with the cost of the tendons and reinforcing bars as designed in this example.

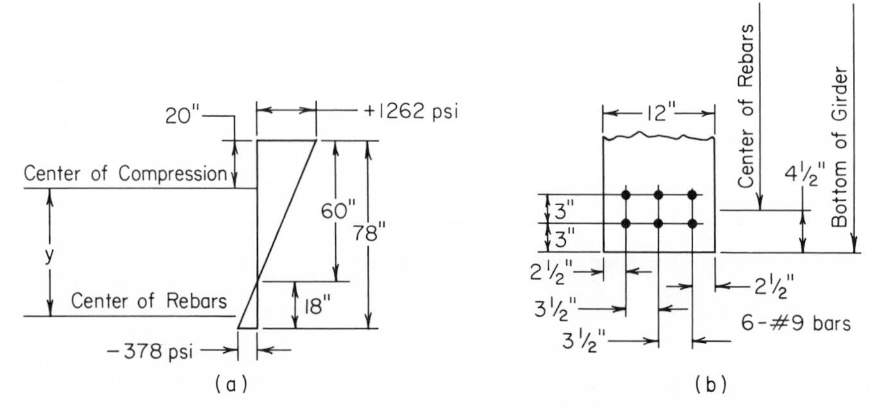

FIG. 14-7. Stress diagram at station *H* under initial prestress plus dead load. Reinforcing bars used to take the tension in the bottom.

Step 12. *a*. From the shear and moment diagrams for a two-span elastic structure as found in any handbook the reactions under uniform load are

$$R_G = R_K = 0.375wR \qquad \text{and} \qquad R_H = 1.25wR$$

which give, for dead load,

$$R_G = R_K = 0.375 \times 1{,}625 \times 125 = 76{,}200 \text{ lb}$$
$$R_H = 1.25 \times 1{,}625 \times 125 = 254{,}000 \text{ lb}$$

b. Using the formula from Fig. 14-3,

$$w_T = \frac{8BF}{R^2} = \frac{8(50.55/12)987{,}000}{125^2} = 2{,}130 \text{ lb per ft}$$

Applying the standard formula used in *a*,

$$R_G = R_K = 0.375 \times 2{,}130 \times 125 = -99{,}800 \text{ lb}$$
$$R_H = 1.25 \times 2{,}130 \times 125 = -333{,}000 \text{ lb}$$

c. Use formulas in Fig. 14-3 to find concentrated downward loads applied to piers by tendons.

At H, $\tan \phi = \dfrac{4B + M}{R} = \dfrac{(4 \times 50.55) + 14.9}{125 \times 12} = 0.1447$

At K, $\tan \phi = \dfrac{M - 4B}{R} = \dfrac{14.9 - (4 \times 50.55)}{125 \times 12} = -0.1249$

Then

$$R_H = F \tan \phi = 987{,}000 \times 0.1447 = 142{,}800 \text{ lb}$$

This is the reaction at H from the tendons between H and K. There is an equal reaction at H from the tendons between H and G. Thus the total reaction at H is $142{,}800 \times 2 = 285{,}600$ lb.

$$R_G = R_K = 987{,}000 \times 0.1249 = 123{,}300 \text{ lb}$$

This is a positive downward reaction. The negative sign for $\tan \phi$ at K simply indicates that the slope of the tendons is in the opposite direction to that for which the equation was set up.

d. Adding the values in *a*, *b*, and *c*, we get the following net reactions on the piers under final prestress plus dead load:

$$R_H = 254{,}000 - 333{,}000 + 285{,}600 = 206{,}600 \text{ lb}$$
$$R_G = R_K = 76{,}200 - 99{,}800 + 123{,}300 = 99{,}700 \text{ lb}$$

The sum of the reactions $R_G + R_H + R_K$ is 406,000 lb, which checks the total dead weight of 1,625 lb per ft $\times$ 125 $\times$ 2.

When a uniform live load is distributed over the portions of the span where it will cause the most shear, the maximum reactions which can be developed are

$$R_G = R_K = 0.438wR \qquad \text{and} \qquad R_H = 1.25wR$$

Using numerical values

$$R_G = R_K = 0.438 \times 1{,}000 \times 125 = 54{,}750 \text{ lb}$$
$$R_H = 1.25 \times 1{,}000 \times 125 = 156{,}250 \text{ lb}$$

Step 13. From AASHO, Sec. 1.13.6 the required ultimate strength for a bridge is $1.5D + 2.5L$. This is not necessarily the most severe condition in a continuous structure. At some points on the structure the maximum ultimate moment may be due to $1.0D$ on part of the structure and $1.5D$ plus $2.5L$ on the rest of the structure.

The first problem, therefore, is to plot the curve of maximum ultimate moment. For greater simplicity in calculations this can be expressed as $1.0D$ for the full length of the structure and $0.5D$ plus $2.5L$ on those parts of the structure which create maximum moments.

The dead-load moments have already been computed in Step 8.

The additional load will be $(0.5 \times 1{,}625) + (2.5 \times 1{,}000) = 3{,}312$ lb per ft. Using $w = 3{,}312$ and the coefficients in the table in Fig. 14-4, compute maximum positive and maximum negative moments. For this computation $w(2R)^2$ is $(3{,}312/12)(2 \times 1{,}500)^2$. Combine each of these values with the dead-load moment to find the critical values for ultimate moment as follows (moments are given in inch-kips):

	Station on girder					
	G or K	1	2	3	4	H
A = dead-load moment. .	0	+17,060	+21,930	+14,620	−6,090	−37,800
B = max pos moment. . .	0	+42,200	+59,600	+52,180	+19,870	
C = max neg moment. . .	0	−7,450	−14,900	−22,350	−32,300	−77,000
$D = A + B$.	0	+59,260*	+81,530*	+66,800*	+13,780*	−37,800
$E = A + C$.	0	+9,610	+7,030	−7,730*	−38,390*	−114,800*

The possible critical ultimate moments are those marked with an asterisk (*). We must now compute the ultimate capacity of the structure at these points.

In checking ultimate moments we need values for A_s and f'_s. In Step 11 we set F = 987,000 lb, using 0.196-in.-diameter wire at a final stress of 150,000 psi. From this $A_s = 987{,}000 \div 150{,}000 = 6.58$ sq in. The ultimate strength of 0.196-in.-diameter wire is $f'_s = 250{,}000$ psi.

First check ultimate negative-moment capacity at H. Under negative moment this section would be considered as a rectangular section, since only the rectangular section is in compression. Use the formula from AASHO, Sec. 1.13.10(A). From Fig. 14-1 the width b of the compression area at H is 12 in. However, if we use a value of 12 in., we shall be completely ignoring the reinforcing steel shown in Fig. 14-7*b*. Since the AASHO Specifications do not outline a design procedure including reinforcing steel in compression under ultimate conditions, the following transformed section method is suggested.

Under design load conditions the addition to the section provided by the reinforcing steel is in the ratio of the E of the steel to the E of the concrete, since both materials are working at stresses below their yield points. Under ultimate moment conditions each material will have reached its yield point. If one material reaches its yield point ahead of the other, it will continue to change length without an appreciable increase in stress while the other material is subjected to additional stress by the continuing change in length.

In this problem the concrete $f'_c = 5{,}000$ psi and concrete under ultimate bending conditions is assumed to yield at 85 per cent f'_c, so its yield point is $5{,}000 \times 85\% = 4{,}250$ psi. The minimum yield point of the intermediate-grade bars is 40,000 psi. Thus adding the steel bars is equivalent to adding an area of concrete equal to $(40{,}000/4{,}250 - 1)A'_s = 8.4A'_s$ in which A'_s is the area of reinforcing steel. Since the center of gravity of the bars is 4½ in. from the bottom of the girder, the center of the additional area will act at this point. We shall assume that the additional area is a uniform width 9 in. deep. The additional area is 8.4 × 6 sq in. = 50.4 sq in., and its width is 50.4 ÷ 9 = 5.6 in. Our transformed section now has a flange with a width b of 12 + 5.6 = 17.6 in. and a depth of 9 in. as shown in Fig. 14-8.

Refer to AASHO Specifications for procedure in computing ultimate strength.

From Fig. 14-1 and Step 1, $d = 78 - 10 = 68$ in.

$$p = \frac{A_s}{bd} = \frac{6.58}{17.6 \times 68} = 0.0055$$

From Sec. 1.13.10(C)

$$f_{su} = f'_s\left(1 - 0.5\frac{pf'_s}{f'_c}\right)$$

$$= 250{,}000\left(1 - \frac{0.5 \times 0.0055 \times 250{,}000}{5{,}000}\right) = 216{,}000 \text{ psi}$$

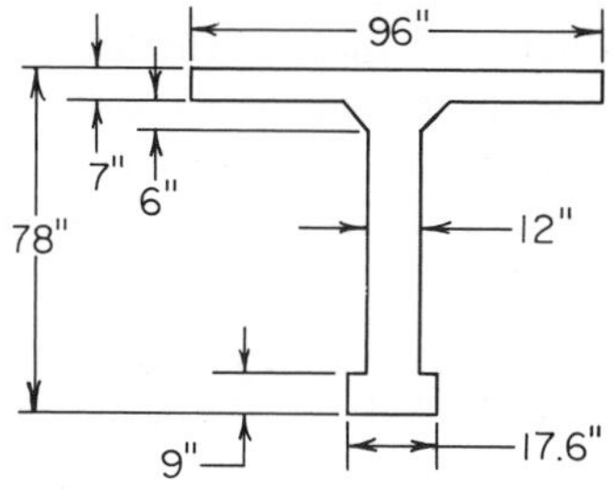

FIG. 14-8. Transformed section at station H resulting from reinforcing bars shown in Fig. 14-7.

From Sec. 1.13.10(B)

$$\frac{1.4dpf_{su}}{f'_c} = \frac{1.4 \times 68 \times 0.0055 \times 216{,}000}{5{,}000} = 22.6 \text{ in.}$$

Since the flange thickness of 9 in. is less than this, the formula from Sec. 1.13.10(B) will be used. For use in this formula

$$A_{sf} = \frac{0.85f'_c(b - b')t}{f_{su}}$$

$$= \frac{0.85 \times 5{,}000(17.6 - 12)9}{216{,}000} = 0.98 \text{ sq in.}$$

$$A_{sr} = A_s - A_{sf} = 6.58 - 0.98 = 5.60 \text{ sq in.}$$

The formula is

$$M_u = A_{sr}f_{su}d\left(1 - \frac{0.6A_{sr}f_{su}}{b'df'_c}\right) + 0.85f'_c(b - b')t(d - 0.5t)$$

Substituting numerical values,

$$M_u = 5.60 \times 216{,}000 \times 68\left(1 - \frac{0.6 \times 5.6 \times 216{,}000}{12 \times 68 \times 5{,}000}\right)$$
$$+ 0.85 \times 5{,}000(17.6 - 12)9(68 - 4.5) = 81{,}250{,}000 \text{ in.-lb}$$

This is only 70.7 per cent of the required 114,800,000 in.-lb, which means that additional tensile steel is required.

Section 1.13.12 says we may use conventional steel at its yield point. We shall use intermediate-grade bars with $f'_y = 40{,}000$ psi. The tensile force in the tendons is $216{,}000 \times 6.58 = 1{,}421{,}000$ lb. Since this is 70.7 per cent of the required force, the amount still needed is

$$\left(\frac{100 - 70.7}{70.7}\right)(1{,}421{,}000) = 588{,}000 \text{ lb}$$
$$588{,}000 \div 40{,}000 = 14.7 \text{ sq in.}$$

Actually we shall need more than this because the additional steel will raise the neutral axis, which will shorten the lever arm of the steel. We shall try sixteen #9 bars with an area of 16 sq in. We shall assume that the center of gravity of the reinforcing steel is at the same elevation as the center of gravity of the tendons. Check the percentage of steel in accordance with Sec. 1.13.11.

$$\frac{pf_{su}}{f'_c} + \frac{p'f'_y}{f'_c} = \frac{0.0055 \times 216{,}000}{5{,}000} + \frac{0.01337 \times 40{,}000}{5{,}000}$$
$$= 0.2375 + 0.107 = 0.3445$$

Since this is greater than 0.30, the percentage of steel is too high. The best solution is to increase the width of the concrete section at H and for a sufficient distance each side to cover the area of critical negative moment. Try an 18-in. width as shown in Fig. 14-9. This will give a transformed section with a flange 23.6 in. wide by 9 in. deep. Now the formula in Sec. 1.13.10(B), $1.4dpf_{su}/f'_c$, must be rewritten to include the reinforcing steel. It will be $1.4d[(pf_{su}/f'_c) + (p'f'_y/f'_c)]$, and the numerical values to be used in it will be $p = 6.58/(23.6 \times 68) = 0.00413$

$$p' = 16/(23.6 \times 68) = 0.00997$$

The value of f_{su} for the new section at H is

$$f_{su} = 250{,}000\left(1 - 0.5\,\frac{0.00413 \times 250{,}000}{5{,}000}\right) = 224{,}000 \text{ psi}$$

Substituting in the formula,

$$1.4 \times 68\left(\frac{0.00413 \times 224{,}000}{5{,}000} + \frac{0.00997 \times 40{,}000}{5{,}000}\right) = 25.2 \text{ in.}$$

Since this is greater than 9 in., the formula from Sec. 1.13.10(B) will apply, but it will be revised as follows to include the reinforcing steel.

$$M_u = (A_{sr}f_{su} + A'_s f'_y)d\left(1 - 0.6\,\frac{A_{sr}f_{su} + A'_s f'_y}{b'df'_c}\right) + 0.85f'_c(b - b')t(d - 0.5t)$$

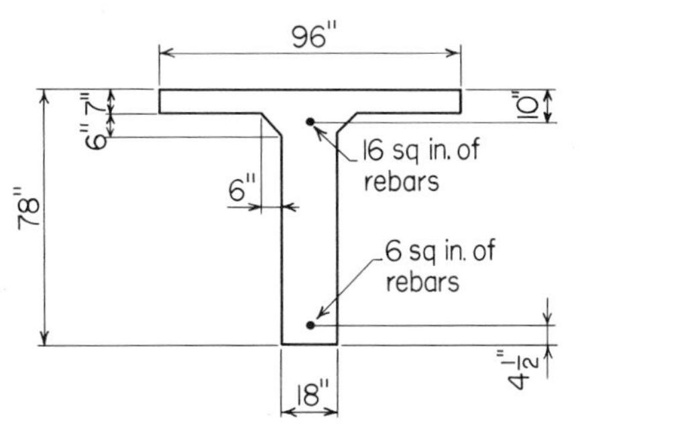

n-1 = 6 See text for computation

Section Properties — Taking Moments About Top

	A, sq in.	y, in	Ay	Ay^2	I_o
96 x 7 =	672	3.5"	2,350	8,220	2,745
18 x 71 =	1,278	42.5	54,300	2,308,000	537,000
6 x 6 x $\frac{2}{2}$ =	36	9.	324	2,915	72
Bars 16 x 6 =	96	10.	960	9,600	
Bars 6 x 6 =	36	73.5	2,650	194,500	
	2,118		60,584	2,523,235 539,817 3,063,052	539,817

$y_t = 60{,}584 \div 2{,}118 = 28.6$

$y_b = 78 - 28.6 = 49.4$

$I = 3{,}063{,}052 - 2{,}118\,(28.6)^2 = 1{,}330{,}000$

$Z_t = 1{,}330{,}000 \div 28.6 = 46{,}500$

$Z_b = 1{,}330{,}000 \div 49.4 = 26{,}920$

FIG. 14-9. Cross section and section properties at H with 18-in.-wide stem.

$$A_{sf} = \frac{0.85 \times 5{,}000(23.6 - 18)9}{224{,}000} = 0.96 \text{ sq in.}$$

$$A_{sr} = 6.58 - 0.96 = 5.62 \text{ sq in.}$$

Substituting numerical values in the formula,

$$M_u = (5.62 \times 224{,}000 + 16 \times 40{,}000) \times 68\left[1 - \frac{0.6(5.62 \times 224{,}000 + 16 \times 40{,}000)}{18 \times 68 \times 5{,}000}\right] + 0.85 \times 5{,}000(23.6 - 18)9(68 - 4.5)$$

$$= 105{,}200{,}000 + 13{,}600{,}000 = 118{,}800{,}000 \text{ in.-lb}$$

which is more than the 114,800,000 in.-lb required.

In summary the cross section at H will be as shown in Fig. 14-9 with 16 sq in. of reinforcing steel in the top and 6 in the bottom. In this example the center of gravity of the reinforcing bars in the top was placed at the same elevation as the center of gravity of the tendons in order to simplify calculations. When the design is completed, the details should be checked to see that a pattern of tendons and bars can be worked out to meet this requirement. If it cannot, the best possible pattern should be worked out and the resulting ultimate strength computed.

When the centers of gravity of the bars and tendons are not the same, an equivalent one for the two groups must be computed for use in determining d. The distance to the equivalent center of gravity y_{cg} can be found by the following equation, taking moments about any desired point:

$$y_{cg} = \frac{(A_s f_{su})y + (A'_s f'_y)y'}{A_s f_{su} + A'_s f'_y} \qquad (14\text{-}2)$$

in which y and y' represent the distances from the point about which moments are being taken to the center of gravity of the tendons and the center of gravity of the bars. Equation (14-2) applies only to calculations for ultimate moment.

The foregoing calculations for ultimate moment at H are based on the earlier decision in Step 13 of Design Procedure that "moment redistribution not be considered in design at the present time." The resultant cross section at H is sufficient to carry the full ultimate moment before extensive yielding takes place, which means that there will be no appreciable redistribution until the computed ultimate is passed.

Use of moment redistribution would probably provide a more economical structure. It would decrease the moment at H, where it is maximum and where the cross section is least efficient. It would increase the positive moments, but the section is more efficient under positive moment, and the maximum addition to the original design, if any, would probably be a small amount of reinforcing steel with no change in the concrete cross section. It is the author's feeling that use of moment redistribution in continuous prestressed concrete structures is permissible for engineers who thoroughly understand their subject.[1-4]*

* Superscript numbers indicate references listed in the Bibliography at the end of the chapter.

Since the new section at H has a larger I than the section throughout the rest of the structure, the moment developed at H will be larger than that found in the foregoing calculations. In the design of an actual structure it would be necessary to establish the length of the heavier cross section, compute bending moments based on the nonuniform I, and check stresses. Our purpose, however, is to illustrate prestressed concrete design procedure rather than to conduct an exercise in elastic analysis. We shall assume that the new bending moments at H have been computed and were found to be

$M_D = -38{,}800{,}000$ in.-lb
$M_L = -23{,}830{,}000$ in.-lb
$M_T = +50{,}650{,}000$ in.-lb
Required $M_u = -118{,}000{,}000$ in.-lb
F still $= 987{,}000$ lb

Because of their considerable area the reinforcing bars were included in computing section properties in Fig. 14-9. In this computation $E_c = 1{,}800{,}000 + 500(5{,}000) = 4{,}300{,}000$ and $E_s = 30{,}000{,}000$. Thus

$$n = \frac{30{,}000{,}000}{4{,}300{,}000} = 7 \qquad \text{and} \qquad n - 1 = 6$$

Using these moments and the section properties computed in Fig. 14-9, the stresses at H under the various combinations of loading are

$$\frac{F}{A_c} = 987{,}000 \div 2{,}118 = +466 \text{ psi}$$

$$f^t{}_F = \frac{50{,}650{,}000}{46{,}500} + 466 = +1{,}555 \text{ psi}$$

$$f^b{}_F = -\frac{50{,}650{,}000}{26{,}920} + 466 = -1{,}414 \text{ psi}$$

$$f^t{}_D = \frac{-38{,}800{,}000}{46{,}500} = -835 \text{ psi}$$

$$f^b{}_D = \frac{38{,}800{,}000}{26{,}920} = +1{,}440 \text{ psi}$$

$$f^t{}_{F+D} = +720$$

$$f^b{}_{F+D} = +26$$

$$f^t{}_L = \frac{-23{,}830{,}000}{46{,}500} = -512 \text{ psi}$$

$$f^b{}_L = \frac{23{,}830{,}000}{26{,}920} = +885 \text{ psi}$$

$$f^t{}_{F+D+L} = +208 \text{ psi}$$

$$f^b{}_{F+D+L} = +911 \text{ psi}$$

We have already computed the ultimate moment capacity of the new section and found it to be 118,800,000 in.-lb, which is more than the new required value, so the new cross section is satisfactory.

It should be noted that use of moment redistribution for ultimate design not only would save the extra material used in the vicinity of H but would eliminate the design complications which go with a variable moment of inertia in a continuous structure.

Next check ultimate positive-moment capacity at stations 1, 2, 3, and 4. First we must evaluate the term $1.4pf_{su}/f'_c$ for each station to see whether to apply the formula for rectangular or flanged sections. From Fig. 14-3

$$d = D + y_B + y$$

$$p = \frac{A_s}{bd}$$

$$f_{su} = f'_s\left(1 - 0.5\,\frac{pf'_s}{f'_c}\right)$$

Substituting known numerical values in these formulas we get the value of f_{su} and use the appropriate formulas for ultimate moment to get

	Station on girder			
	1	2	3	4
d	54.27	67.49	64.51	45.33
p	0.00126	0.00102	0.00106	0.00151
f_{su}	242,000	243,000	243,000	240,500
$1.4dpf_{su}/f'_c$	4.63	4.68	4.65	4.61
M_u	83,200,000	105,000,000	100,000,000	68,500,000
Required M_u	59,260,000	81,530,000	66,800,000	13,780,000

Computations for station 4 are as follows:

$D = 10$ in.

$X = 25$ ft 0 in. $= 300$ in.

Values of M, B, etc., are given in Step 1.

$$y_B = \frac{MX}{R} = \frac{14.9 \times 300}{1{,}500} = 2.98 \text{ in.}$$

$$y = 4B\frac{X}{R}\left(1 - \frac{X}{R}\right)$$

$$y = 4 \times 50.55\,\frac{300}{1{,}500}\left(1 - \frac{300}{1{,}500}\right) = 32.35$$

$$d = 10 + 2.98 + 32.35 = 45.33$$

$$p = \frac{A_s}{bd} = \frac{6.58}{96 \times 45.33} = 0.00151$$

$$f_{su} = 250{,}000\left(1 - \frac{0.5 \times 0.00151 \times 250{,}000}{5{,}000}\right) = 240{,}500$$

$$\frac{1.4dpf_{su}}{f'_c} = \frac{1.4 \times 45.33 \times 0.00151 \times 240{,}500}{5{,}000} = 4.61$$

Since this is less than the 7-in. flange thickness, the formula from Sec. 1.13.10(A) will apply.

$$M_u = 6.58 \times 240{,}500 \times 45.33\left(1 - \frac{0.6 \times 0.00151 \times 240{,}500}{5{,}000}\right)$$

$$= 68{,}500{,}000$$

Each of the positive-moment capacities is greater than the required ultimate at that point, so no revisions are necessary. It is also apparent that the increase in positive moments due to moment redistribution would have to be rather large before additional steel became necessary.

Check the ultimate negative-moment capacity at stations 3 and 4. In computing the ultimate positive-moment capacity we found d from the top to the center of gravity of the tendons. Subtracting this value of d from the depth of the member will give the value of d for the negative-moment computation. At 4, $d = 78 - 45.33 = 32.67$ in. and at 3, $d = 78 - 64.51 = 13.49$ in. In both cases the tendons are on the compression side of the neutral axis. Before the load in the tendons could increase appreciably beyond F, cracks would develop in the tensile side and extend through the neutral axis to the tendons. Such long cracks would greatly reduce the moment of inertia of the section and would result in a redistribution of moment.

Development of these cracks can be kept within reasonable limits by placing a proper amount of reinforcing steel near the tensile surface of the concrete. This gives us a condition where the centers of gravity of the bars and tendons are not at the same elevation. In this case, however, Eq. (14-2) is not applicable. It can be applied only when the tendons and bars are in the same general vicinity. When the tendons are on the compression side of the neutral axis, they are usually too far from the bars for combined action. As the bending moment increases, the stress in the bars will increase but there will be practically no change in stress in the tendons. When the bars reach their yield point, they will elongate rapidly, cracks on the concrete will approach the tendons, and their tension will finally begin to increase. Here again we cannot take advantage of an increase in stress in the tendons because, before this can take place, the reinforcing bars must yield, which would result in a redistribution of moments.

The foregoing discussion indicates that ultimate moment design at points in a structure where the tendons are on the compression side of the neutral axis and/or the tendons and reinforcing bars are too far apart to work together presents a problem entirely different from the normal condition. Reinforcing bars must be placed near the tensile surface of the concrete. These bars must be selected so that the section will carry the required ultimate moment before the bars yield. The following analysis is suggested:

1. Let T' be the total tension in reinforcing bars at their yield point.
2. The total force in the compression side of the member will be $T' + F$.
3. Since the concrete stress at ultimate moment is assumed to be $0.85f'_c$, the concrete area required to carry the total force will be $(T' + F)/0.85f'_c$.

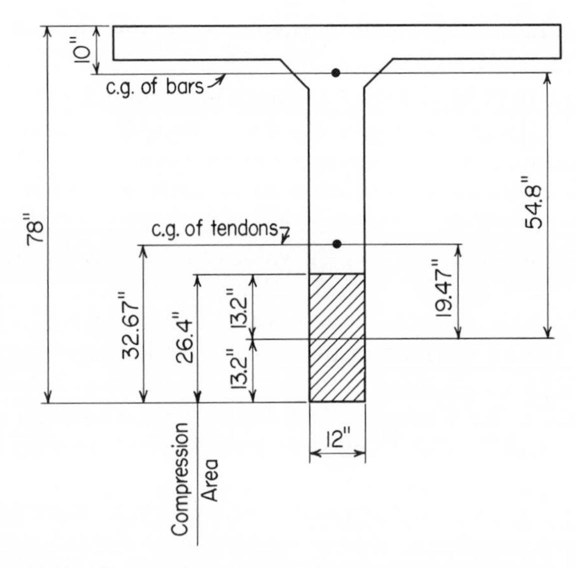

FIG. 14-10. Cross section at station 4 under ultimate moment conditions.

4. The ultimate negative moment the member can carry will be ($F \times$ lever arm to center of compression force) plus ($T' \times$ lever arm to center of compression force). The simplest design procedure is to assume a value for T', determine a pattern of bars and their center of gravity, and follow the foregoing procedure to find M_u. If the result is too large or too small, the value of T' can be revised and the new M_u computed.

The required ultimate negative moment at station 4 is 38,390,000 in.-lb. Assume nine #9 bars with their center of gravity 10 in. from the top surface as shown in Fig. 14-10. Then $T' = 9 \times 40{,}000 = 360{,}000$ lb. $T' + F = 360{,}000 + 987{,}000 = 1{,}347{,}000$ lb. Required concrete area to resist this force is $1{,}347{,}000/(0.85 \times 5{,}000) = 317$ sq in. Since the concrete section is 12 in. wide, its depth will be $317 \div 12 = 26.4$ in. and the distance from the bottom to its center of gravity is 13.2. Using the lever arms shown on Fig. 14-10,

$$M_u = 360{,}000 \times 54.8 + 987{,}000 \times 19.47 = 38{,}950{,}000 \text{ in.-lb}$$

which is satisfactory. The designer should draw the ultimate negative-moment diagram from station 3 to H and compute the required reinforcing at points along the diagram.

Step 14. In Step 12 we found the dead-load reactions on the piers and the reactions on the piers from the tendons. The shear in the concrete section at a pier is the reaction on the pier minus the vertical load carried by the tendon. Thus under the dead-load condition, using values computed in Step 12,

$$V_G = V_K = 99{,}700 - 123{,}300 = -23{,}600 \text{ lb}$$

This means that the vertical component of the tension in the tendons is greater than the reaction on the pier and the concrete section has a negative shear of 23,600 lb.

The dead-load shear on either side of the pier at H is

$$V_H = \left(\frac{206{,}600 - 285{,}600}{2}\right) = -39{,}500 \text{ lb}$$

which is also a negative shear.

The shear in the concrete section is negative for the full span because the vertical upward load applied to the concrete by the tendons is 2,130 lb per ft and the dead load is only 1,625 lb per ft, leaving a net upward load of 505 lb per ft.

The maximum live-load shear at G and K is 54,750 lb, giving a net maximum of

$$V_G = V_K = 54{,}750 - 23{,}600 = 31{,}150 \text{ lb positive shear}$$

At H the maximum live-load shear is $156{,}250 \div 2 = 78{,}125$, giving a net of $V_H = 78{,}125 - 39{,}500 = 38{,}625$ lb positive shear.

It is apparent from these figures that design load shear is not critical. We shall not go through the calculations here, but ultimate load shear must be computed. The ultimate shear requirement should be computed in the same manner as the ultimate moment was computed in Step 13. Thus the applied loading for ultimate shear would be $D + (0.5D + 2.5L)$ with D applied the full length of the structure and ($0.5D$ plus $2.5L$) applied in those areas which produce maximum shear. We have already computed the net shear under dead load and found it to be negative. Shear due to $0.5D + 2.5L$ applied to produce maximum shear must be added to the net dead-load shear to get ultimate shear. Ultimate shear will be positive at all points. Once ultimate shear is established, the design of stirrups will follow the procedure used in Chap. 5. Note that the shear carried by the tendons has already been included in computing dead-load shear; it should not be included again.

Step 15. Camber computations are simple, since the structure is fully elastic under dead plus live loads. The camber under dead load plus initial prestress is due to the net uniform upward load of the tendons minus the dead weight of the concrete member. In Step 11 initial prestress was found to be $1.166F$. Thus the upward load from the tendons under initial prestress will be $1.166 \times 2{,}130 = 2{,}485$ lb per ft, and the net upward load including dead load will be $2{,}485 - 1{,}625 = 860$ lb per ft. The camber under dead load plus initial prestress will be that due to a uniform upward load of 860 lb per ft applied to the elastic structure. Camber growth will be the same as in previous examples. Deflection from the dead-load position under live load will be that for the elastic structure subjected to the live load.

Step 16. In addition to details around the tendon anchors, bearing plates, etc., similar to those worked out in Chap. 11 there are several details peculiar to a continuous structure.

When the prestressing load is applied, the structure will shorten. If it is firmly attached to its piers, the piers will bend as the structure shortens and part of the prestressing force will be used to bend the piers rather than to prestress the structure. One means of eliminating this problem in a two-span structure is to anchor the structure to the center pier and support it on rockers at each end pier. The amount of shortening due to elastic compression, closing of shrinkage cracks, etc., can be computed, and the rockers can be tilted so they will be at the desired angle after the prestressing operation.

From the parabolic path of tendons shown in Fig. 14-3 it appears that the tendons come to a point at H. While this assumption is usually sufficiently accurate for computations, the path of tendons over an intermediate support should be detailed as shown in Fig. 14-11. There are so many factors influencing the value of r that it is impossible to write an equation for it that will cover all conditions. The suppliers of the tendons chosen should be consulted concerning the most efficient radius. Until the specific tendons are chosen, it is suggested that the designer estimate the number of tendons that will be used and determine the ultimate strength F_u of one tendon. Then the minimum value of r should be

$$r = 0.085\sqrt{F_u} \tag{14-3}$$

If details permit, a larger value of r will reduce the unit pressure between the tendon and the concrete. For a round post-tensioned tendon in a cored hole or in a metal hose, the unit pressure between tendon and concrete is not p divided by the diameter of the tendon. This would be true if the tendon were the same diameter as the hole so that they were in contact for the full 180° of the underside of the tendon. Actually the hole is larger than the tendon, and the contact surface, for a round tendon, approaches line contact. A group of parallel wires or strands in one hole will spread across the full width, and their unit pressure will approach p divided by the diameter of the hole.

Each tendon should be tensioned from both ends at one time. The tension at H will be that applied at the jack less that lost by friction between the jack and H. The tendons chosen must be large enough to give the required tension at H after friction losses. Friction is sometimes reduced for long tendons by using galvanized tendons or galvanized hose.

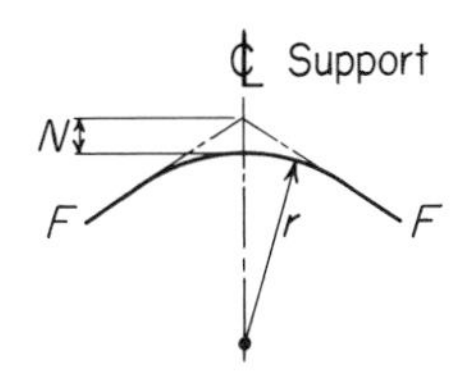

p = pressure of tendon against concrete in pounds per linear inch where radius of curvature of tendon is r expressed in inches.

$$p = \frac{F}{r}$$

See text for recommended values of r

FIG. 14-11. Actual path of tendons over an intermediate pier. See text for comments on values of r, etc.

STRUCTURES WITH PRECAST SECTIONS

14-4. Characteristics. Precast prestressed concrete members can readily be designed and detailed for assembly into continuous and rigid frame structures. Precast prestressed members are joined to other precast prestressed members or to reinforced members. In either case, although the joint is not prestressed, full continuity can be developed. See discussion and example in Art. 15-3.

14-5. Design Procedure. Steps in the design procedure for a structure of this type are presented herewith. They refer specifically to a multispan bridge in which precast prestressed I beams are erected on the piers, joined over the piers with reinforcing bars, and covered with a poured-in-place slab. The bridge cross section is of the type shown in Figs. 5-1 and 5-2, and the joint as shown in Fig. 14-12. The concrete around the reinforcing bars at the piers would be poured at the same time as the deck slab, which means that the weight of the deck slab would be carried by the precast beam acting as a simple span member rather than as a continuous member. The live load would be carried by the composite section acting as a continuous elastic member. If the erection schedule and details permitted, the reinforcing bars and concrete over the piers could be placed and the concrete allowed to cure before the slab was placed. With this method the beam would act as a continuous member under the dead weight of the slab. For this method of erection or other variations from the example presented here the design procedure should be altered to suit the revised erection procedure.

Many of the computations will be similar to those shown in Chap. 5.

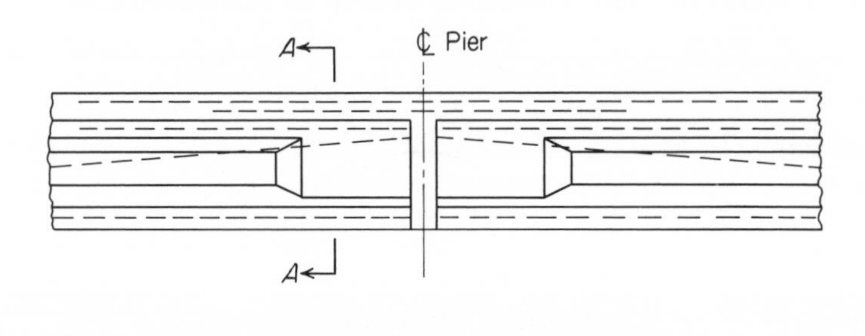

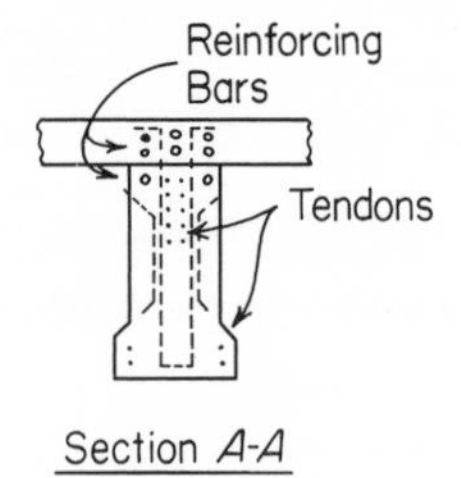

FIG. 14-12. Joint of reinforcing bars and concrete creating continuity at support between precast prestressed members.

Step 1. Compute the stresses in the beam as a simple span member under its own dead weight.

Step 2. Compute the stresses in the beam as a simple span member under the weight of the poured-in-place slab.

Step 3. Compute the stresses in composite section composed of beam and slab acting as a continuous elastic member under live load.

Step 4. Compute the total stresses in the beam by adding stresses in Steps 1, 2, and 3.

Step 5. Compute the required prestressing force and location of the tendons to offset tensile stresses found in Step 4. Tendons will extend for the length of the precast member. They will not be continuous over the supports.

Step 6. Design the connection between the beams to carry live-load negative moment over the piers. This is designed as a reinforced concrete member. Reinforcing bars to carry the tension are placed in the slab over the beam or in a cavity provided in the top of the beam. These bars must be long enough to transfer their load through the concrete to the steel in the precast beams. Provide concrete between the ends of the beams to carry the compression in the bottom of the beams and to bond the bars in the top.

Step 7. Check the structure as now designed for the critical loading conditions and for ultimate strength.

Step 8. Compute the reactions on the piers. Since the dead weight of the beam and slab is carried with the beam acting as a simple span member, the pier reactions for these loads will be computed on a simple span basis. The pier reactions due to live load will be computed as for a continuous elastic structure.

Step 9. Design the stirrups, connection between the beam and slab, and other details in the same manner as in Chap. 5.

BIBLIOGRAPHY

1. Lin, T. Y.: Strength of Continuous Prestressed Concrete Beams under Static and Repeated Loads, *J. Am. Concrete Inst.,* June, 1955, pp. 1037–1059.
2. Janney, Jack, and W. J. Eney: Full Scale Test of Bridge on Northern Illinois Toll Highway, *Proc. World Conf. on Prestressed Concrete,* 1957.
3. Ozell, A. M.: Behavior of Simple Span and Continuous Composite Prestressed Concrete Beams, Part 2, *J. Prestressed Concrete Inst.,* June, 1957, pp. 42–74.
4. Lin, T. Y.: "Design of Prestressed Concrete Structures," chap. 10, Continuous Beams, John Wiley & Sons, Inc., New York, 1955.
5. Parme, A. L., and G. H. Paris: Designing for Continuity in Prestressed Concrete Structures, *J. Am. Concrete Inst.*, September, 1951, pp. 45–64.
6. Moorman, B. B.: Continuous Prestressing, *Proc. Am. Soc. Civil Engrs. Structural Div.,* Separate No. 588.
7. Morice, P. B., and H. E. Lewis: Prestressed Continuous Beams and Frames, *Proc. Am. Soc. Civil Engrs. Journal Structural Div.,* Paper 1055.
8. Fiesenheiser, E. I.: Rapid Design of Continuous Prestressed Members, *J. Am. Concrete Inst.,* April, 1954, pp. 669–676.
9. Leonhardt, Fritz: Continuous Prestressed Concrete Beams, *J. Am. Concrete Inst.,* March, 1953, pp. 617–634.

CHAPTER 15

Typical Structures and Their Components

15-1. Scope. Prestressed concrete is an extremely versatile material. It can perform, economically, most of the functions performed by other structural materials and also some functions which are unique to it. Engineers and architects can establish basic requirements for their structures and then select members to suit. Joints can be pinned or rigid. Poured-in-place concrete becomes a composite part of the structure to help carry live load.

This chapter illustrates the standard members available and a few special applications. These or simple adaptations of them may fit your structure. If not, local fabricators, suppliers of materials for prestressed concrete, and/or engineers experienced in the field are always available to help make your own special ideas economically feasible.

15-2. Highway Bridges of Standard Precast Members. Standards for prestressed concrete members that are used in bridges have been established and are kept up to date by a Joint Committee composed of members from the American Association of State Highway Officials Committee on Bridges and Structures and members from the Prestressed Concrete Institute. The most frequently used of these standard members are I beams, box beams, slabs, and piles. Since each set of standards is brought up to date periodically by the AASHO-PCI Joint Committee, the designer should make sure that he is working from a current drawing. Copies are available at a nominal charge from PCI.

The AASHO-PCI Standard Sections were set up by men who had considerable experience with similar members in the prestressed concrete bridge field. Hundreds of bridges have been built in accordance with these standards. The bridges are structurally sound, fast to erect, practically maintenance-free, and often appreciably lower in first cost than other materials. See Art. 15-7 for cost comparison on one specific project.

As this is written the standards for piles have just been revised and are presented herewith and in Figs. 15-1*a* to *d* and 15-2*a* and *b*. A discussion of prestressed concrete piles is presented in Chap. 12. When L/D exceeds 25 the design approach shown in Fig. 12-4 is recommended.

General Notes—Prestressed Concrete Piles AASHO-PCI Standards

Purpose: These standards have been revised to show the latest developments in materials and uses of prestressed concrete piles. In addition to square and octagonal piles, details for pretensioned cylinder piles have been added to these standards.

Any other shape of pretensioned concrete piles, or post-tensioned cylinder piles, with similar properties, may be used subject to approval of the Engineer.

Specifications: AASHO Standard Specifications for Highway Bridges, current edition.

Concrete: Concrete in the precast prestressed piles and build-ups with driving shall have a minimum compressive cylinder strength (f'_c) of 5,000 psi at 28 days. Concrete in build-ups without driving shall have a minimum compressive cylinder strength (f'_c) of 3,000 psi. Compressive cylinder strength at transfer of prestressing force shall be not less than 4,000 psi.

Higher concrete strengths may be used and advantage may be taken of such greater strength for handling and driving stresses and column loading, subject to approval of Engineer.

Prestressing Reinforcement: Seven-wire stress relieved strand shall conform to the general requirements of ASTM Designation A416, and may be either regular or high strength, in accordance with strand manufacturer's published tables. Subject to the approval of the Engineer, prestressing may be increased as required for handling or driving by increasing the number or size of strands. In general the unit prestress after losses should not exceed $0.2f'_c$, unless special conditions warrant and appropriate adjustment is made in allowable pile capacity. Broken wires within individual strands will be permitted up to 2 per cent of the total number of wires in each pile, providing that there is not more than one broken wire per strand. Two or more broken wires per strand will be cause for replacement of the strand, even though the two broken wires are within the 2 per cent limitation.

Build-ups and Splices: Build-ups, precast or cast-in-place, may be used if specified or authorized by the Engineer. Two prestressed pile sections may be spliced by the use of dowels extending from the tip of the upper prestressed section into cored or drilled holes in the lower prestressed section. For hollow core or cylinder piles, this splice connection may be made in the walls of the pile or in a solid plug at the head and tip of the spliced sections. The dowels shall have an area equal to 1½ per cent of the gross cross-section of pile and shall be adequately bonded into both sections. The dowel holes and space between spliced sections shall be filled with a material having properties fully equal to that of the concrete and adhesive strength equal to the shear and tensile strength of the concrete. Such properties shall be obtained within a time limit consistent with the driving requirements of the pile.

Any alternate method of splicing providing equal results may be considered for approval.

Forms: For forming the exterior of piles, the use of steel forms on concrete founded casting beds is required, unless otherwise approved by the Engineer. Side forms for square and octagonal piles, may have a maximum draft on each side not exceeding ¼ in. per ft.

For forming the interior of piles with hollow cores, forms shall be constructed of an approved material which will not deform or break during prestressing operations.

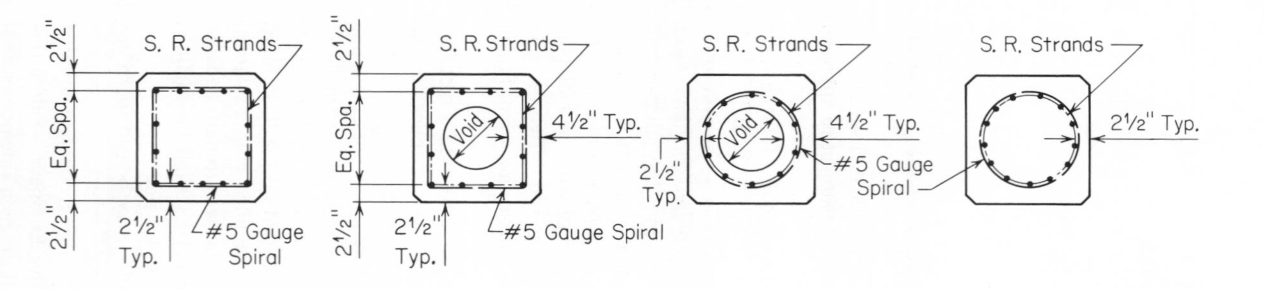

Pile Properties

Pile size diam, in. (1)	Area Ac, sq in.	Approx wt, lb per lin ft (2)	Min prestress force, kips (3)	Strands per pile diam (4)		Section modulus, in.^{3}	Perimeter, in.	Design bearing capacity concrete strength, tons (5)	
				7/16 in.	½ in.			5,000 psi	6,000 psi
10	100	105	70	4	4	167	40	50	60
12	144	150	101	6	5	288	48	72	86
14	196	205	138	8	6	457	56	98	117
16	256	265	180	11	8	683	64	128	153
18	324	335	227	13	10	972	72	162	194
20	400	415	280	16	12	1,333	80	200	240
22	484	505	339	20	15	1,775	88	242	290
24	576	600	404	23	18	2,304	96	288	345
20 HC	305	320	214	13	10	1,261	80	152	183
22 HC	351	365	246	14	11	1,647	88	175	210
24 HC	399	415	280	16	12	2,097	96	200	240

FIG. 15-1*a*. AASHO-PCI Standards. Square piles. See also Figs. 15-1*c* and *d* and General Notes in the text.

S. R. Strands at Equal Spacing
#5 Gauge Spiral
See Note 6
2½" Typ.
2½" Typ.
4½" Typ.
Void

NOTES (for both square and octagonal piles):

(1) Voids in 20 in., 22 in. and 24 in. diameter hollow-core (HC) piles are 11 in., 13 in. and 15 in. diameter, respectively, providing a minimum 4½ in. wall thickness. If a greater wall thickness is desired, properties should be increased accordingly.

(2) Weights based on 150 lb per cu ft of regular concrete.

(3) Minimum prestress force based on unit prestress of 700 psi after losses.

(4) Based on 7/16 in. and ½ in. high strength strand with an ultimate strength of 31,000 and 41,300 lb respectively. If regular strength strand is used, the number of strands per pile should be increased accordingly in conformance with the strand manufacturer's tables.

(5) Design bearing capacity based on 5,000 psi and 6,000 psi concrete and an allowable unit stress on the tip of the pile of $0.2f'cAc$. These bearing capacity values may be increased if higher-strength concrete is used.

(6) Circular piles of the same diameter may be used in lieu of octagonal subject to engineer's approval.

Pile Properties

Pile size diam, in. (1)	Area *Ac*, sq in.	Approx wt, lb per lin ft (2)	Min prestress force, kips (3)	Strands per pile diam (4)		Section modulus, in.^{3}	Perimeter, in.	Design bearing capacity concrete strength, tons (5)	
				7/16 in.	½ in.			5,000 psi	6,000 psi
10	83	85	59	4	4	111	34	41	50
12	119	125	84	5	4	189	40	59	71
14	162	170	114	7	5	301	46	81	97
16	212	220	149	9	7	449	54	106	127
18	268	280	188	11	8	639	60	134	160
20	331	345	232	14	10	877	66	165	198
22	401	420	281	16	12	1,167	72	200	240
24	477	495	334	19	15	1,515	80	238	286
20 HC	236	245	166	10	8	805	66	118	141
22 HC	268	280	188	11	8	1,040	72	134	160
24 HC	300	315	210	12	9	1,308	80	150	180

FIG. 15-1*b*. AASHO-PCI Standards. Octagonal piles. See also Figs. 15-1*c* and *d* and General Notes in the text.

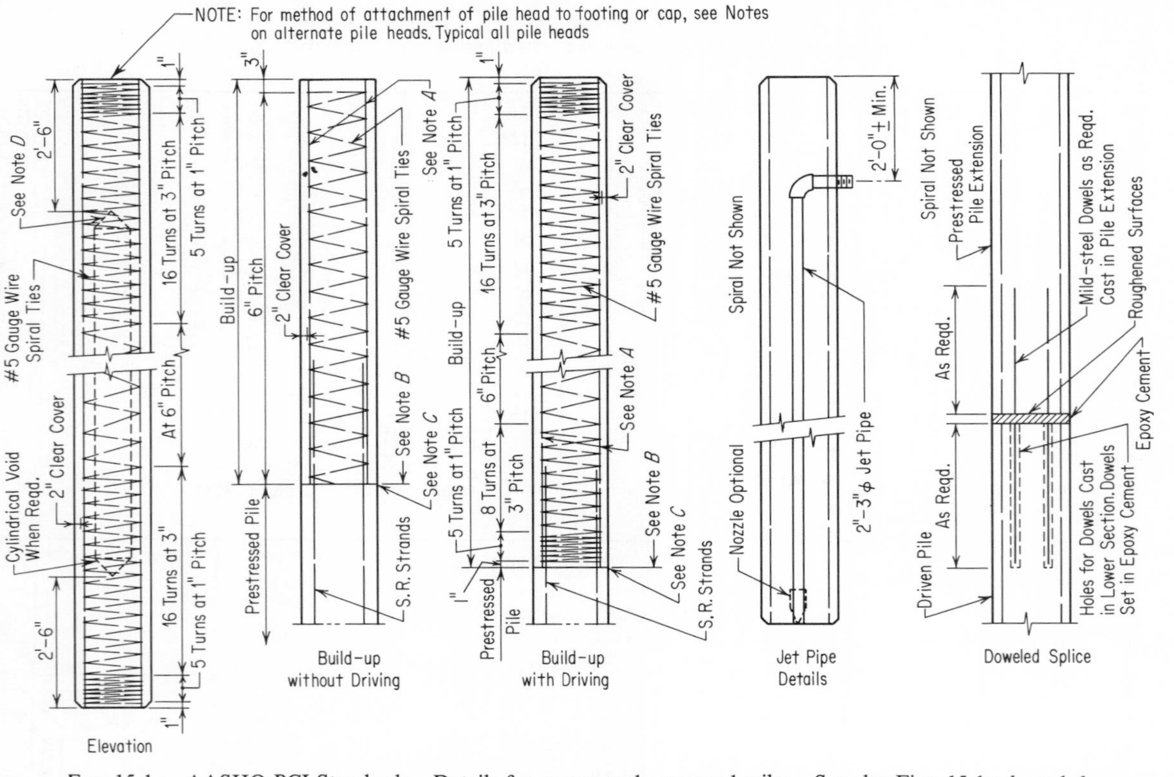

FIG. 15-1*c*. AASHO-PCI Standards. Details for square and octagonal piles. See also Figs. 15-1*a*, *b*, and *d*.

NOTES ON PILE HEADS:

NOTE *A*. The minimum area of reinforcing steel shall be 1½ per cent of the gross cross section of concrete. Placement of bars shall be in a symmetrical pattern of not less than four bars.

NOTE *B*. Method of attachment of pile to build-up may be by any of the methods given in the notes on Alternate Pile Heads. If mild reinforcing steel is used for attachment, the area shall be no less than that used in the build-up.

NOTE *C*. Concrete around top half of pile shall be bush-hammered to prevent feather edges.

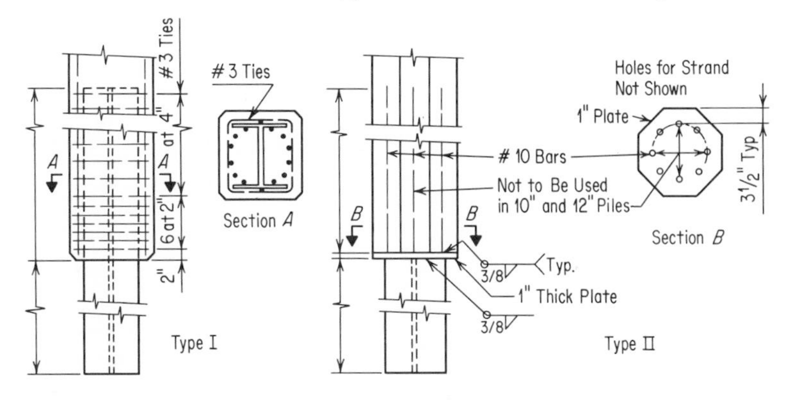

Alternate Pile Tips

FIG. 15-1*d*. AASHO-PCI Standards. Details for square and octagonal piles. See also Figs. 15-1*a*, *b*, and *c*.

Alternate Pile Tips

When driving into rock or hard strata, either type I or type II alternate tips may be used in lieu of the standard flat tip. Size and length of steel section used shall be as determined by engineer for adequate penetration. Type I or type II tips may be used for either square or octagonal piles.

Alternate Pile Heads

Reinforcement may be specified to project from the pile into the cap or footing. If so required, attachment of the pile to the cap or footing may be made by any one of the following methods unless otherwise specified:

1. Allow all strands to project a minimum of 24 in.
2. Cast mild reinforcing steel in pile head with bars projecting for anchorage.
3. Provide cored holes in pile head for subsequent use of grouted dowel bars.
4. Drill holes in pile head for installation of grouted dowel bars. Special care must be taken to prevent damage to the pile head.

If mild reinforcing steel is used for projection into cap or footing, the minimum area of steel required shall be twice the area of the prestressing strands with not less than four bars being used. Arrangement of bars shall be in a symmetrical pattern with bars as close as practical to the sides of the pile. Anchorage of bars shall be sufficient to develop strength of bar but not less than 20 bar diameters.

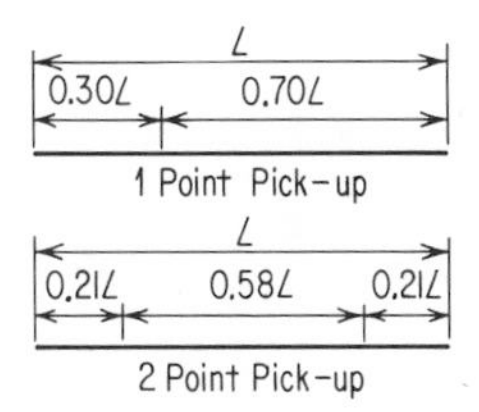

Picking Points

NOTE: Unless special lifting devices are attached for pickup, pickup points shall be plainly marked on all piles after removal of the forms and all lifting shall be done at these points.

The use of special embedded or attached lifting devices, the employment of other pickup points or any other method of pickup shall be subject to approval by the engineer.

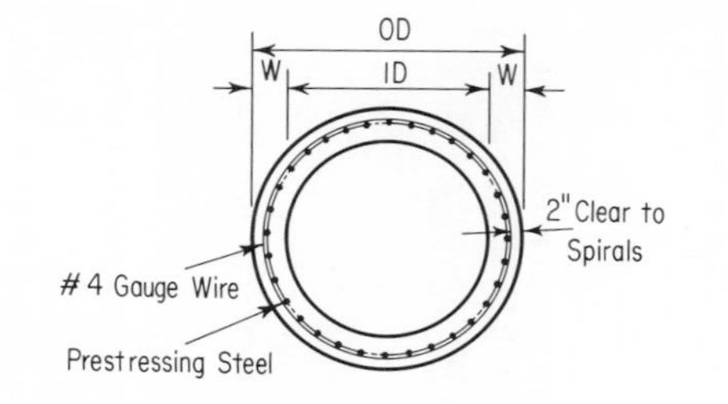

Pile Properties

Pile size, in.			Area Ac, sq in.	Approx wt, lb per lin ft (1)	Min prestress force, kips (2)	Strands per pile diam (3)		I, in.^{4}	Section modulus, in.^{3}	Perimeter, in.	Design bearing capacity concrete strength, tons (4)	
OD	ID	W				7/16 in.	½ in.				5,000 psi	6,000 psi
36	26	5	487	508#	414	24	18	60,000	3334	113	242	292
	24	6	565	590#	481	28	21	66,100	3676	113	282	339
48	38	5	675	703#	574	33	25	158,200	6593	151	337	405
	36	6	792	826#	674	39	29	178,100	7422	151	396	475
54	44	5	770	802#	655	38	28	233,400	8645	170	385	462
	42	6	904	940#	769	44	33	264,600	9802	170	452	542

NOTES:

(1) Weights are based on 150 lb per cu ft of regular concrete.

(2) Minimum prestress force based on unit prestress of 850 psi after losses.

(3) Based on 7/16 in. and ½ in. high-strength strand with an ultimate strength of 31,000 lb and 41,300 lb, respectively. If regular-strength strand is used, the number of strands per pile should be increased accordingly in conformance with strand manufacturer's tables.

(4) Design bearing capacity based on 5,000 psi and 6,000 psi concrete and an allowable unit stress on the tip of the pile of $0.2f'cAc$. These bearing capacity values may be increased if higher-strength concrete is used.

FIG. 15-2*a*. AASHO-PCI Standards. Pretensioned cylinder piles. See also Fig. 15-2*b* and General Notes in the text.

If a moving mandrel is used for forming the inner void, special precautions shall be taken to prevent fallout of inner surfaces, tensile cracks and separation of concrete from strands.

Pick-up and Handling: Maximum lengths for pick-up are determined using the following stress assumptions.

Loading: 1½ times full dead load. Allowable tensile stress equals $6.0\sqrt{f'_c}$. These stress and loading criteria are based on normal care in handling the pile. If handling is such that damage to the pile becomes evident, the Engineer may require a higher load factor or lower allowable stress as necessary to insure no damage to piles.

Driving: Pile heads shall be protected from direct impact of the hammer by cushion blocks consisting of several plies of soft compressible wood or other approved material.

Jetting will be permitted and/or required when necessary to obtain the required penetration. Internal jets may be installed provided they are securely anchored to the pile and are imbedded in the concrete.

The driving head (helmet) shall be sufficiently large and shallow so as not to bind the head of the pile if it twists slightly during driving. Hollow piles which are open-ended at the tip shall have vents to relieve internal hydraulic pressure.

Tolerances: Voids, when used, shall be located within ½ in. of position shown on plans.

Pile ends shall be plane surfaces and perpendicular to axis of pile with a maximum tolerance of ⅛ in. per ft transversely.

The maximum sweep (deviation from straightness measured along two perpendicular faces of the pile, while not subject to bending forces) shall not exceed ⅛ in. in any 10 ft of its length.

Chamfers and Corners: All corners of square piles shall be chamfered to at least ¾ in. or rounded to approximately 1 in. radius.

General: When piles are ordered in accordance with this standard plan, the standard pile details shall be used. Alternate pile heads, pile tips, splices, build-ups or other alternates shall be used only if specified or authorized by the Engineer.

Where specific methods are indicated for achieving a result, other methods which will insure equal results may be considered for approval by the Engineer.

Small areas of honeycomb which are purely surface in nature extending to a depth of no more than 1 in. may be repaired in a manner satisfactory to the Engineer. Honeycomb extending to the plane of reinforcing will be cause for rejection.

Pile Capacities and Loading:

(a) Loads on Piles as Columns—The maximum compressive stress on prestressed concrete piles in excess of the effective prestress shall not exceed the following:

$$f_{pc} = 1{,}000 \text{ for } L/D = 0 \text{ to } 12$$
$$f_{pc} = 1{,}240 - 20\,L/D \text{ for } L/D = 12 \text{ to } 25$$

where L = effective length of pile
D = diameter or width of pile
f_{pc} = allowable unit compressive stress in psi exclusive of effective prestress

For piles considered hinged at both ends, L shall be taken as the actual length of pile. For piles considered hinged at one end and fully fixed at the other end, L shall be taken as 0.7 of the length between hinge and assumed location of fixity. For piles considered fully fixed at both ends, L shall be considered as 0.5 of the length between

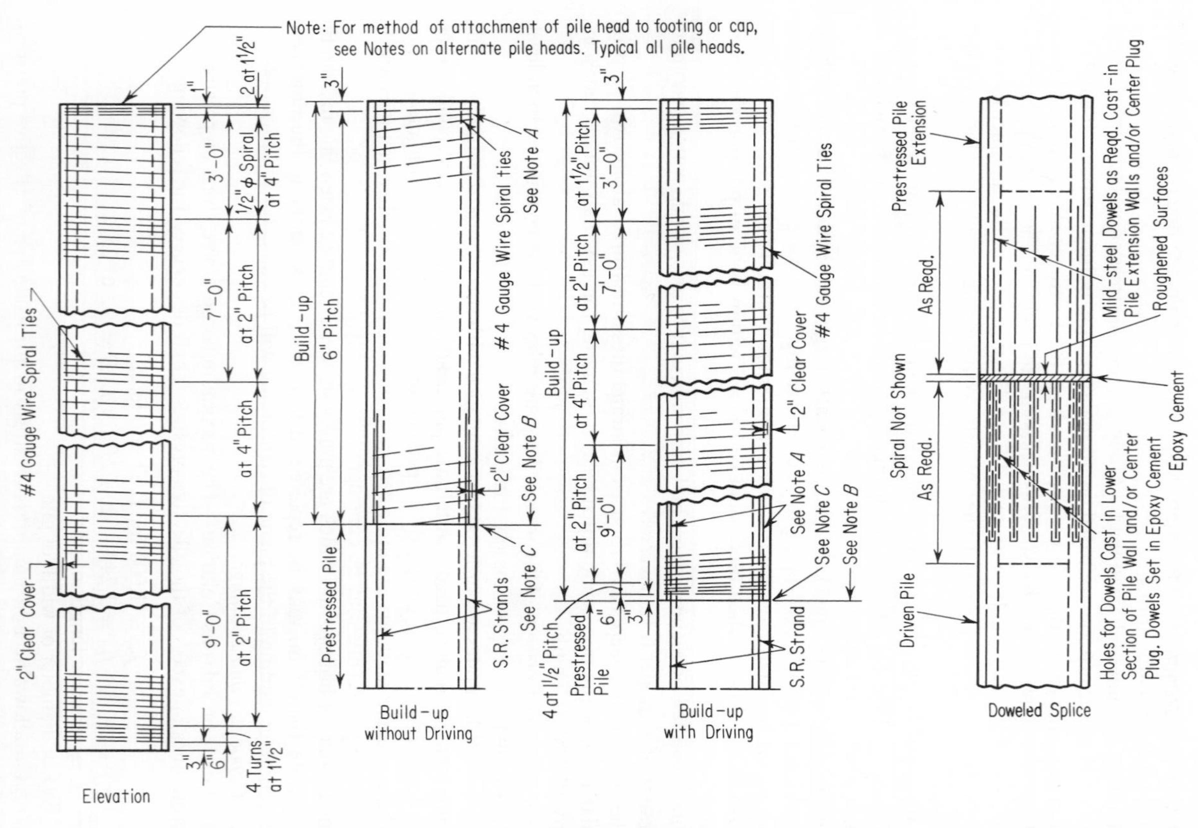

FIG. 15-2*b*. AASHO-PCI Standards. Details for pretensioned cylinder piles. See also Fig. 15-2*a* and General Notes in the text.

NOTES ON PILE HEADS:

NOTE *A*. The minimum area of reinforcing steel shall be 1½ per cent of the gross cross section of concrete. Placement of bars shall be in a symmetrical pattern of not less than eight bars.

NOTE *B*. Method of attachment of pile to build-up may be by any of the methods given in the notes on Alternate Pile Heads. If mild reinforcing steel is used for attachment, the area shall be no less than that used in the build-up.

NOTE *C*. Concrete around top half of pile shall be bush-hammered to prevent feather edges.

Alternate Pile Heads

Reinforcement may be specified to project from the pile into the cap or footing. If so required, attachment of the pile to the cap or footing may be made by any one of the following methods, unless otherwise specified:

1. Allow all strands to project a minimum of 24 in.
2. Cast mild reinforcing steel in plug poured in top of pile head, after driving, with bars projecting for anchorage.
3. If pile is driven with a plug in pile head, cored holes may be provided in plug for subsequent use of grouted dowel bars.

Plug in pile head may be cast in place or precast. If precast, plug shall be adequately bonded to inner wall of pile by an epoxy, a nonshrink grout or other means acceptable to the engineer. Depth of plug in pile head shall not be less than the outside diameter of the pile.

If mild reinforcing steel is used for projection into the cap or footing, the minimum area of steel required shall be 1½ per cent of the gross cross section of concrete with not less than eight bars being used. Arrangement of bars shall be in a symmetrical pattern with bars as close as practical to the sides of the pile. Anchorage of bars shall be sufficient to develop strength of bar but not less than 20 bar diameters.

the two assumed fixed ends. When pile is designed as a cantilever, L shall be considered as 2 times the length of the cantilever.

Where it is necessary to use piles with an L/D greater than 25, they shall be thoroughly investigated for elastic stability using recognized formulae and applying a minimum factor of safety of two. [Author's note: See Chap. 12.]

The above factor shall apply either for direct axial load or combination of axial load and bending. Where bending stresses occur, the maximum allowable tensile stress in concrete shall not exceed 250 psi.

The above factors are based upon a design concrete strength (f'_c) of 5,000 psi. For any higher values, these stresses may be increased in direct proportion to design concrete strengths.

If an effective prestress in excess of $0.2 f'_c$ is used, the stresses as permitted above shall be reduced accordingly.

(b) Capacity in End Bearing of Prestressed Concrete Piles—The capacity of prestressed concrete piles in end bearing shall not exceed $0.2 f'_c$.

(c) Capacity of Prestressed Concrete Piles as Friction Piles—The capacity of prestressed concrete piles shall preferably be determined by loading tests. The capacity shall be limited to the lesser of either (a) above or the safe capacity resulting from test loading.

Standards for prestressed concrete beams for I-beam and slab bridges (Fig. 15-3) were published in 1957 and revised in 1965. A complete design example of a bridge using these standard sections is presented in Chap. 5.

GENERAL NOTES FOR AASHO-PCI STANDARD I BEAMS

Specifications: AASHO Standard Specifications for Highway Bridges, latest edition, together with any tentative or supplemental specifications approved by the AASHO Committee on Bridges and Structures.

Live Load: All Highway Live Loads as specified in AASHO Standard Specifications for Highway Bridges.

Purpose: The purpose of the beams shown on this sheet shall be to establish a limited number of simple, practical sections leading to uniformity and simplicity of practice, forming, and production methods, and which are applicable to all conditions of highway bridge loading and to all spans within the approximate limits shown. The purpose is specifically not to disrupt or supplant established prestressed concrete beam practice utilizing present plants and forms. Beams of similar cross sections but with minor dimensional differences, manufactured with presently established plant facilities and which comply with structural and geometric requirements for any particular project may be substituted upon submission by the Producer of data necessary to show compliance with job requirements and approval by the Engineer.

Span Limits: The span limits shown on this sheet are approximate only and are not mandatory at either limit. Lateral spacing of beams shall be varied in keeping with the requirements of span and loading. The span limits shown contemplate the use of concrete weighing 150 lb per cu ft, HS 20-44 live load, simple span construction, cast-in-place deck slabs 6 to 8 in. thick composed of concrete with f'_c not less than 3,500 psi and having elastic properties approximately equal to those of the beam concrete. All dead load is assumed to be carried by the beam alone with live load carried by the beam and slab composite section.

By using light weight concrete, continuous construction or live loadings lighter than HS 20-44, span limits may be increased.

Concrete: Recommended minimum strengths for concrete in beams are $f'_c =$ 5,000 psi; at transfer of stressing force, $f'_{ci} = 4{,}000$ psi. Concrete of greater com-

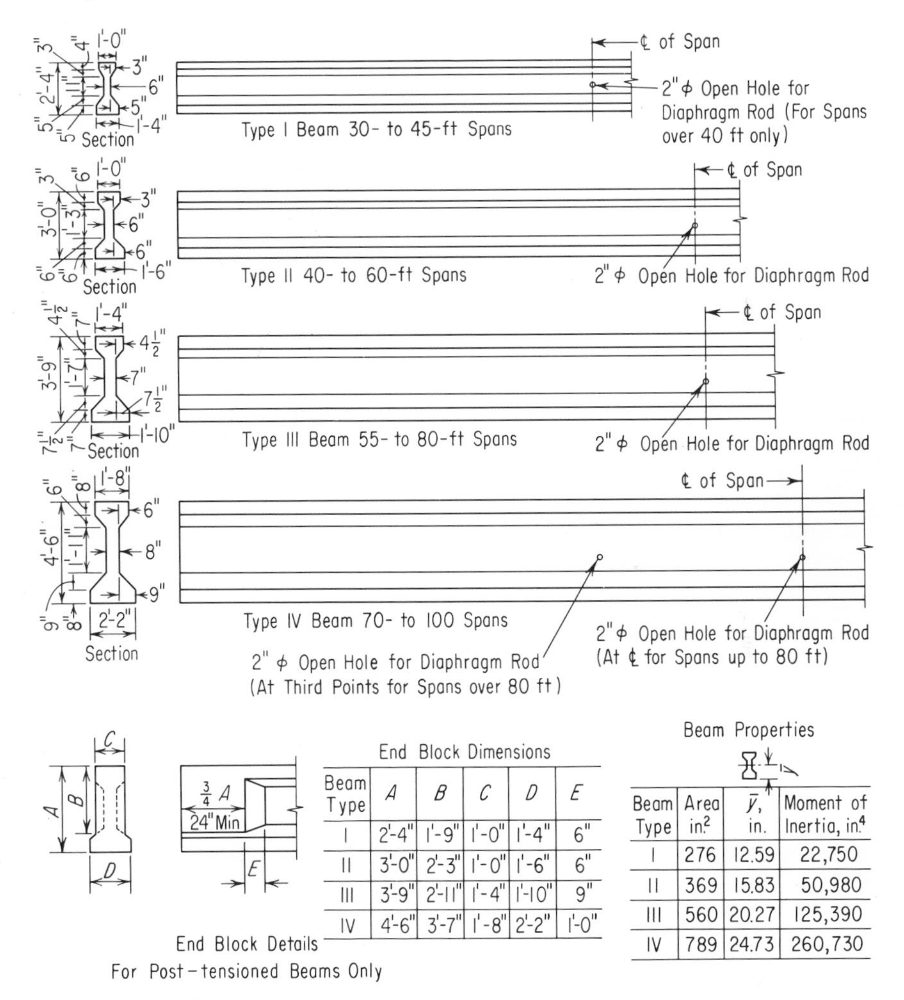

End Block Dimensions

Beam Type	A	B	C	D	E
I	2'-4"	1'-9"	1'-0"	1'-4"	6"
II	3'-0"	2'-3"	1'-0"	1'-6"	6"
III	3'-9"	2'-11"	1'-4"	1'-10"	9"
IV	4'-6"	3'-7"	1'-8"	2'-2"	1'-0"

Beam Properties

Beam Type	Area in.^{2}	$\bar{y}$, in.	Moment of Inertia, in.^{4}
I	276	12.59	22,750
II	369	15.83	50,980
III	560	20.27	125,390
IV	789	24.73	260,730

FIG. 15-3. AASHO-PCI Standard Prestressed Concrete Beams for Highway Bridge Spans of 30 to 100 ft. These beams are designed for use in beam and slab bridges of the type illustrated in Fig. 5-1. See text for General Notes for AASHO-PCI Standard Beams.

pressive strength may be used in which case allowable working stresses and resulting utility of the beams will be based upon the actual concrete specifications for the particular project.

Prestressing Reinforcement: Prestressing reinforcement shall generally be designed for particular projects or for prevailing bridge practices and available manufacturing facilities.

The beams are applicable for use with any acceptable type of prestressing in current practice; namely, pretensioning with straight or deflected strands, post-tensioning or a combination of pretensioning and post-tensioning. Placing of a portion of the stressing force on a draped or deflected profile will result in economy of beam section for all sizes and spans but not, necessarily, economy of manufacture or overall job cost. Use of draped reinforcement will generally be required for the longer spans in each beam series.

Materials for prestressing reinforcement shall be in accordance with the latest applicable designations of ASTM specifications for the particular type of tendons or subsequent developments by manufacturers which have generally been approved and accepted by member departments of AASHO.

Broken wires within individual strands will be permitted up to 2 per cent of the total number of wires in each beam, providing that there is not more than one broken wire per strand. Two or more broken wires per strand will be cause for replacement of the strand even though the two broken wires are within the 2 per cent limitation.

End Zones: Pretensioned: Where all tendons are pretensioned 7-wire strands, the use of end blocks will not be required.

In pretensioned beams, vertical stirrups acting at a unit stress of 20,000 psi to resist at least 4 per cent of the total prestressing force shall be placed within the distance of $d/4$ of the end of the beam, the end stirrups to be as close to the end of the beam as practicable. These vertical stirrups are to be provided in addition to those required as shear steel.

Post-tensioned: For beams with post-tensioning tendons, end blocks shall be used to distribute the concentrated prestressing forces of the anchorage.

End blocks shall have sufficient area to allow the spacing of the prestressing steel. Preferably, they shall be as wide as the narrow flange of the beam. Their length shall be at least three-fourths of the beam depth or 24 in. minimum. In post-tensioned members, a closely spaced grid of vertical and horizontal bars shall be placed near the face of the end block to resist crushing. Closely spaced horizontal and vertical bars shall be placed through the length of block.

Diaphragms: Diaphragms of precast or cast-in-place construction using prestressed or non-stressed reinforcement are recommended at span ends. Intermediate diaphragms are not required in spans up to 40 ft; are recommended at mid-span for spans above 40 ft and to 80 ft; and are recommended at span third points for spans in excess of 80 ft.

Forms: For standard plant manufacturing, the use of steel forms on concrete floored casting beds is recommended.

Chamfers and Corners: All exposed corners shall be chamfered not less than ¾-in. or rounded to ¾-in. radius. Angles of intersection between webs and flanges shall be rounded to not less than ¾-in. radius.

Finish of Tops: Tops of all beams shall be left rough. At approximately the time of initial set, all laitance shall be removed with a coarse wire brush.

Handling: In the handling of beams, they must be maintained in an upright position at all times and must be picked up only by means of approved devices anchored within the end zones. Disregard of this requirement may result in collapse of the member.

Mild Steel Reinforcing, Elastomeric Bearing Pads, Shoes and Miscellaneous Details:

All details not shown or specified here-on shall be designed for particular job requirements and shall be in accordance with applicable job specifications.

Standard AASHO-PCI box beams are illustrated in Figs. 15-4 and 15-5. When used adjacent to each other as shown in these figures, box beams provide bridges of minimum depth and maximum clearance above the roadway or body of water they are crossing. Adjacent box beams can be covered with any desired wearing surface. If they are covered with poured-in-place concrete this concrete forms a composite section with the beam for carrying live load.

Although it is not shown on the standard drawings, the standard box beams are frequently used as stringers to build beam and slab bridges. These structures are referred to as spread-box bridges. For example, in a structure that required 33-in.-deep adjacent boxes the designer might elect to use 39-in. or 42-in. boxes spaced some distance apart with a poured-in-place concrete slab on top and spanning between the boxes. The slab and box would function as a composite section to carry live load. Design analysis for such a structure would be similar to the example in Chap. 5.

General Notes—AASHO-PCI Prestressed Concrete Box Beams

*Specifications:** AASHO Standard Specifications for Highway Bridges, current edition; Criteria for Prestressed Concrete Bridges, U.S. Bureau of Public Roads, 1954; Tentative Recommendations for Prestressed Concrete, ACI-ASCE Joint Committee 323; and any subsequent revisions as approved by the Committee on Bridges and Structures of the AASHO.

Live Load: All highway live loads as specified by the AASHO Standard Specifications for Highway Bridges. Live loads shall be distributed in accordance with Article 13.2(c) of the AASHO Standard Specifications for Highway Bridges.

Purpose: The purpose of the standards shown on this sheet shall be to establish a limited number of simple, practical sections leading to uniformity and simplicity of forming and production methods. These standards shall be applicable to all conditions of highway bridge loading and usage within the approximate span limits shown. The purpose is specifically not to disrupt or supplant established prestressed concrete box beam practice utilizing present plant and forms. Box beams of similar cross section but with minor dimensional variations, manufactured with presently established plant facilities, which comply with structural and geometrical requirements for any particular project may be substituted upon submission by the producer of the data necessary to show compliance with the requirements of the job and upon approval of the substitution by the Engineer. Further, the purpose is not to supersede other standard sections adopted by the AASHO and PCI, but rather to complement those standards.

*When these standards are revised the notes on Specifications and Prestressing Reinforcement will probably be changed to be similar to the notes under these headings as shown for Prestressed Concrete Piles, pp. 245–254.

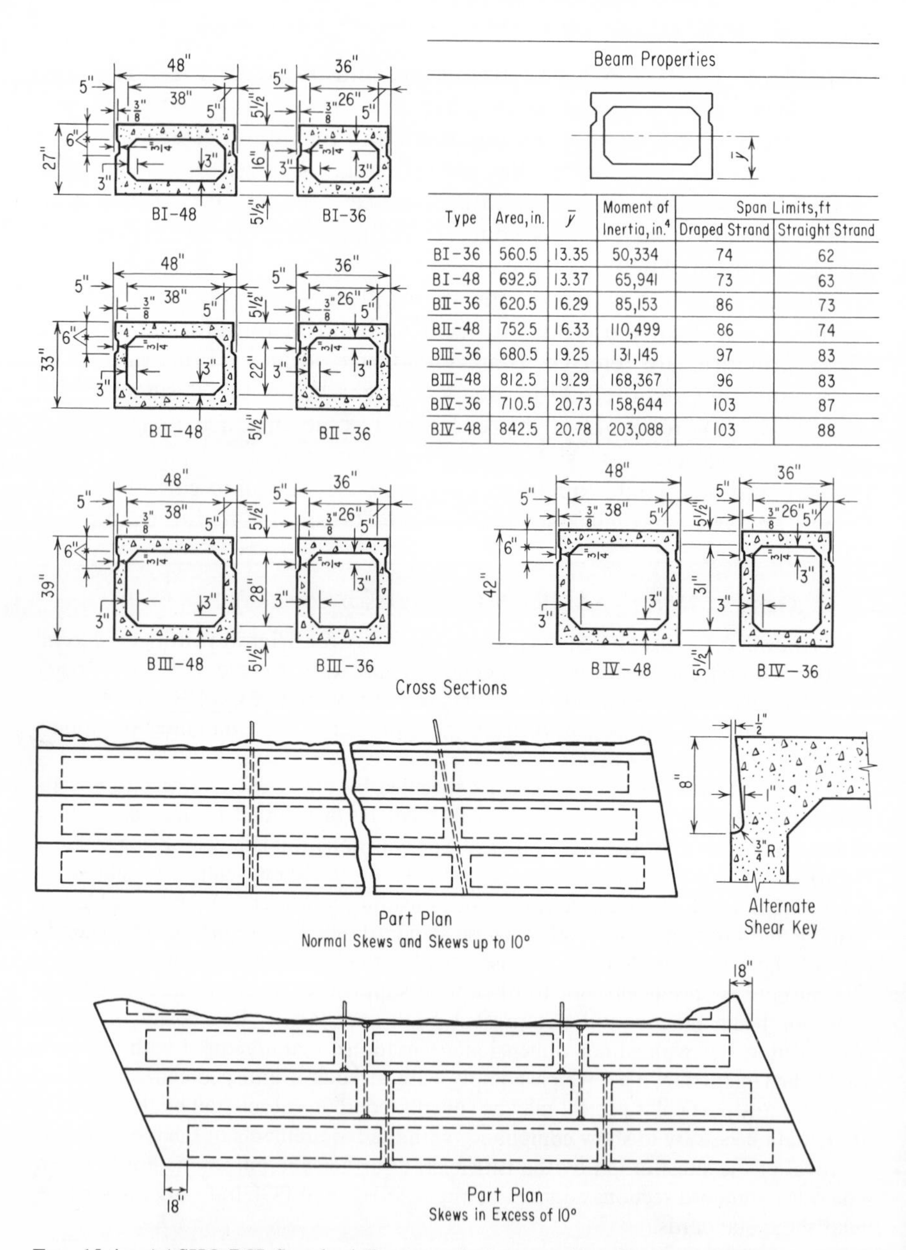

Beam Properties

Type	Area, in.	$\bar{y}$	Moment of Inertia, in.[4]	Span Limits, ft	
				Draped Strand	Straight Strand
BI-36	560.5	13.35	50,334	74	62
BI-48	692.5	13.37	65,941	73	63
BII-36	620.5	16.29	85,153	86	73
BII-48	752.5	16.33	110,499	86	74
BIII-36	680.5	19.25	131,145	97	83
BIII-48	812.5	19.29	168,367	96	83
BIV-36	710.5	20.73	158,644	103	87
BIV-48	842.5	20.78	203,088	103	88

FIG. 15-4. AASHO-PCI Standard Prestressed Concrete Box Beams for Highway Bridge Spans to 103 ft. See also Fig. 15-5.

Span Limits: The span limits shown for the various sections on this sheet are based upon the following design conditions: H20-S16-44 live load, 28-ft roadway, live load distribution as specified above, concrete weighing 150 lb per cu ft with $f'_c =$ 5,000 psi and $f'_{ci} = 4{,}000$ psi, an allowance of 30 lb per sq ft for wearing surface and allowable stresses as given in the design specifications.

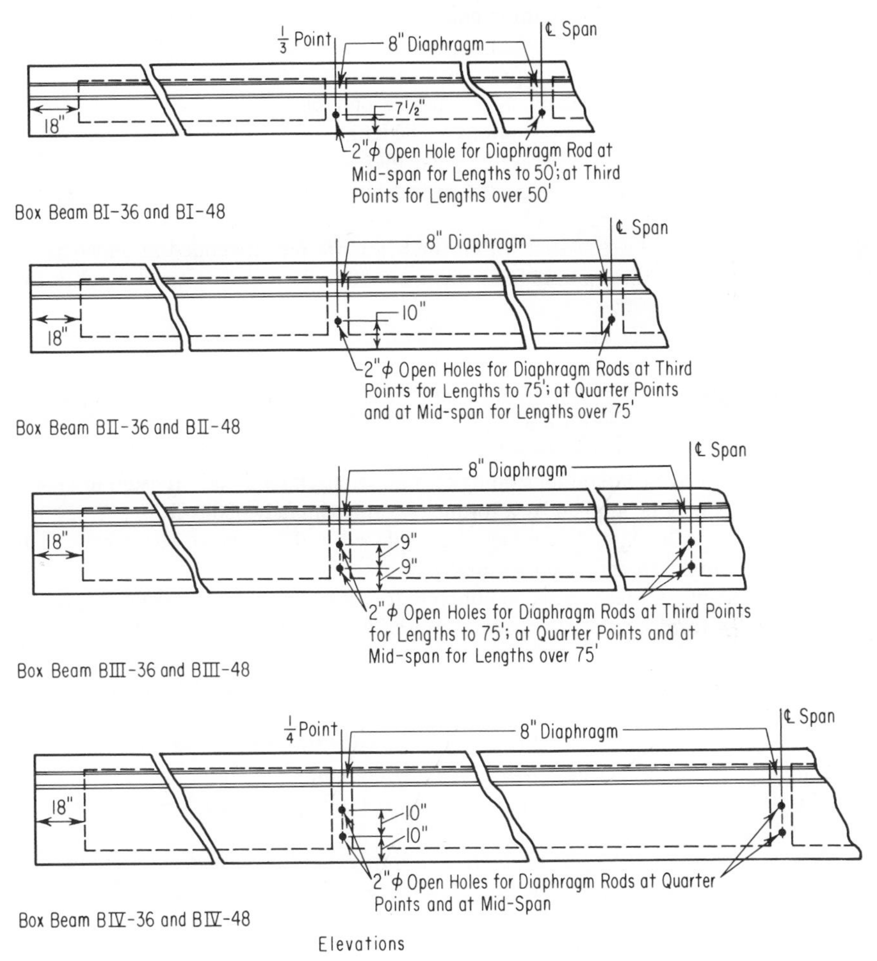

FIG. 15-5. Elevations of AASHO-PCI Box Beams. See also Fig. 15-4.

Span limits shown are approximate only and are not a rigid limitation of the sections. The upper span limits may be extended by reduced loading, increased concrete strength, use of lightweight concrete, or other approved means within the limits of design specifications.

Concrete: Recommended minimum strengths for concrete in box beams are $f'_c = 5{,}000$ psi; at transfer of stressing force $f'_{ci} = 4{,}000$ psi. Concrete of greater or less compressive strength, but not less than $f'_c = 4{,}000$ psi, may be used, in which case allowable working stresses and resulting utility of the box beams will be based upon the actual concrete specifications for the particular project.

*Prestressing Reinforcement:** Prestressing reinforcement shall generally be designed for particular projects or prevailing bridge practices and available manufacturing facilities.

Materials for prestressing reinforcement may be any of the materials specified in the governing specifications or subsequent developments by manufacturers which have generally been accepted in prestressed practice.

End Blocks: The box beams shown utilize end blocks 18-in. long which have proven satisfactory in many installations. The length of end blocks may be increased to accomodate local plant facilities or particular job requirements. Sufficient mild steel reinforcement should be provided in end blocks to resist the tensile forces due to concentrated prestressing loads.

Diaphragms: Diaphragms cast within the beam are recommended at midspan for spans up to 50 ft, at third points for spans from 50 to 75 ft and at quarter points for spans over 75 ft.

Lateral Ties: Lateral ties shall be provided through the diaphragms in the positions indicated except that for the 39- and 42-in. deep sections, when adjacent units are tied in pairs, one tie at diaphragm locations centered between bottom of key and bottom of beam will be permitted.

Each tie shall be equivalent to a 1¼-in. mild steel bar tensioned to 30,000 lb or an equal force applied by lateral tensioning of high strength tendons. Tension in 1¼-in. mild steel bars may be applied by a torque of approximately 600 ft-lb.

Shear Keys: After lateral ties have been placed and tightened, shear keys shall be filled with high strength, non-shrinking mortar.

Forms: The use of steel forms on concrete founded casting beds is recommended. Voids may be formed of any approved material and be vented during the curing period.

Chamfers and Corners: All exposed corners shall be chamfered ¾-in. or rounded to ¾-in. radius.

Finish: Tops shall be given a broom finish normal to L of roadway.

Handling: In handling, the box beams must be maintained in an upright position at all times and must be picked up only by means of approved devices near the ends of the beams.

Mild Steel Reinforcing, Bearing Pads, Anchorages and Miscellaneous Details: All details not shown or specified hereon shall be designed for particular job requirements and shall be in accordance with applicable job specifications.

Standards for AASHO-PCI Prestressed Concrete Slabs are shown in Figs. 15-6 and 15-7. These members are designed for the shorter span bridges.

General Notes—AASHO-PCI Prestressed Concrete Slabs

*Specifications:** AASHO Standard Specifications for Highway Bridges, current edition; Criteria for Prestressed Concrete Bridges, U.S. Bureau of Public Roads,

*When these standards are revised the notes on Specifications and Prestressing Reinforcement will probably be changed to be similar to the notes under these headings as shown for Prestressed Concrete Piles, pp. 245–254.

1954; Tentative Recommendations for Prestressed Concrete by ACI-ASCE Joint Committee 323, and any subsequent revisions approved by the Committee on Bridges and Structures of the AASHO.

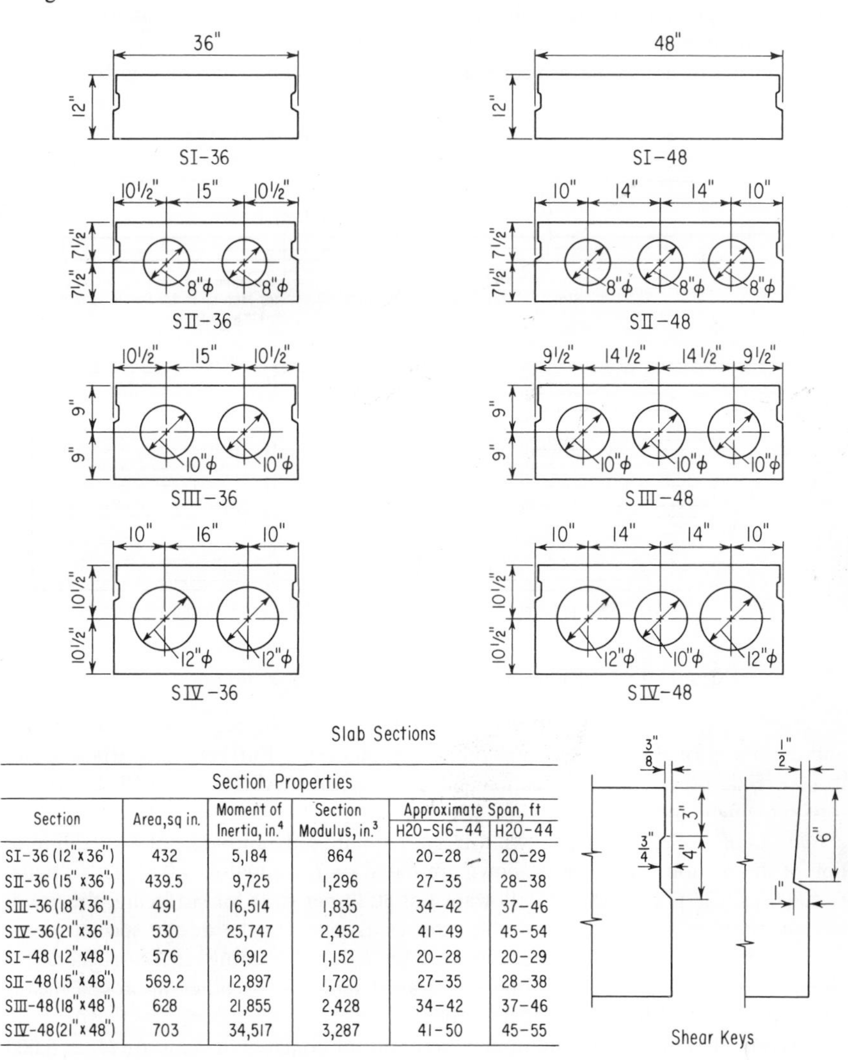

Section Properties

Section	Area, sq in.	Moment of Inertia, in.^{4}	Section Modulus, in.^{3}	Approximate Span, ft	
				H20-S16-44	H20-44
SI-36 (12"x36")	432	5,184	864	20-28	20-29
SII-36 (15"x36")	439.5	9,725	1,296	27-35	28-38
SIII-36 (18"x36")	491	16,514	1,835	34-42	37-46
SIV-36 (21"x36")	530	25,747	2,452	41-49	45-54
SI-48 (12"x48")	576	6,912	1,152	20-28	20-29
SII-48 (15"x48")	569.2	12,897	1,720	27-35	28-38
SIII-48 (18"x48")	628	21,855	2,428	34-42	37-46
SIV-48 (21"x48")	703	34,517	3,287	41-50	45-55

FIG. 15-6. AASHO-PCI Standard Prestressed Concrete Slabs for Highway Bridge Spans up to 55 ft. See also Fig. 15-7.

Live Load: All highway live loads as specified by AASHO Standard Specifications for Highway Bridges. Live loads shall be distributed in accordance with Section 13.2(c) of the AASHO Standard Specifications for Highway Bridges.

Purpose: The purpose of the standards shown on this sheet shall be to establish a limited number of simple, practical sections leading to uniformity and simplicity of forming and production methods. These standards shall be applicable to all conditions of highway bridge loading and usage within the approximate span limits indicated in the SUMMARY OF SECTIONS on this sheet. The purpose is specifically not to disrupt or supplant established prestressed concrete slab practice utilizing present plant or forms. Similar sections with minor dimensional variations, manufactured with established plant facilities, which meet structural and geometrical requirements of a specific project may be substituted upon submission by the pro-

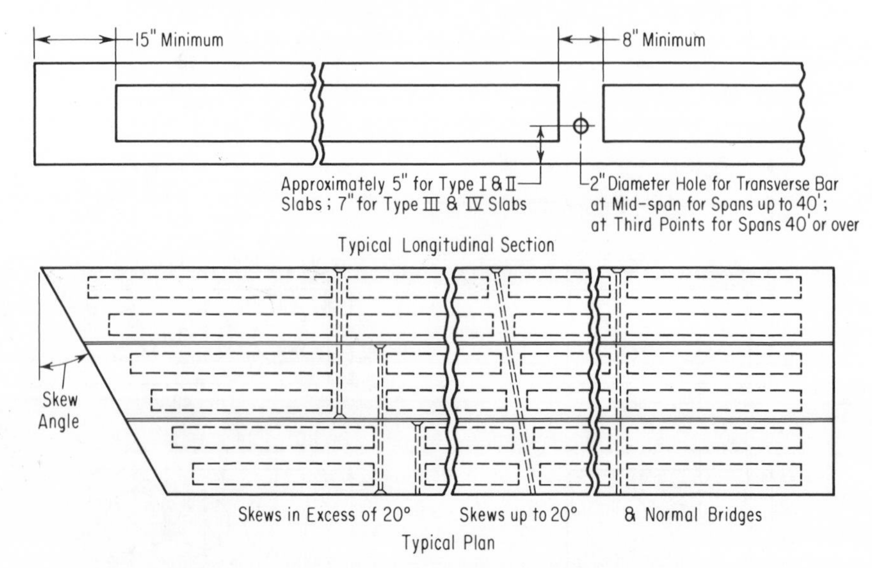

FIG. 15-7. Elevations of AASHO-PCI Slabs. See also Fig. 15-6.

ducer of the data necessary to show compliance with the requirements of the job and upon approval of the substitution by the engineer. Further, the purpose is not to supercede other standard sections adopted by the AASHO and PCI, but rather to complement these standards.

Span Limits: Span limits shown are approximate only and are not a rigid limitation of the section. The limits shown are based on f'_{ci} equal to 4,000 psi; f'_c equal to 5,000 psi; 28-ft roadway, an allowance of 30 lb per sq ft for surfacing, the use of straight pretensioning, and allowable stresses as given in the design specifications. The upper span limits may be extended by reduced loading, increased concrete strength, use of light weight concrete, draped tendons, or other approved means within the limits of the design specifications.

Concrete: Recommended minimum strengths for concrete in slabs are $f'_c = 5{,}000$ psi; at transfer of stressing force, $f'_{ci} = 4{,}000$ psi. Concrete of greater or less compressive strength, but not less than $f'_c = 4{,}000$ psi, may be used, in which case allowable working stresses and resulting utility of the slabs will be based upon the actual concrete specifications for the particular project.

*Prestressing Reinforcement:** Prestressing reinforcement shall generally be de-

*When these standards are revised the notes on Specifications and Prestressing Reinforcement will probably be changed to be similar to the notes under these headings as shown for Prestressed Concrete Piles, pp. 245–254.

signed for particular projects or for prevailing bridge practices and available manufacturing facilities.

Materials for prestressing reinforcement may be any of the materials specified by governing specifications or subsequent developments by manufacturers which have generally been accepted in common prestressed practice.

End Blocks: The slabs shown utilize end blocks 15-in. long which have proven satisfactory in many installations. The length of end blocks may be increased to accomodate local plant facilities or particular job requirements. Sufficient mild steel reinforcement should be provided in end blocks to resist the tensile forces due to concentrated prestressing loads.

Diaphragms: Diaphragms cast within the slab are recommended at midspan for spans up to 40 ft and at third points for spans 40 ft or over.

Lateral Ties: Lateral ties shall be provided through the diaphragms in the positions indicated. Each tie shall be equivalent to a 1¼-in. mild steel bar tensioned to 30,000 lb or an equal force applied by lateral tensioning of high strength tendons. Tension in 1¼-in. mild steel may be applied by torquing to approximately 600 ft-lb.

Shear Keys: After lateral ties have been placed and tightened, shear keys shall be filled with high strength, non-shrinking mortar.

Forms: The use of steel forms on concrete founded casting beds is recommended.

Chamfers and Corners: All exposed corners shall be chamfered ¾ in. or rounded to ¾-in. radius.

Finish: Tops shall be given a broom finish, normal to centerline of roadway.

Handling: In handling, the slabs must be maintained in an upright position at all times and must be picked up only by means of approved devices near the ends of the slabs.

Mild Steel Reinforcing, Bearing Pads, Anchorages and Miscellaneous Details: All details not shown or specified hereon shall be designed for particular job requirements and shall be in accordance with applicable job specifications.

15-3. Continuous Bridges of Precast Members. Bridges of the type described in the preceeding section can easily be made continuous by placing reinforcing steel bars in the poured-in-place slab to carry the negative moment in the region of the pier. The prestressed beams will carry their own dead weight and the dead weight of the poured-in-place slab as simple span members. After the slab has cured, all additional loads will be carried by the continuous structure. The fact that the negative moment is being carried by a reinforced concrete member and the positive moment is being carried by a prestressed concrete member does not alter the moment diagram. Test data indicate that the precompression due to the pretensioned strands near the bottom of a precast member does not influence the negative-bending strength of the section.

Florida's Sebastian Inlet Bridge designed by Howard, Needles, Tammen and Bergendoff of Orlando, Florida, is composed of precast pretensioned members yet takes full advantage of continuity. As shown in Figs. 15-8 to 15-10, although it has two 100-ft side spans and a 180-ft main span, the longest precast section required was only 120 ft. Four prestressed I-beam stringers support the 28-ft roadway and two 2-ft-wide safety walks.

15-4. Railway Bridges. The American Railway Engineering Association has prepared and adopted a complete code for prestressed concrete railroad bridges entitled "Specifications for Design, Materials and Construction of Prestressed Concrete Structures." In its tentative form this was printed in *AREA* Bulletin 554, December, 1959.

Complete drawings and specifications for a 28-ft span standard railway trestle are included in a report entitled Prestressed Concrete for Railway Structures, which appeared in *AREA* Bulletin 583 for January, 1964. Details presented cover 24-in.-sq, hollow precast prestressed concrete piles, precast or cast-in-place reinforced concrete pile caps, elastomeric bearing pads, and prestressed concrete box beams. The box beams are designed so that four 3-ft-wide beams or three 4-ft-wide beams will carry Cooper E72 loading.

FIG. 15-8. All precast pretensioned prestressed concrete bridge over Sebastian Inlet in Florida. Main span = 180 ft; two side spans = 100 ft; approach spans = 73 ft. For further details see text and Figs. 15-9 and 15-10.

Reprints of both these items from the *AREA* Bulletin are available from Portland Cement Association.

15-5. Post-tensioned Hollow-box Bridges. The longest prestressed concrete bridges are the hollow-box type illustrated in Figs. 15-11 to 15-13. Two styles have been used. The three-span continuous structure shown in Fig. 15-11 is one. The other has a long main span with cantilever ends that are from one-sixth to one-fourth the length of the main span. All the weight of the cantilever ends is carried by the two main piers, thus creating a large negative moment over the piers which in turn reduces the positive moment in the main span.

Figure 15-12 shows details of a hollow-box bridge at mid-span. This type of cross-section makes maximum use of the concrete by putting the majority of it in the top and bottom flanges with a minimum amount in the webs. The structure is supported on falsework until it has cured and the galvanized strands have been pulled through the hollow boxes into position and post-tensioned. The galvanized strands are sometimes given the additional protection of a coat of bitumastic paint after they have

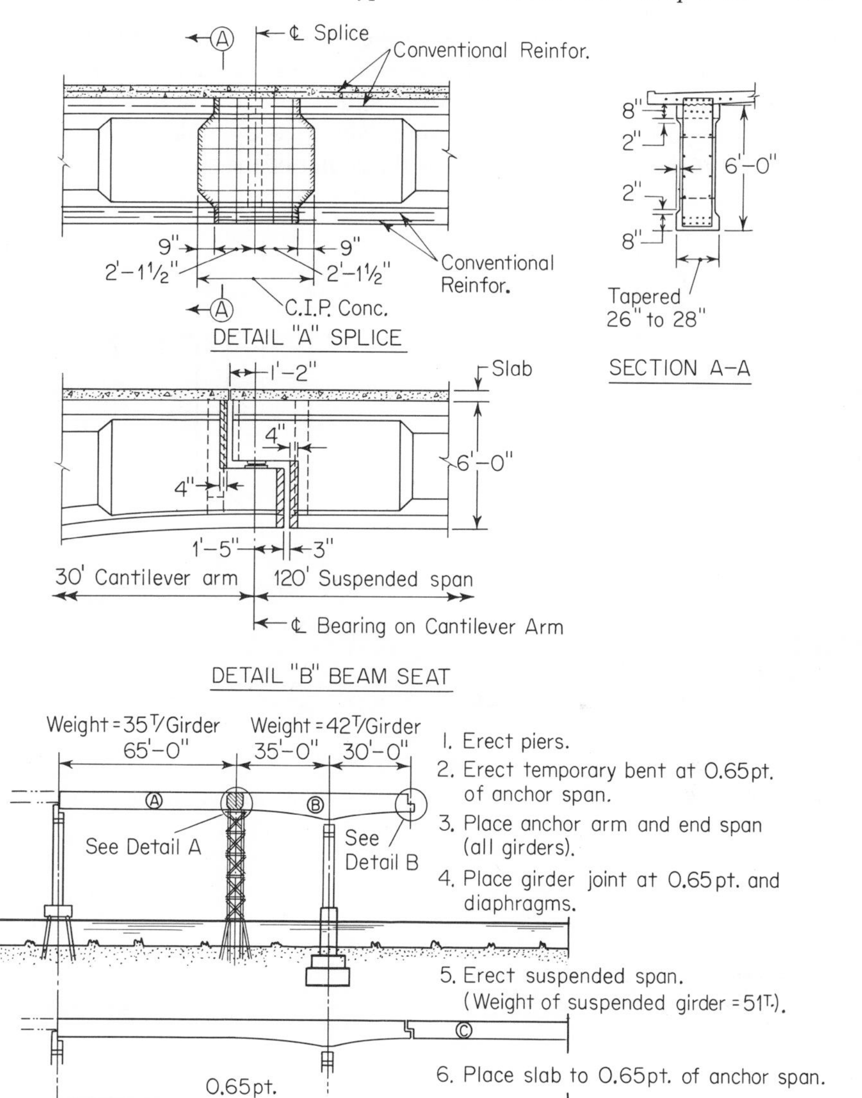

FIG. 15-9. Joint details and erection plan for Sebastian Inlet Bridge. See text and Figs. 15-8 and 15-10.

been tensioned. Strands can be inspected at any time during the life of the structure. If the designer desires, the details can readily be worked out so that any strand can be removed and replaced by a new one.

A pleasing architectural effect can be achieved by making these structures deep at the piers and shallow at the center of the span. If this differ-

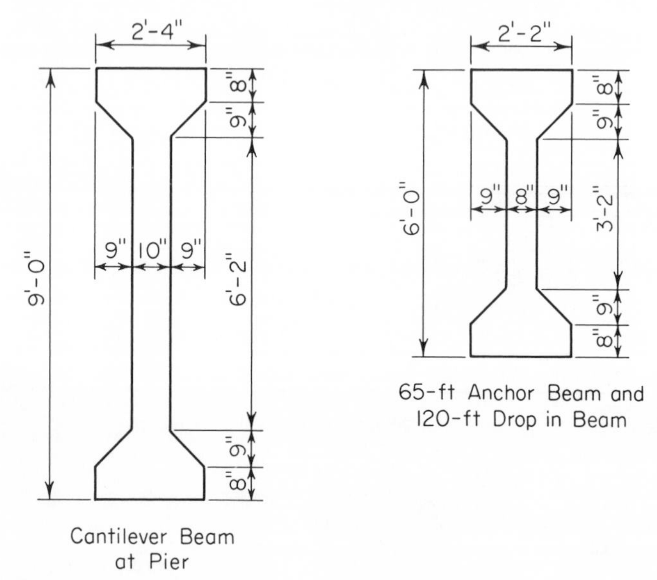

FIG. 15-10. Precast pretensioned girder sections for Sebastian Inlet Bridge. Fabricator: Juno Prestressors, Inc. See text and Figs. 15-8 and 15-9.

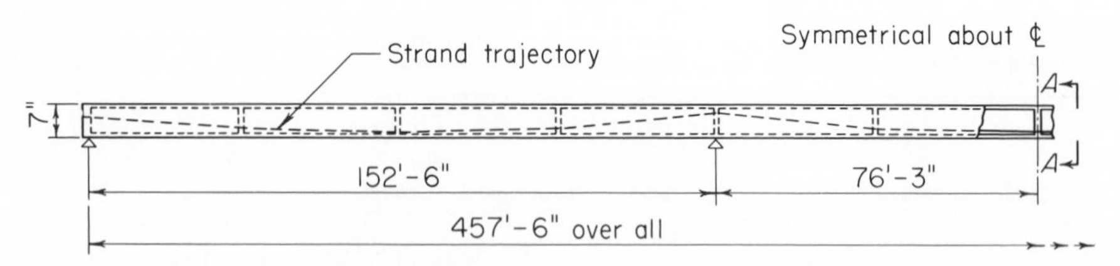

FIG. 15-11. Elevation of three-span post-tensioned hollow-box bridge. (*From Eng. News-Record, Dec.* 27, 1956.)

ence in depth is kept within reasonable limits, the stresses remain reasonable also.

When checking ultimate strength of a structure of this type with unbonded tendons, refer to Sec. 1.13.10(c) of the American Association of State Highway Officials Standard Specifications for Highway Bridges for 1961 or the corresponding section in the latest edition of these specifications.

15-6. Components of Buildings. A major portion of the prestressed concrete members used in building construction are of the precast pretensioned type which are made in a local fabricating plant and hauled to the job site where they can be lifted directly from the truck to their

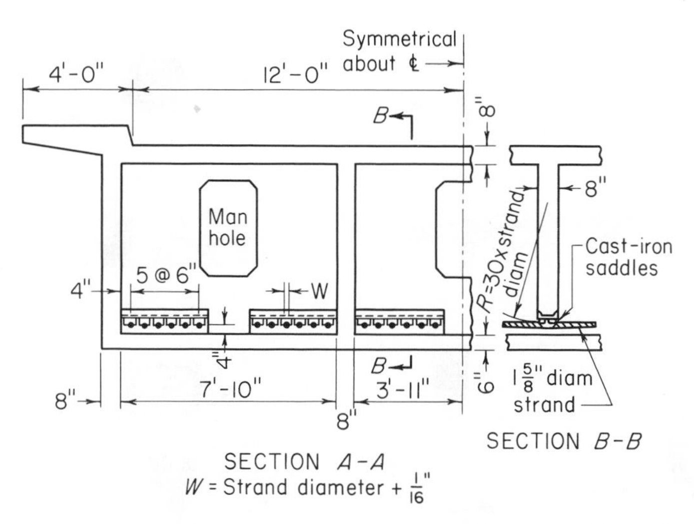

FIG. 15-12. Details at mid-span of post-tensioned hollow-box bridge shown in Fig. 15-11. (*From Eng. News-Record, Dec.* 27, 1956.)

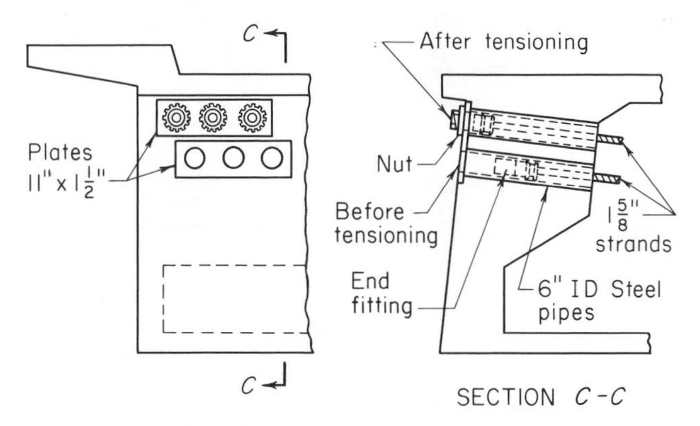

FIG. 15-13. Details at end of post-tensioned hollow-box bridge shown in Fig. 15-11. (*From Eng. News-Record, Dec.* 27, 1956.)

final position in the structure. Figure 15-14 shows the most common precast pretensioned members and indicates their uses and capacities. Since the functions of some of these members overlap each other, one fabricator seldom has forms for all of them. Before designing a structure

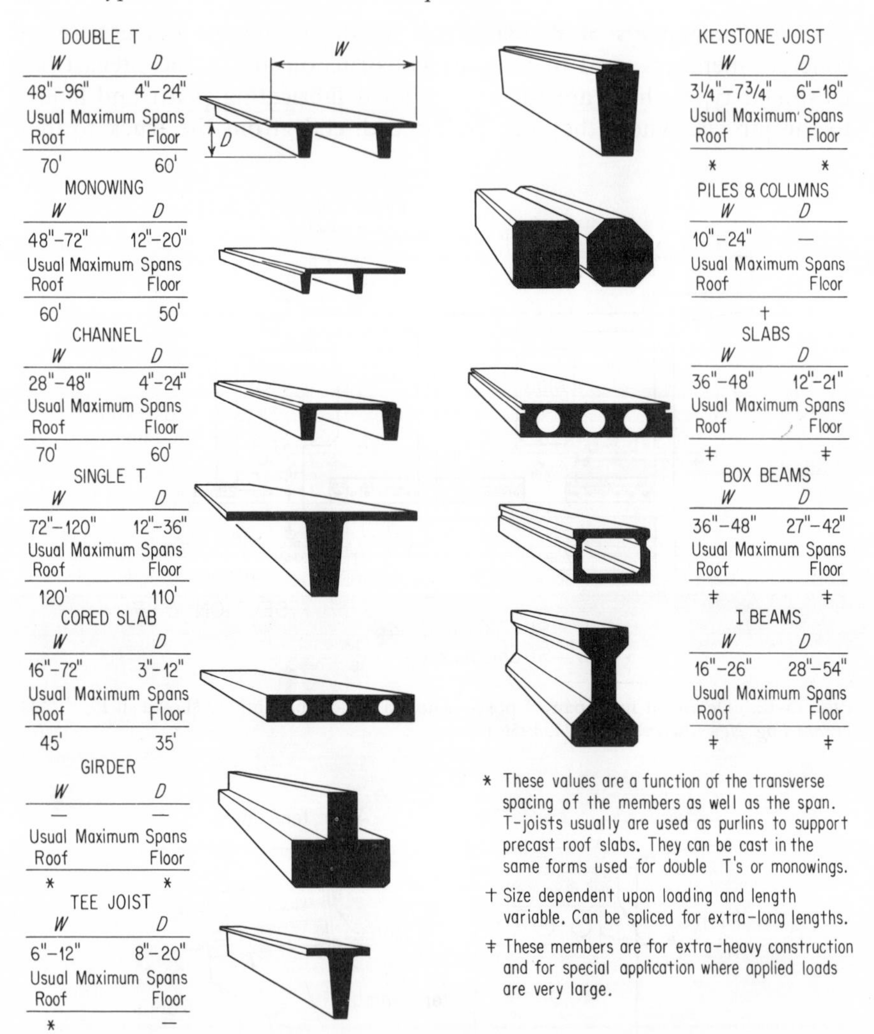

FIG. 15-14. Typical precast prestressed concrete sections. The average casting yard is equipped to produce several typical sections as standard items from forms on hand. When used as a floor, the area-covering members (double T's, monowings, single T's, slabs, and channels) are usually covered with a 2- to 4-in. layer of poured-in-place concrete. When used as a roof, a cover of roofing material usually is sufficient. The most common details of these members are shown above. (*Courtesy of CF&I Steel Corporation.*)

it is advisable to determine which sections are available within reasonable shipping distance.

Floors and roofs are made from double T's, monowings, channels, single T's, and slabs as illustrated in the left-hand column of Fig. 15-14. These members are placed with flanges or sides touching so that they completely cover the area. Flanged members are tied together at intervals by welding inserts cast into the edges of the flanges. Slabs often have shear keys along the sides which are filled with concrete after they are in place in the structure.

Floors are completed by pouring a 2- to 4-in. concrete topping on the precast members. This creates a composite section for carrying live load. Forms are not needed except at expansion joints and the outside edges of the floor. Utilities such as electrical conduits are easily placed on the precast deck before the top slab is poured.

Roofs can be completed with poured-in-place topping or simply with a roofing compound. One popular detail is the use of sheets of insulation to cover the deck and then a topping of roofing material.

When the bays in a building are rectangular rather then square and the members being used for floor or roof are single T's, double T's, or channels, it is often more economical to use these members to span the long direction with girders in the short direction.

The "girder" illustrated in Fig. 15-14 is called an inverted T or a ledger beam. It is used to support single T's, double T's, channels, or slabs which rest on the ledges and are flush top with the stem. Similar members with only one ledge called L beams are used where the T's frame into one side only.

Joists of the T, I, or keystone type are normally used as roof purlins with precast lightweight slabs spanning between them. The slabs are relatively short span members of Insulrock, Tectum, or similar materials.

I joists are sometimes used with a composite poured-in-place slab to form a floor or roof system.

Precast prestressed concrete piles are discussed in Chap. 12. When precast prestressed beams, T's, etc., are used for a building it is only logical to precast the columns too and eliminate the need for forms at the job site. Reinforced concrete columns are precast in single units several stories high and set on footings in much the same manner as steel columns. One such detail is illustrated in Art. 15-7. Some fabricators find it more economical to make and handle precast prestressed columns than precast reinforced columns, and the former are entirely adequate structural members.

Precast pretensioned bridges are used throughout the country and the slabs, box beams, and I beams in the lower right-hand side of Fig. 15-14 are standard bridge members. Complete details on these sections are presented in Art. 15-2. They are readily available and are sometimes useful in buildings that have heavy loadings.

15-7. Framing Building Members. Both pinned and rigid connections between precast members can be designed and fabricated economically. Details for rigid connections include reinforcing bars lapped and covered with poured-in-place concrete, reinforcing bars welded together and covered with concrete, embedded steel plates or shapes welded together after erection, and post-tensioned joints. Pinned connections are usually the bearing type where steel members partially embedded in one member

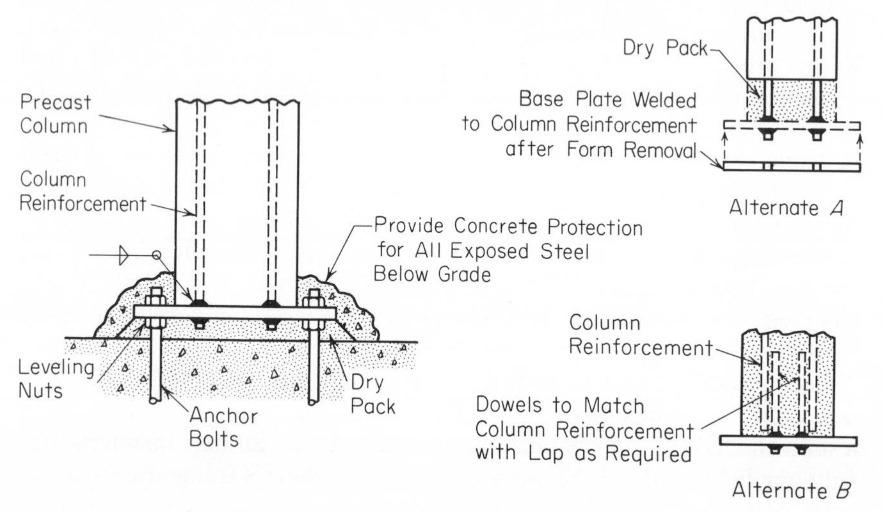

CB-1 Extended Base Plate with Anchor Bolts

This is the most popular column-base connection, since it provides immediate stability upon erection, and it is easy to adjust the column for plumb and elevation. With properly sized base plate, bolts and reinforcing, a limited amount of moment transfer may be assumed. When large moments occur at the column base, other types of connections are preferred.

1. If the columns are to be cast in continuous pile forms, it is advisable to weld the base plate to the column reinforcement after the column is removed from the forms as shown in Alternate *A*. The space above the base plate is then dry-packed as shown in the sketch, using a mix having strength equal to column concrete.
2. The base plate may be welded to dowels which in turn lap with main column bars as shown in Alternate *B*. This allows more careful welding, since bench welding is substituted for job welding.
3. Care must be taken to check conditions that exist prior to grouting such as axial construction loads and bending due to wind or other lateral loads.
4. For heavy erection loads which are imposed prior to grouting beneath column, steel shims or 6″ × 6″ grout pads may be placed at the center of the column and set to proper elevation prior to erection of column, thus making plumbing of column easier and faster, and preventing deflection of base plate due to dead loads and erection loads.

FIG. 15-15. Detail of connection at base of precast column where moment transfer is small. (*Reprinted by permission of the Prestressed Concrete Institute.*)

bear on steel members in another member or where one member bears on an elastomeric pad which in turn bears on the other member.

Criteria for the design of connections are set forth in Suggested Design of Joints and Connections in Precast Structural Concrete, Title No. 61-51, pages 921 to 937 of the *Journal of the American Concrete Institute* for

August, 1964. Excerpts from this that apply to design examples in Chap. 13 are reproduced in Appendix B. Details are illustrated in the handbook Connection Details for Precast-Prestressed Concrete Buildings published by the Prestressed Concrete Institute. This handbook shows 46 connection details, a few of which are reproduced here in Figs. 15-15 to 15-20.

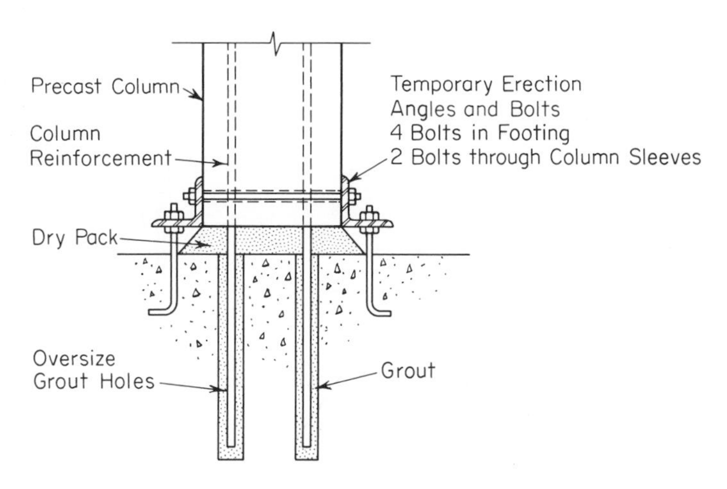

CB-4 Dowelled Connection

This connection eliminates all extraneous materials. Providing sufficient embedment length is available, the detail is capable of resisting high moments at the base of the column.

1. The oversize grout holes may be cast or drilled into the supporting member.
2. Temporary erection angles and bolts may be used for positioning the column. When the grout has set, the angles and through bolts are removed for use on other columns. Steel wedges will serve the same function if the column is properly braced.
3. Holes are filled with grout just prior to placing the column. Grout consistency to permit displacement of some grout when the column bars are inserted. Nonshrinking or high-adhesive (epoxy, etc.) grouts are preferred.
4. Steel shims can be placed to proper grade at the center of the column to prevent deflection of the angles and to facilitate plumbing. Dead loads can be applied immediately, since shims will carry the loads.
5. For short columns, threaded inserts may be used in lieu of the pipe sleeves shown in the detail. For very long columns, guying or some form of bracing should be used to provide stability during erection.

FIG. 15-16. Detail of connection at base of precast column where moment transfer is large. (*Reprinted by permission of the Prestressed Concrete Institute.*)

Precast reinforced or precast prestressed columns shorten erection time and add to the economy of buildings made of precast members. Two details for the bases of such columns are illustrated in Figs. 15-15 and 15-16.

In the design of prestressed concrete flexural members we make allowance for stress losses which reduce the tension in the prestressing tendons.

In designing and detailing the joints and connections for these members we should not overlook the effect of the stress losses. (See Sec. 301.3.5 of Suggested Design of Joints and Connections in Precast Structural Concrete.)

Nonrigid connections are usually the simple bearing type. Figure 15-17

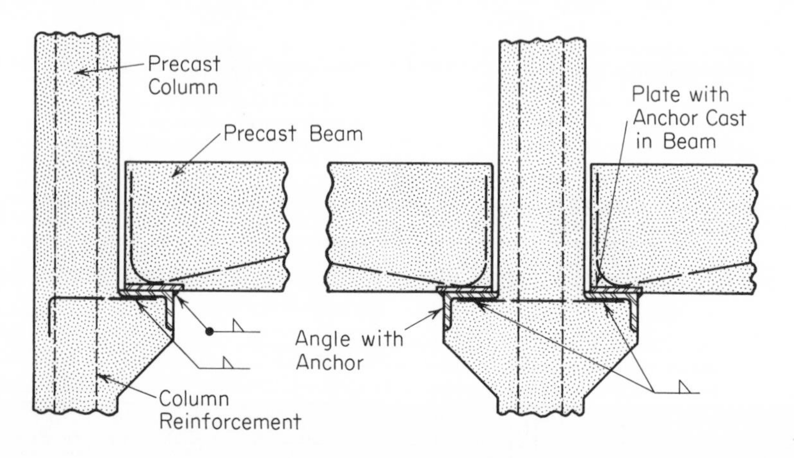

BC-2 Welded, Simple Spans

This connection should be considered when it is desirable to keep the columns free of continuity moments. However, eccentricity moments should be computed for large differences in beam spans or loadings. Stability of the frame must be furnished by other means such as shear walls.

1. A fireproof connection may be obtained by recessing the column steel, then welding and grouting over.
2. It is essential that horizontal anchors are welded to steel which is embedded in the column and beam. These anchors must be adequate to resist axial tensions due to temperature drop and creep shortening of prestressed beams which occur after erection.
3. If heavy dead loads are to be placed on the beam, make tack welds during erection and provide full fillet welds after all dead loads are in place. This reduces stresses in the welds.
4. Consideration may also be given to a welded connection at the top of the beam, with a flexible bearing pad where the beam bears on the haunch. This may result in partial restraint.

FIG. 15-17. Simple-span connection of precast beam to column corbel. See Sec. 301.3 of ACI Title No. 61-51 (Appendix B) for criteria on corbel design. See Art. 13-7 for discussion of high allowable bearing pressures where corner of concrete is confined by a steel angle. (*Reprinted by permission of the Prestressed Concrete Institute.*)

shows a steel plate cast into the beam bearing on a steel angle cast into the corbel of the column. This detail transmits shear, but there is no transfer of moment between beam and column. Criteria for the design of corbels, based on an extensive series of tests, have been developed and are presented in Sec. 301.3 of ACI Title No. 61-51 discussed earlier in this section. These criteria should be reviewed and followed.

Nonrigid connections between beams cantilevering over columns or other beams can be detailed using elastomeric bearing pads as shown in Fig. 15-18.

Precast pretensioned members joined with welded or lapped reinforcing bars and poured-in-place concrete provide a continuous or rigid-frame structure. Numerous load tests conducted throughout the country have proved that full composite action and continuity are obtained.[1]* One

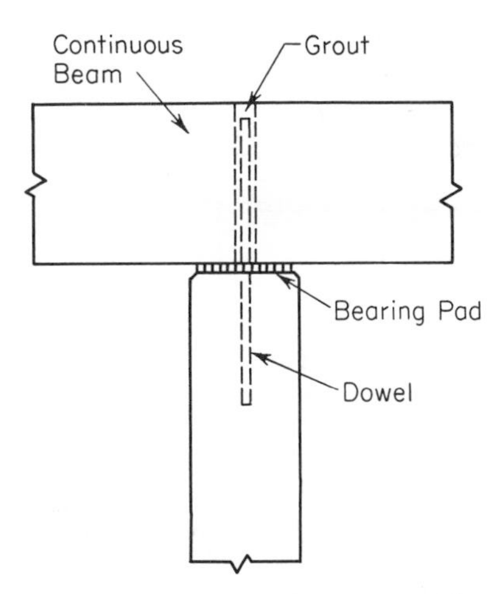

BC-8 Dowelled, Continuous Spans (Top of Column Only)

This detail is common where the beam cantilevers over the top of a column, or at an end span where continuity is achieved by some other means such as post-tensioning.

1. Where the beam is not continuous with the column, it is wise to chamfer the edges of the column to prevent chipping.
2. The dowels may be cast integrally with the column, or they may be grouted into oversize holes.
3. The size of the dowel hole should be adequate to allow for beam tolerances.
4. Where an immediate connection is desired, anchor bolts may be substituted for the dowels.
5. Where small movements are desirable, the lower portion of the dowel hole may be grouted with mastic, or anchor bolts may be left entirely ungrouted. Bolts should never be left ungrouted if subject to deterioration or freezing.

FIG. 15-18. Nonrigid connection for beam cantilevering over column or supporting beam. (*Reprinted by permission of the Prestressed Concrete Institute.*)

example is a report by Jack R. Janney and John F. Wiss entitled Load-deflection and Vibration Characteristics of a Multi-Story Precast Concrete Building which appeared in the *Journal of the American Concrete Institute,* April, 1961. Although post-tensioned joints can be and are used, they are not normally necessary to achieve continuity.

Two economical details for developing continuity between precast

*Superscript numbers indicate references listed in the Bibliography at the end of the chapter.

beams are illustrated in Figs. 15-19 and 15-20. In both cases reinforcing bars near the top of the beam carry the tensile load across the joint, poured-in-place concrete between the beams carries the compressive force, and bearing pads transfer the shear from beam to support. In Fig. 15-19 the top reinforcing bars project into pockets in the ends of the beams, and the tensile load is transferred from one set of bars to the other by the lap bar in the cast-in-place concrete. In Fig. 15-20 the tensile load is transferred from one set of bars to the other through a structural steel angle which

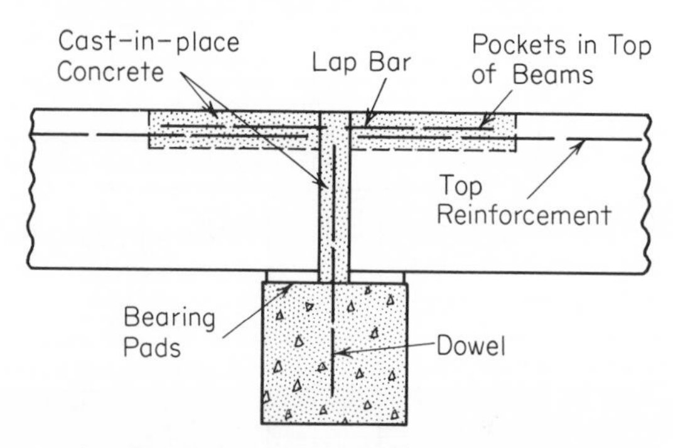

BG-5 Lap Splice, Continuous Spans

This is the simplest form of continuous beam connection.

1. If a compressible bearing pad is used, it should extend continuously across the grouted joint.

2. In cramped locations the grout may not be sound enough to produce a full bond. If in doubt, use type BG-4 or BG-6.

3. Exterior span should be type BG-1 or similar connection. It is usually not desirable to put excessive torque in the main beam to achieve a fixing moment in the secondary.

FIG. 15-19. Negative-moment connection between two precast beams using lapped reinforcing bars. (*Reprinted by permission of the Prestressed Concrete Institute.*)

is welded to both sets of bars. Data on transferring the load from one reinforcing bar to another are given in Fig. 15-21. Mild-steel bars can be welded without problems. Intermediate-grade bars can be welded with proper electrodes and procedure. High-carbon bars, usually those with yield strengths above 40,000 psi, should be avoided where welding is required.

Figs. 15-22 and 15-23 show precast prestressed soffit beams supporting double-T roof and floor sections. The precast prestressed soffit beam is set in place in the structure and supported at short intervals. It will not carry any dead- or live-load moment or shear until its composite top has been cast and cured. Double T's, with top flanges cut back flush with the side of the soffit beam, are set in place and concrete is cast in place to the top of the T's for roofs or to the top of the floor slab for floors. When the concrete has cured, shores are removed and the composite beam carries the dead and live loads.

There are many ways of connecting beams to columns with reinforcing bars so that the joint is completely rigid. One such detail is shown in Fig. 15-24.

Post-tensioned joints are sometimes advantageous. Figure 15-25 shows a detail which provided a completely rigid joint and still permitted the use of a precast column which was several stories high.

Expansion joints are required in prestressed concrete buildings as they

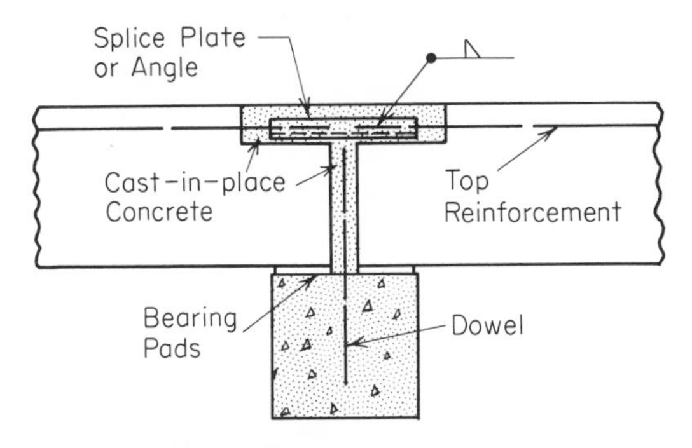

BG-4 Welded, Continuous Spans

This connection relies on welding to connect the tension steel. As a result the pocket length is less than that required for lap splices, and the connection is immediate. However, it is more costly than lap splices.

1. Splices made with angles are preferred when space permits, since they offer a concentric load transfer from bar to bar.

2. It is advisable to inspect welds closely to see that the ultimate strength of the negative reinforcement is obtained.

Fig. 15-20. Negative-moment connection between two precast beams using reinforcing bars spliced by welding. (*Reprinted by permission of the Prestressed Concrete Institute.*)

are in buildings of other materials. Fig. 15-26 shows a split ledger beam in the Marine Plaza Parking Garage. This multistory structure was constructed of precast reinforced concrete columns, 60-ft-span ledger beams 29 ft 7½ in. center to center supporting double T's with a cast-in-place topping. Three bays of double T's, totaling about 90 ft, were made continuous with an expansion joint at the end of each three-bay group. Split columns and split ledger beams form the expansion joint.

When single T's, double T's, or similar members are used for a floor or roof deck it is standard practice to tie the flanges of adjacent members together at reasonable intervals. This is accomplished by casting reinforcing bars in the edges of the flanges at desired intervals and welding a connecting bar between the cast-in bars as shown in Fig. 15-27.

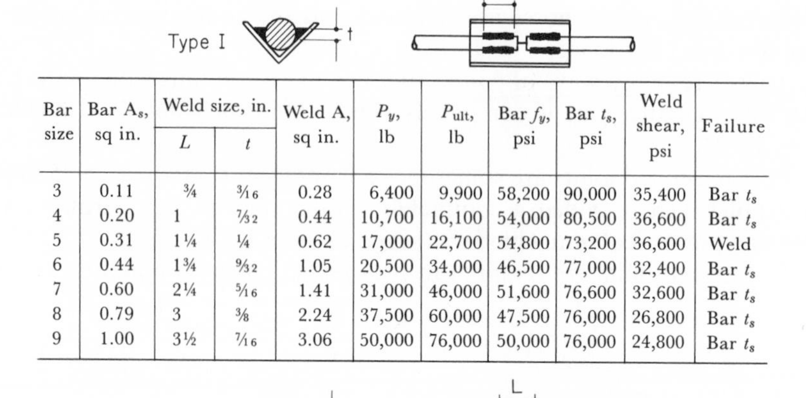

Bar size	Bar A_s, sq in.	Weld size, in.		Weld A, sq in.	P_y, lb	P_{ult}, lb	Bar f_y, psi	Bar t_s, psi	Weld shear, psi	Failure
		L	t							
3	0.11	¾	3⁄16	0.28	6,400	9,900	58,200	90,000	35,400	Bar t_s
4	0.20	1	7⁄32	0.44	10,700	16,100	54,000	80,500	36,600	Bar t_s
5	0.31	1¼	¼	0.62	17,000	22,700	54,800	73,200	36,600	Weld
6	0.44	1¾	9⁄32	1.05	20,500	34,000	46,500	77,000	32,400	Bar t_s
7	0.60	2¼	5⁄16	1.41	31,000	46,000	51,600	76,600	32,600	Bar t_s
8	0.79	3	⅜	2.24	37,500	60,000	47,500	76,000	26,800	Bar t_s
9	1.00	3½	7⁄16	3.06	50,000	76,000	50,000	76,000	24,800	Bar t_s

L

Type II

t

t

Bar size	Bar A_s, sq in.	Weld size, in.		Weld A, sq in.	P_y, lb	P_{ult}, lb	Bar f_y, psi	Bar t_s, psi	Weld shear, psi	Failure
		L	t							
3	0.11	¾	3⁄16	0.28	6,030	6,950	54,800	63,200	24,800	rotation of joint due to eccentricity
4	0.20	⅞	7⁄32	0.38	10,100	11,800	50,500	59,000	31,100	
5	0.31	1¼	¼	0.62	ND	19,500	ND	63,000	31,400	
6	0.44	1⅝	9⁄32	0.92	ND	23,500	ND	53,500	25,600	
7	0.60	2	5⁄16	1.25	ND	28,500	ND	47,500	22,800	
8	0.79	2¾	11⁄32	1.89	ND	42,800	ND	54,200	22,600	
9	1.00	3⅜	⅜	2.54	ND	58,100	ND	58,100	22,800	

ND = not determined.

L

Type III

t

Bar size	Bar A_s, sq in.	Weld size, in.		Weld A, sq in.	P_y, lb	P_{ult}, lb	Bar f_y, psi	Bar t_s, psi	Weld shear, psi	Failure
		L	t							
3	0.11	1⅜	3⁄16	0.26	6,200	9,500	56,500	85,500	37,500	rotation of joint due to eccentricity
4	0.20	1⅜	7⁄32	0.30	10,200	12,300	51,000	61,500	41,000	
5	0.31	2½	¼	0.62	16,300	19,200	53,500	62,000	31,000	
6	0.44	3⅛	9⁄32	0.88	20,800	29,000	47,200	66,000	23,600	
7	0.60	4	5⁄16	1.25	32,000	37,000	53,400	61,600	29,600	
8	0.79	5⅜	11⁄32	1.85	ND	53,700	ND	68,000	29,000	
9	1.00	6⅜	⅜	2.39	ND	71,800	ND	71,800	30,000	

ND = not determined.

FIG. 15-21. Tensile tests on three types of welded connections using intermediate-grade reinforcing bars. These data are part of a report by Arthur R. Anderson presented before the Structural Division Session on Composite Design in Building Construction at an ASCE convention in Washington, D. C., and later reprinted in the September, 1960, issue of *Progressive Architecture,* pages 172 to 179.

All bars are intermediate-grade billet steel, ASTM A15, rolled to ASTM A305.

All welding electric arc with low-hydrogen electrodes are AWS class E7015 or E7016. On multiple passes, surface is thoroughly cleaned prior to welding the next pass.

On type I joints, the cross-section area of structural angle bar should be at least 1.5 times the reinforcing-bar cross section.

The Pascack Hills High School in Montvale, New Jersey, (Figs. 15-28 to 15-30) utilizes many of the advantages of precast prestressed concrete construction. The exterior load-bearing wall panels are precast prestressed single T's with flanges blocked out for windows at first- and second-floor levels. Single-T floor members are supported on structural steel T sections cast into the wall panels as shown in Fig. 15-30*a*. The top of each wall panel was cast with an opening into which the single-T roof member was set so that the outside end of the roof member was flush with the outside face of the stem of the wall panel. Maximum roof T span is 88 ft and maximum floor T span is 58 ft.

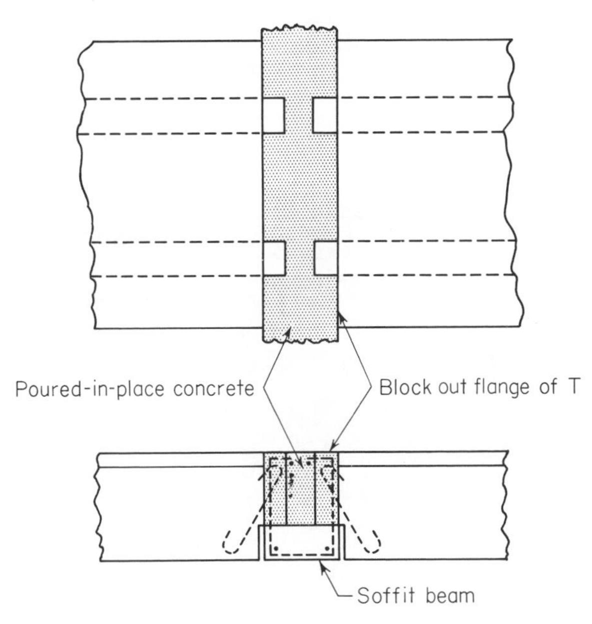

FIG. 15-22. Composite roof beam composed of precast prestressed soffit beam, double T's, and poured-in-place concrete.

Single T's and double T's are most economical when the underside is painted or sprayed with a coat of acoustical material to serve as a ceiling for the room below. Where necessary, however, ceilings can easily be hung from T's as shown in Fig. 15-31.

Precast pretensioned members provide excellent maintenance-free fire-resistant stadiums at relatively low cost. Fig. 15-32 shows one of many possible details which was adopted for Municipal Stadium, Bridgeport, Connecticut.

15-8. Large Special Projects. Precast pretensioned prestressed concrete members are well adapted to single projects which can be designed so that members of identical cross section are used dozens or hundreds of times. Numerous repetitions of use also make it economical to fabricate special sections such as the insulated wall panel shown in Fig. 15-34. Construction of a casting yard at or close to the project makes it feasible to deliver precast members on schedule and lift them directly from the truck to their final place in the structure thus obviating the need for storage space and rehandling at the job site. The best way to demonstrate what can be achieved is to illustrate two specific cases.

One of the first such projects in the United States was the Illinois Toll

FIG. 15-23. Soffit beam supporting double-T floor members. Note stirrups projecting from soffit beam and reinforcing bars for negative moment passing through precast column. Cast-in-place concrete will cover beam and T's to provide finished floor surface.

Road built during 1957. Of a total of 289 bridges on the entire project, the superstructures of 224 are carried wholly or in part by precast prestressed concrete I-beam stringers. At the time this project was designed there was not a great deal of experience with precast pretensioned I-beam and slab bridges, but the design engineers, Joseph K. Knoerle and Associates, Inc., introduced two new concepts which appreciably reduced the cost of the precast members and developed into standard practice. These were the elimination of end blocks for I beams and the use of deflected pretensioned strands. Engineers estimates indicated that the cost of these bridges, with supporting structure, would be about 21 per cent less than that of competitive materials; this was confirmed by actual bid prices.

Building projects also find definite advantages in precast prestressed concrete members. The Great Atlantic & Pacific Tea Company's food processing plant at Horseheads, New York, which covers 35 acres under one roof is a good example. Before selecting the materials to be used for the structural roof system the designer, the Rust Engineering Company, made a study of ten structural systems in which different materials were

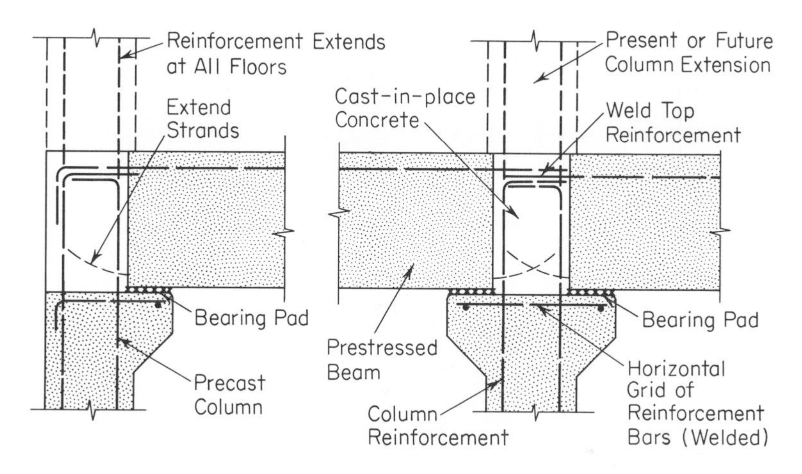

BC-3 Welded, Continuous Spans

This detail provides monolithic behavior between the beams and columns and easily accommodates present or future extensions of the column to floors above.

1. The welding of top reinforcement may be by lap splices or angle splices.
2. The prestressing strands should extend far enough into the cast-in-place concrete to resist moment reversals and axial shortening of the beams.
3. The column must be designed for its portion of the continuity moments.
4. This detail can be used for precast or prestressed beams. Bottom mild steel in the beam should extend into the connection for precast beams.
5. A small chamfer at the outer edge of the concrete haunches will reduce spalling. Also, ¼-in. bearing plates will help prevent cracking of outer corners.

FIG. 15-24. Rigid connection of beams to column using reinforcing bars. See Sec. 301.3 of ACI Title No. 61–51 (Appendix B) for corbel design. (*Reprinted by permission of the Prestressed Concrete Institute.*)

used in the arrangements most advantageous to each but conforming to one typical module (30 by 50 ft). Each system was rated for its qualities of initial cost, durability, maintenance cost, fire resistance, alterability, and suitability in connection with the food processing, packaging, and storing functions of the structure. After a comparison of the systems, prestressed concrete double T's were selected for the roof. Similar studies resulted in the choice of prestressed double T's for floors and exterior wall panels.

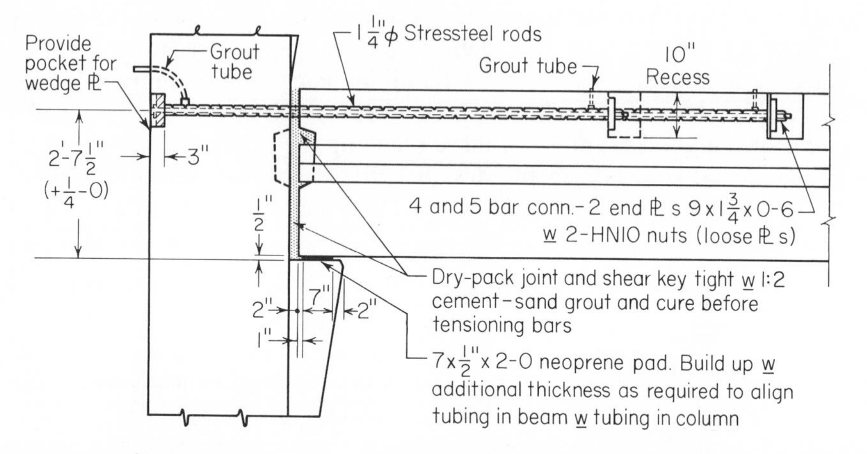

FIG. 15-25. Marine Plaza Parking Garage, Milwaukee, Wisconsin. Typical post-tensioned rigid-frame connection of ledger beam to precast column. (*Courtesy Ross H. Bryan, Consulting Engineer.*)

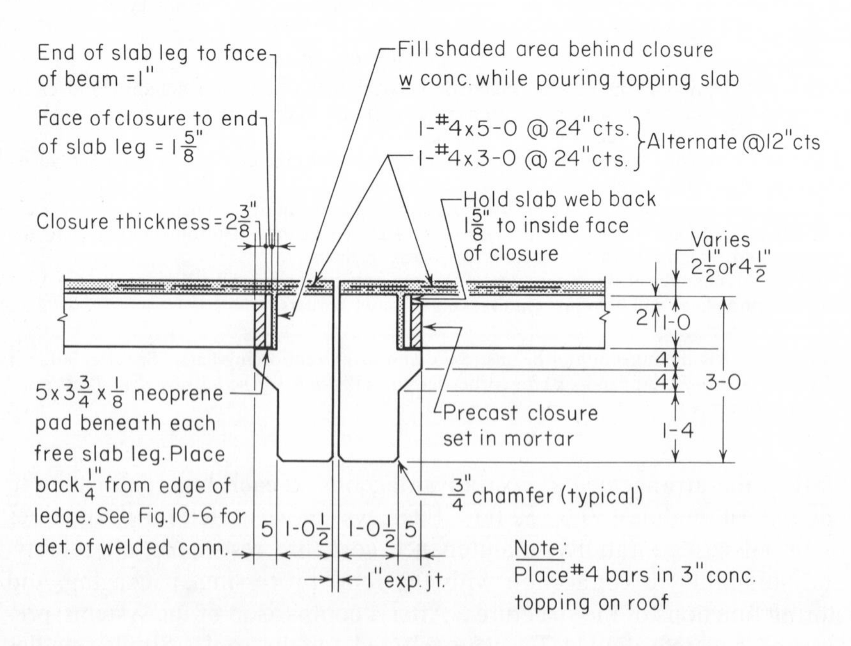

FIG. 15-26. Marine Plaza Parking Garage. Cross section through split ledger beams at expansion joint. (*Courtesy Ross H. Bryan, Consulting Engineer.*)

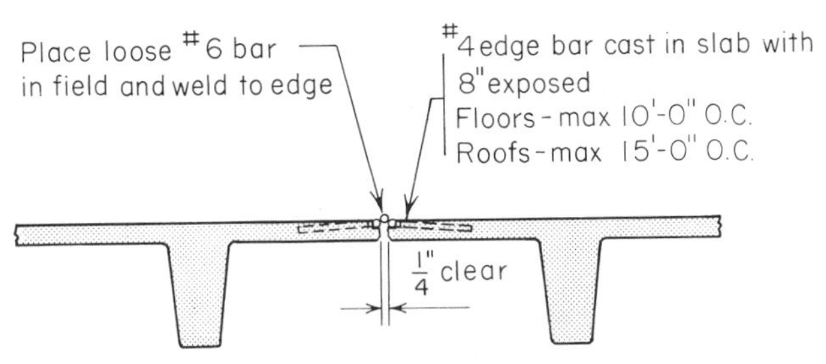

FIG. 15-27. Detail used by Nebraska Prestressed Concrete Co. for connecting flanges of adjacent double T's.

Vital statistics of the A & P plant, illustrated in Figs. 15-33 to 15-36, are:

Poured-in-place reinforced concrete columns.

Precast post-tensioned girders 30 ft long resting on neoprene pads at columns except at expansion joints where they rest on a combination of neoprene and Teflon to permit more motion.

Precast pretensioned double-T floor members with 3-in. poured-in-place topping. Floor members span 25 ft and are designed for 300 lb per sq ft uniform live load or 1,000 lb concentrated load.

Precast pretensioned double-T roof members on 50-ft spans.

Precast pretensioned double-T wall panels insulated with 2-in. thick sheets of Styrofoam.

FIG. 15-28. Pascack Hills High School. One of the completed two-story buildings composed of precast single-T wall panels and precast single-T floors and roofs. (*Courtesy of C. W. Blakeslee and Sons, Inc., of New Haven, Connecticut, fabricators of the prestressed concrete members.*)

FIG. 15-29*a*. Pascack Hills High School. Erecting precast pretensioned single-T wall panels on poured foundation wall. (*Courtesy of C. W. Blakeslee and Sons, Inc.*)

FIG. 15-29*b*. Pascack Hills High School. Single-T floor members span from exterior wall panels to ledger beam. Single-T roof members have clear span from wall to wall. Note cutout in flange of roof T near center of span. (*Courtesy of C. W. Blakeslee and Sons, Inc.*)

For fabricating the prestressed members on the A & P job, Dickerson Structural Concrete Corporation set up a temporary plant about one-half mile from the job site. This plant, which began producing precast prestressed members ten weeks after award of the contract, included six pretensioning beds with a total length of 1,750 ft plus a 630-ft line for the

FIG. 15-30*a*. Pascack Hills High School. Single-T floor member framed to wall panel. Adjacent wall panel with projecting inverted structural steel T waiting to receive floor T. The floor T's have two structural steel plates projecting from their ends and the plates straddle the stem of the structural steel T. Note steel plate tie between flanges of precast wall panels. (*Courtesy of C. W. Blakeslee and Sons, Inc.*)

post-tensioned beams. Casting all of the members for the project took 26 weeks and used 25,000 cu yd of concrete and 325 miles of prestressing strand. Thirty-eight weeks after award of the contract the precast members were all in place in the structure, the plant was dismantled, and the plant site was cleared. One interesting statistic out of many: on a record day one erection crew set more than an acre of concrete roof.

FIG. 15-30*b*. Pascack Hills High School. Single-T roof member framing into precast opening at top of precast single-T wall panel. See Fig. 15-28 for external appearance of this joint in completed structure. (*Courtesy of C. W. Blakeslee and Sons, Inc.*)

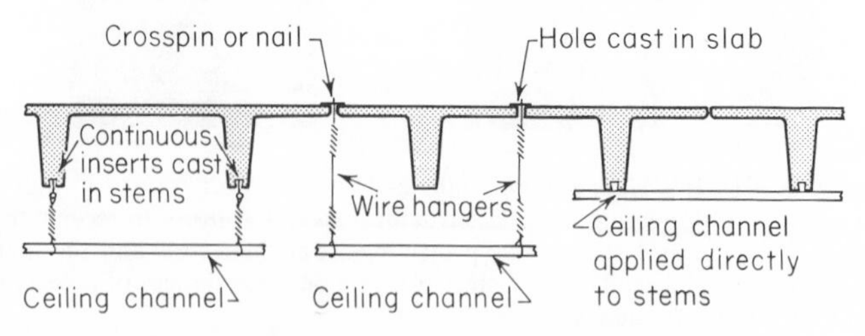

FIG. 15-31. Some of the many methods of supporting a hung ceiling from precast prestressed members. (*Courtesy Nebraska Prestressed Concrete Co.*)

15-9. Post-tensioning in Buildings. There are many applications of post-tensioning in building construction. Practically all of them involve tensioning of the tendons with the prestressed member in its final position in the structure. In many cases post-tensioning is chosen when the shipment of pretensioned members is not feasible because of weight or size or when it is desirable to tie together several members of the structure with the

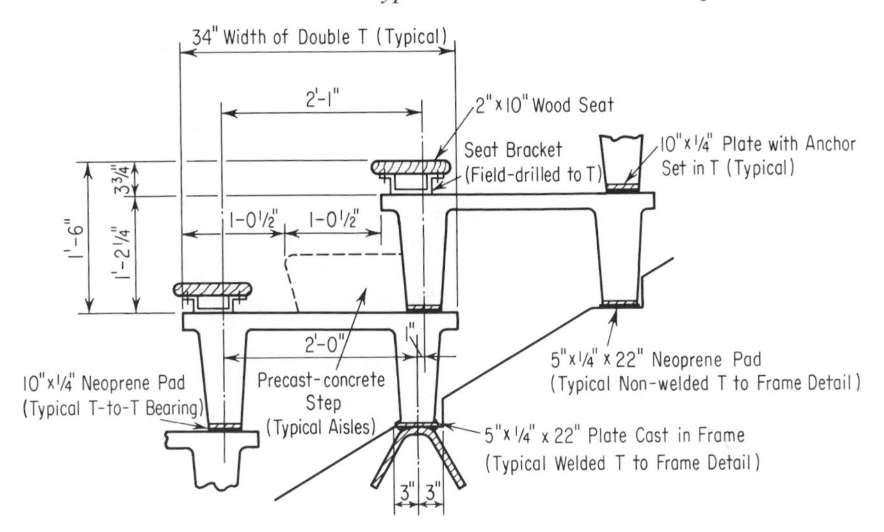

FIG. 15-32. Precast pretensioned stadium members made in double-T forms with most of the overhanging flanges blocked off. (*Courtesy of C. W. Blakeslee & Sons, Inc., New Haven, Connecticut.*)

FIG. 15-33*a*. Casting wall panels for A & P project. Stems and bottom 2 in. of slab have been cast. Sheets of Styrofoam insulation are being placed. Vertical bars will be bent over above insulation, hinged form sides will be turned up to add 4 in. to total depth, and 2 in. of top slab will be poured. (Project designer and general contractor—the Rust Engineering Company. Prestressed concrete by Dickerson Structural Concrete Corporation.)

FIG. 15-33*b*. Wall panels after removal from forms. Note large cutout between stems. Holes in stems where strands are recessed will be filled and finished.

tendons. In addition to the structure covered in this section, post-tensioned applications are included in Arts. 15-5, 15-8, 15-10, and 15-11.

Figures 15-37 to 15-41 illustrate a 143-ft-diameter post-tensioned roof over a college gymnasium.

The structural elements in the roof are precast, pie-shaped, single-T sections which are pretensioned to withstand stresses resulting from shipping and erection. Each T also has two cored holes each of which will accommodate a Freyssinet cable composed of twelve ½-in.-diameter seven-wire strands.

The entire roof is supported on a ring of columns which form a 124-ft-diameter circle. At the job site each T was erected as a radius of the circle

FIG. 15-33*c*. Wall panels in place are sturdy and provide insulation and pleasing architectural effect.

with its outer end resting on a column (and cantilevering about 9 ft 6 in.) and its inner end resting on false work. The inner ends of the T's left an open circle about 13 ft in diameter at the center of the structure. In this

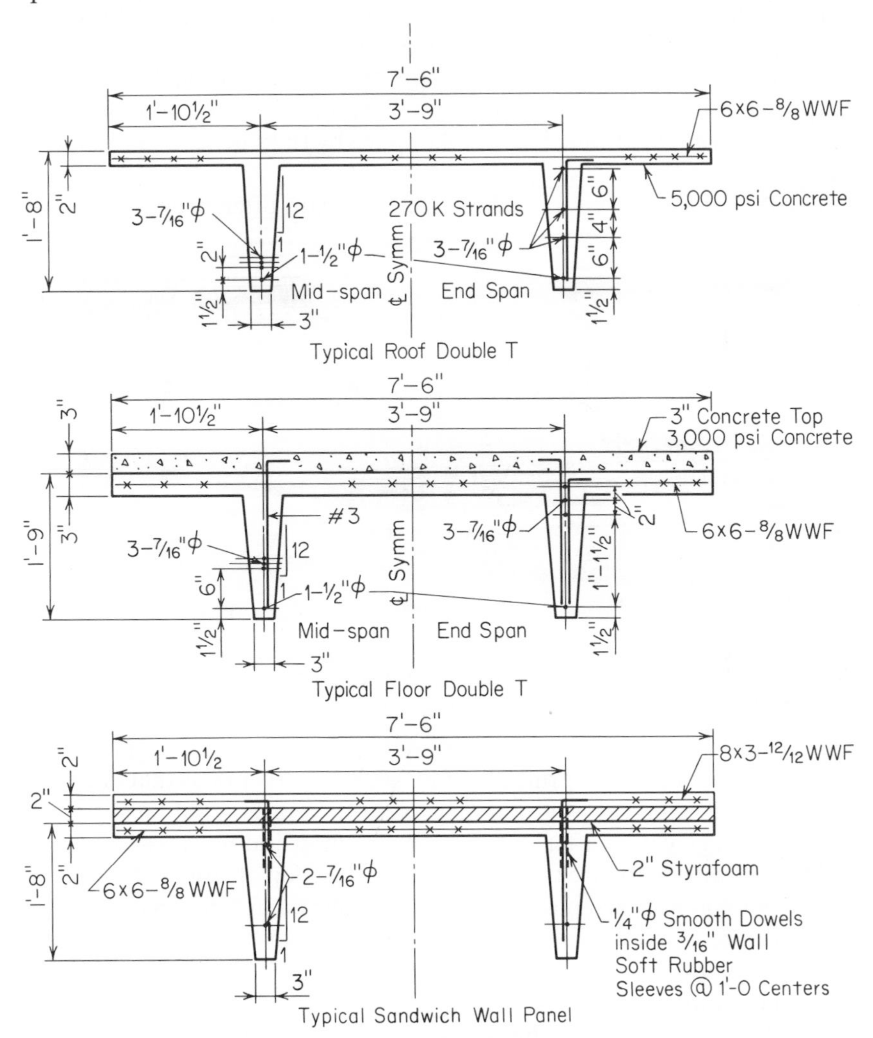

FIG. 15-34. Cross section of all double T's was the same except for thickness of top slab. Number of strands varied with span and loading. (A & P project.)

circle a steel tension ring was placed flush bottom with the T's, the cables were threaded through it, and post-tensioned and anchored against it. A concrete compression ring was poured flush with the top of the T's. A concrete slab was poured on top of and between the T's to complete the roof.

In effect the two T sections on one diameter of the circle span 124 ft from column to column. The tensile force in the bottom of the T at its inner end is delivered to the tension ring by the anchors of the post-tensioned cables. This force passes around the ring and into the bottom of the opposite T. Similarly, the compressive forces in the top flanges pass around the concrete compression ring. In summary, each 124-ft-span T is spliced at mid-span in tension at the bottom and in compression at the top. Forces are distributed across the splice by rings instead of by straight links.

15-10. Lift Slabs. Prestressed concrete lift slabs often prove economical

FIG. 15-35. A double-T floor section was taken from stock pile, covered with 3-in. poured-in-place slab and loaded to failure, demonstrating that actual ultimate capacity exceeded design requirements for ultimate strength. (A & P project.)

for buildings. The basic principle of lift-slab construction is that of casting all floor slabs and roof at ground level and then raising the finished slabs to their final elevation after they have been cured and prestressed.[2-4]

The following steps are typical in the construction of a lift-slab building:

1. Steel columns are placed on their footings and anchored. Usually there are two rows of columns. The slabs will span between the two rows and cantilever beyond each row as shown in Fig. 15-42.

2. The bottom slab of the building is cast on grade in its final position.

3. When the bottom slab has set, it is covered with paper or other bond-preventing material.

4. A steel collar for the first-floor slab is placed around each column. The function of these collars is to transfer the load from the floor slab to the column when the slab has been lifted to its final position.

5. Prestressing tendons, electrical conduits, etc., are placed, and the first-floor is poured.

6. When the first-floor slab has set sufficiently, steps 3, 4, and 5 are

FIG. 15-36. This 32-acre prestressed concrete plant was installed in six weeks by Dickerson Structural Concrete Corporation to manufacture members for A & P project. It includes office, batching plant, carpenter shop, and maintenance, inspection, and testing facilities. Most of the equipment was selected so it could be salvaged for use elsewhere after completion of this project.

FIG. 15-37. Student Activities-Physical Education Building at State University of New York Agricultural and Technical College, Farmingdale, New York. High circular roof is 143 ft in diameter supported on 124-ft-diameter circle of columns. See text and Figs. 15-38 to 15-41. (*Courtesy Braenstress, Inc., Wyckoff, New Jersey. Architect: Max O. Urbahn; Engineers: Summers & Molke.*)

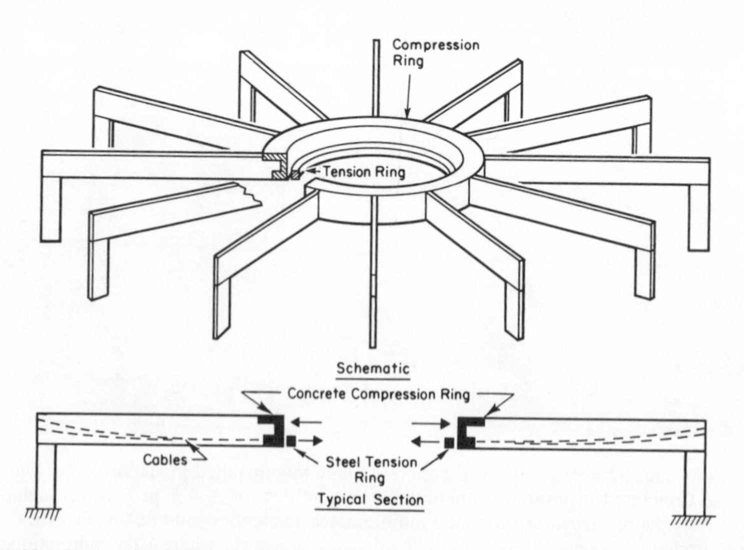

FIG. 15-38. Basic scheme for circular precast post-tensioned roof. See text and Figs. 15-37, 15-39, 15-40, and 15-41. (*Courtesy Braenstress, Inc.*)

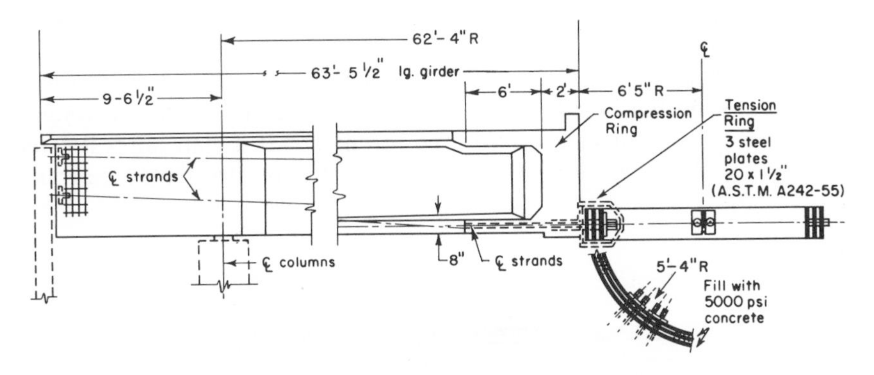

FIG. 15-39. Details of precast post-tensioned roof. See text and Figs. 15-37, 15-38, 15-40, and 15-41. (*Courtesy Braenstress, Inc.*)

FIG. 15-40. Cables threaded through steel tension ring but not yet tensioned. See text and Figs. 15-37, 15-38, 15-39, and 15-41 for other details on circular roof. (*Courtesy Braenstress, Inc.*)

carried out for the second-floor slab. This procedure is repeated for each slab until all floors have been cast.

7. Each slab is prestressed before it is lifted. The exact time at which the tendons are tensioned can be chosen to suit the contractor as long as the concrete has reached its required strength. If a slab is prestressed while other slabs are resting on it, some of the prestress may be trans-

FIG. 15-41. Erecting precast T for circular roof. Post-tensioned cables are in place but not tensioned. See text and Figs. 15-37 to 15-40. (*Courtesy Braenstress, Inc.*)

ferred by friction to other slabs. However, as each slab is lifted, this friction is lost and all the tension in its tendons is carried by the slab itself, giving it the proper prestress.

8. The top slab is lifted to its final elevation, and each collar is permanently welded or bolted to its column.

9. The next slab is lifted to its final elevation and permanently attached to the columns, etc., until all slabs are in place.

The lifting operation is the most precise part of the entire construction procedure. Jacks are mounted on top of each column and attached to the slab by long threaded rods. The jacks are operated simultaneously from a central control panel so that the slab is kept level as it is raised. Maintenance of the necessary level at each column requires a complicated control system and an experienced operator. The jacking equipment, controls, and operators are usually rented to the general contractor by a firm specializing in lifting slabs.

In the taller buildings the columns may be too slender to permit raising

FIG. 15-42. Prestressed concrete lift slabs in Litchfield County Hospital, Winsted, Connecticut. Each of the five large slabs is 182 by 43 ft by 8½ in. The slabs were post-tensioned with Freyssinet cables and the lifting operation was carried out by New England Lift Slab Corp., a licensee of the Youtz-Slick Lift Slab Method. The Freyssinet cables run the full length of the slab in both the 182- and 43-ft directions. They are near the bottom of the slabs in the center of the bays and near the top of the slabs in the column lines.

the top slab to its final position without lateral bracing. For a structure of this type with six slabs to be lifted, one erection procedure would be as follows:

1. Erect columns that extend slightly above the third-floor level and mount jacks on them.
2. Raise the top slab to the top of the columns and bolt it temporarily in place.
3. Guy the top slab to the ground.
4. Raise each of the next two slabs and bolt them temporarily in place.

5. Raise the third-floor slab and fasten it permanently to the columns.
6. Raise the second- and first-floor slabs and fasten them permanently to the columns.
7. Provide the necessary bracing between columns up to the third floor or guy the third floor to the ground.
8. Release guys from the top slab.

FIG. 15-43. North Carolina Mutual Life Insurance Co. building, Durham, North Carolina. Precast segmental trusses by Concrete Materials, Inc., Charlotte, North Carolina. Architects: Welton Becket and Associates. Associate architects: M. A. Ham, Associates, Inc. Consulting engineers: Seelye, Stevenson, Value and Knecht. See text and Figs. 15-44 and 15-45 for details.

9. Remove the jacks, splice on the remainder of the columns, and place the jacks on top.
10. Raise the top slab to the final position, etc.

The economy of lift-slab construction comes from the virtual elimination of forms plus the labor saved by casting all the concrete at ground level. Prestressing the slabs permits wider spacing between columns and keeps the weight of slab to be lifted to a minimum.[5]

The slabs in the structure illustrated in Fig. 15-42 are solid slabs 8½ in.

thick. Another common type of lift slab is the waffle design. These are thinner slabs with ribs running in both directions on the underside of the slab. Each rib has one or more tendons.

15-11. Segmental Members. A segmental prestressed concrete beam is composed of several precast concrete blocks assembled in a line and placed under compression by some type of prestressing tendon to form a beam. A beam of this type is similar to other prestressed concrete beams

FIG. 15-44. North Carolina Mutual Life Insurance Co. building. Erection photograph showing structural steel monorail for handling precast sections. Note completed center core. See text and Figs. 15-43 and 15-45.

except that, as soon as the compressive stress is offset by bending moments, a crack will appear between the precast blocks because there is nothing to carry tensile stress.

Some of the first prestressed concrete members put in service in the United States were of the segmental type. The precast blocks were made on an ordinary building-block machine but were of high-strength concrete. These beams were made by placing the blocks face to face with a mortar joint about ¼ in. thick. Prestressing strands were passed through the holes in the blocks for the full length of the beam and anchored against steel bearing plates in the special end blocks.

The North Carolina Mutual Life Insurance Company building in Durham, North Carolina illustrates a recent more sophisticated application of the use of post-tensioned segmental members (Figs. 15-43 to 15-45). All four faces of the 14-story structure are identical. Each face is composed of seven 20-ft-deep prestressed Vierendeel trusses 108 ft long supported by two columns 40 ft 6 in. on centers and cantilevering 33 ft 9 in. beyond the columns.

FIG. 15-45. North Carolina Mutual Life Insurance Co. building. Erecting a lower chord member. Note lip around end for confining grout. External faces of block are exposed aggregate. See text and Figs. 15-43 and 15-44.

Each truss is made up of precast top and bottom chord T section blocks plus rectangular vertical members all of which are post-tensioned together with Stressteel bars to keep concrete and joints in compression under all loading conditions.

The central core of the building is a 45 by 34 ft rectangle with 12-in.-thick walls which begins below the basement floor and rises 235 ft to a point above the roof. This core was cast in place in one continuous pour of 24 hr per day for 18 days. At every floor level precast prestressed concrete girders span from the core to the trusses to carry the precast prestressed concrete double-T floor members which are covered with a 4½-in. cast-in-place slab. Since each truss is two floors high the prestressed girders at one floor level span in a direction at right angles to those on the next floor so that each truss supports the weight of only one floor. Windows at one floor level are set in openings in the truss and at the next floor level utilize the full width of the building between the top chord of one truss and the bottom chord of the next truss.

Truss sections were moved into place by a structural steel monorail system (Fig. 15-44 and 15-45) and set on tubular shores. The ends of each precast block were cast with a lip around the edge and around each tendon hole (Fig. 15-45) to confine the grout which is to be placed in the joints before post-tensioning. An expanded polyurethane gasket impregnated with a butyl rubber sealer was also attached to the lip to complete the seal. After the grout in the joints had cured, Stressteel bars were post-tensioned and grouted and closure units were set in the corners at the ends of the trusses.

The precast post-tensioned exposed aggregate Vierendeel trusses and columns combined the function of carrying floor loads and providing a desirable architectural effect. Use of precast prestressed concrete girders and double T's between core columns and trusses provides entirely column-free office space.

15-12. Slabs on Grade. Prestressed concrete is an excellent material for slabs on grade which are subjected to heavy load concentrations as in airport runways. Joints, one of the chief sources of maintenance problems in slabs, can be spaced 500 ft or more apart and specially designed.[6] Slabs less than half the thickness required in ordinary concrete are feasible.

The *Journal of the American Concrete Institute*[7] contains a comprehensive report of tests on a 7-in.-thick prestressed slab. The data in this report and the *Highway Research Board Bulletin*[8] can be used as a basis for computing the required prestress in a slab, friction between slab and subbase, etc. In common with the first prestressed concrete structures of other types, this slab was post-tensioned. Later studies indicate that pretensioned bonded strands have several advantages over post-tensioned, at least in the longitudinal direction.

One of the big problems in prestressed slabs on grade is friction. Friction between the slab and subgrade is kept to a minimum by placing a 1-in. layer of round sand on the subbase. The sand is covered with a layer of waterproof building paper, and the slab is poured on the paper. When the slab is prestressed, it shortens and therefore moves with respect to the base. The only friction resisting this motion is the shear developed within the sand layer, which is equivalent to a coefficient of about 0.50.

Where the post-tensioning method is used, friction between the tendon and its enclosure becomes quite large in the long longitudinal tendons. Use of pretensioned strands completely eliminates this friction, since the strands are entirely in the open at the time of tensioning. Pretensioned strands can be tensioned in any length that proves economical on a given project. Where the strands stretch across an expansion joint, the slab can be formed at each side of the joint, and after the concrete has cured to the required strength, the strands can be burned off at the faces of the slab.

Worn or otherwise inadequate runways can be rehabilitated by covering them with a 1-in. layer of sand and then a relatively thin, 3- or 4-in., prestressed concrete overlay. The same principle can be applied to a new slab by casting an unreinforced base and covering it with a thin prestressed slab.[9] Pretensioned strands in the longitudinal direction can be used economically with either of these types because the bottom slab will serve to hold the tension in the strands while the top slab is being poured and cured. When the load of the strands is released into the top slab, it shortens and the bottom slab lengthens.

Use of pretensioned tendons in the transverse direction creates a problem in pouring the slab. The tendons must be placed for the full width before any concrete is poured. Since paving-equipment units are not wide enough to pour the entire width at one pass, the slab must be poured in lanes, which means that one of the rails supporting the paving unit will cross all the transverse tendons. This problem plus the fact that transverse tendons are relatively short usually makes post-tensioned tendons more efficient for transverse prestressing.

Suggested procedure for fabrication of a slab with pretensioned longitudinal strands and post-tensioned transverse tendons is as follows:

1. Prepare the subbase, cover with a sand layer and waterproof paper.

2. Place and tension longitudinal strands for the width of one lane. These can be continuous for the full length of the runway if desired and anchored to abutments at each end.

3. Place transverse tubes for transverse tendons. These tubes can be thin-wall steel or flexible metal hose. In each tube use a pipe just slightly smaller than the tube to keep it straight. The pipe should be a little longer than the lane width.

4. Place any needed reinforcing bars, expansion-joint connections, etc., and pour the lane.

5. Place and tension the longitudinal strands for the second lane.

6. Attach a small-diameter wire rope to the outside end of each of the pipes in the transverse holes. Pull the pipes into the second lane, leaving the end of the pipe in the first lane, and cover the pipe in the second lane with metal hose.

7. Repeat Step 4, etc., until all lanes are poured and have cured to the required strength.

8. For each transverse tendon there is now a cast-in-metal tube with a small-diameter wire rope pulled through it. Attach a post-tensioning tendon to each rope, and pull the tendon into place in the tube.

9. Tension, anchor, and grout all the transverse tendons. (If the longitudinal prestress were applied first, each lane would not necessarily shorten the same amount as the adjacent lane and the transverse tendons would be pinched at the joint between the lanes, thus preventing proper prestressing.)

10. Release the tension in the longitudinal strands from their anchors into the slab. To prevent large motions at the ends of the slab, cut a few strands at each anchor and in each expansion joint, then a few more at each point, etc., until all are released.

Although prestressed concrete runway slabs offer many advantages including economy, especially for heavy planes, they have been slow to gain acceptance. Recent technical periodicals should be checked for articles on the newest developments.[10]

BIBLIOGRAPHY

1. Rostasy, Ferdinand S.: Connections in Precast Concrete Structures—Continuity in Double-T Floor Construction, *J. Prestressed Concrete Inst.,* August, 1962, pp. 18–48.
2. Minges, James S., and Donald S. Wild: Six Stories of Prestressed Slabs Erected by Lift-slab Method, *J. Am. Concrete Inst.,* February, 1957, pp. 751–768.
3. Perry, John P. H.: Office Buildings of 370,000 Square Feet Erected by Lift-slab Method, *Civil Eng.,* June, 1955, pp. 43–47.
4. Brownfield, Allen H.: Growing Pains in Prestressed Concrete Buildings, *Civil Eng.,* February, 1958, pp. 46–49.
5. Rice, Edward K.: Economic Factors in Prestressed Lift-slab Construction, *J. Am. Concrete Inst.,* September, 1958, pp. 347–357.
6. Prestressing Promises Nearly Joint-free Design, *Civil Eng.,* August, 1958, pp. 34–37.
7. Cholnoky, Thomas: Prestressed Concrete Pavement for Airfield, *J. Am. Concrete Inst.,* July, 1956, pp. 59–84.
8. Prestressed Concrete Pavement Research, *Highway Research Board Bull. 179,* Washington, D.C.
9. Jointless Prestressed Floor Resists Heavy Loads in Warehouse, *Eng. News-Record,* Jan. 6, 1949.
10. Prestressed Pavement—A World View of Its Status, Report by Subcommittee VI, ACI Committee 325, *J. Am. Concrete Inst.,* February, 1959, pp. 829–838.

APPENDIX A

ACI Standard
Building Code Requirements for Reinforced Concrete (ACI 318-63)

This code was adopted as a standard by the American Concrete Institute in 1963 and portions of it are reproduced here by permission of ACI. Prestressed concrete is an integral part of this code and Chap. 26 on prestressed concrete is reproduced completely. A few other portions frequently referred to are also reproduced. Copies of the complete code are available from American Concrete Institute at a nominal charge.

The following items are excerpts from ACI 318-63.

104—Approval of special systems of design or construction

(a) The sponsors of any system of design or construction within the scope of this code and which has been in successful use, or the adequacy of which has been shown by analysis or test, and the design of which is either not consistent with, or not covered by this code shall have the right to present the data on which their design is based to a "Board of Examiners for Special Construction"* appointed by the Building Official. This Board shall be composed of competent engineers and shall have the authority to investigate the data so submitted, to require tests, and to formulate rules governing the design and construction of such systems to meet the intent of this code. These rules when approved by the Building Official and promulgated shall be of the same force and effect as the provisions of this code.

301—General definitions

(a) The following terms are defined for general use in this code; specialized definitions appear in individual chapters:

Admixture—A material other than portland cement, aggregate, or water added to concrete to modify its properties.

Aggregate—Inert material which is mixed with portland cement and water to produce concrete.

Aggregate, lightweight—Aggregate having a dry, loose weight of 70 lb per cu ft or less.

* The exact name of this board and its selection should be adapted to the local legal conditions.

Building official—See Section 102(c).

Column—An upright compression member the length of which exceeds three times its least lateral dimension.

Combination column—A column in which a structural steel member, designed to carry the principal part of the load, is encased in concrete of such quality and in such manner that the remaining load may be allowed thereon.

Composite column—A column in which a steel or cast-iron structural member is completely encased in concrete containing spiral and longitudinal reinforcement.

Composite concrete flexural construction—A precast concrete member and cast-in-place reinforced concrete so interconnected that the component elements act together as a flexural unit.

Compressive strength of concrete (f'_c)*—Specified compressive strength of concrete in pounds per square inch (psi). Compressive strength shall be determined by tests of standard 6 x 12-in. cylinders made and tested in accordance with ASTM specifications at 28 days or such earlier age as concrete is to receive its full service load or maximum stress.

Concrete—A mixture of portland cement, fine aggregate, coarse aggregate, and water.

Concrete, structural lightweight—A concrete containing lightweight aggregate conforming to Section 403.

Deformed bar—A reinforcing bar conforming to "Specifications for Minimum Requirements for the Deformations of Deformed Steel Bars for Concrete Reinforcement" (ASTM A 305) or "Specifications for Special Large Size Deformed Billet-Steel Bars for Concrete Reinforcement" (ASTM A 408). Welded wire fabric with welded intersections not farther apart than 12 in. in the direction of the principal reinforcement and with cross wires not more than six gage numbers smaller in size than the principal reinforcement may be considered equivalent to a deformed bar when used in slabs.

Effective area of concrete—The area of a section which lies between the centroid of the tension reinforcement and the compression face of the flexural member.

Effective area of reinforcement—The area obtained by multiplying the right cross-sectional area of the reinforcement by the cosine of the angle between its direction and the direction for which the effectiveness is to be determined.

Pedestal—An upright compression member whose height does not exceed three times its average least lateral dimension.

Plain bar—Reinforcement that does not conform to the definition of deformed bar.

Plain concrete—Concrete that does not conform to the definition for reinforced concrete.

Precast concrete—A plain or reinforced concrete element cast in other than its final position in the structure.

Prestressed concrete—Reinforced concrete in which there have been introduced internal stresses of such magnitude and distribution that the stresses resulting from service loads are counteracted to a desired degree.

* Wherever this quantity appears under a radical sign, the root of only the numerical value is intended; all values are in pounds per square inch (psi).

Reinforced concrete—Concrete containing reinforcement and designed on the assumption that the two materials act together in resisting forces.

Reinforcement—Material that conforms to Section 405, excluding prestressing steel unless specifically included.

Service dead load—The calculated dead weight supported by a member.

Service live load—The live load specified by the general building code of which this code forms a part.

Splitting tensile strength—(see Section 505).

Stress—Intensity of force per unit area.

Surface water—Water carried by an aggregate except that held by absorption within the aggregate particles themselves.

Yield strength or yield point (f_y)—Specified minimum yield strength or yield point of reinforcement in pounds per square inch. Yield strength or yield point shall be determined in tension according to applicable ASTM specifications.

505—Splitting tensile tests of concrete

(a) To determine the splitting ratio, F_{sp}, for a particular aggregate, tests of concrete shall be made as follows:

1. Twenty-four 6 x 12-in. cylinders shall be made in accordance with "Method of Making and Curing Concrete Compression and Flexure Test Specimens in the Laboratory" (ASTM C 192), twelve at a compressive strength level of approximately 3,000 psi and twelve at approximately 4,000 or 5,000 psi. After 7 days moist curing followed by 21 days drying at 73°F and 50 per cent relative humidity, eight of the test cylinders at each of the two strength levels shall be tested for splitting strength and four for compressive strength.
2. The splitting tensile strength shall be determined in accordance with "Method of Test for Splitting Tensile Strength of Molded Concrete Cylinders" (ASTM C 496), and the compressive strength in accordance with "Method of Test for Compressive Strength of Molded Concrete Cylinders" (ASTM C 39).

(b) The ratio, F_{sp}, of splitting tensile strength to the square root of compressive strength shall be obtained by using the average of all 16 splitting tensile tests and all eight compressive tests.

1102—Modulus of elasticity of concrete

(a) The modulus of elasticity, E_c, for concrete may be taken as $w^{1.5}33\sqrt{f'_c}$, in psi, for values of w between 90 and 155 lb per cu ft. For normal weight concrete, w may be considered as 145 lb per cu ft.

1504—Safety provisions*

(a) Strengths shall be computed in accordance with the provisions of Part IV-B.

(b) The coefficient ϕ shall be 0.90 for flexure; 0.85 for diagonal tension, bond,

*The coefficient ϕ provides for the possibility that small adverse variations in material strengths, workmanship, dimensions, control, and degree of supervision, while individually within required tolerances and the limits of good practice, occasionally may combine to result in undercapacity.

and anchorage; 0.75 for spirally reinforced compression members; and 0.70 for tied compression members.

(c) The strength capacities of members so computed shall be at least equal to the total effects of the design loads required by Section 1506.

1506—Design loads*

(a) The design loads shall be computed as follows:

1. For structures in such locations and of such proportions that the effects of wind and earthquake may be neglected the design capacity shall be

$$U = 1.5D + 1.8L \tag{15-1}$$

The loads D, L, W, and E are the loads specified in the general code of which these requirements form a part.

2. For structures in the design of which wind loading must be included, the design capacity shall be

$$U = 1.25(D + L + W) \tag{15-2}$$

or

$$U = 0.9D + 1.1W \tag{15-3}$$

whichever is greater, provided that no member shall have a capacity less than required by Eq. (15-1).

3. For those structures in which earthquake loading must be considered, E shall be substituted for W in Eq. (15-2).
4. In considering the combination of dead, live, and wind loads, the maximum and minimum effects of live loads shall be taken into account.
5. In structures in which it is normal practice to take into account creep, elastic deformation, shrinkage, and temperature, the effects of such items shall be considered on the same basis as the effects of dead load.

CHAPTER 25—COMPOSITE CONCRETE FLEXURAL CONSTRUCTION

2500—Notation

b' = width of area of contact between precast and cast-in-place concretes

d_p = effective depth of the tension reinforcement in precast component

I = moment of inertia of the transformed composite section neglecting area of concrete in tension

M_D = moment due to dead load, produced prior to the time at which the cast-in-place concrete attains 75 per cent of its specified 28-day strength

M_L = moment due to live load and superimposed dead load

Q = statical moment of the transformed area outside of the contact surface about the neutral axis of the composite section

V = total shear

v_h = horizontal shear stress along contact surface

* The provisions of Section 1506 provide for such sources of possible excess load effects as load assumptions, assumptions in structural analysis, simplifications in calculations, and effects of construction sequence and methods.

2501—Definition

(a) Composite concrete flexural construction consists of precast concrete members and cast-in-place reinforced concrete so interconnected that the component elements act together as a unit.

2502—Special design considerations

(a) In regions of negative moment, the bending moment may be assigned to either the composite section or the precast element. When the negative moments are assigned to the composite section, adequate provision for shear transfer must be made throughout the full length of the beam.

2503—Flexural design—Working stress design (Part IV-A)

(a) The design of the composite reinforced concrete member shall be based on allowable stresses, working loads, and the accepted straight-line theory of flexure as given in Part IV-A of this code. The effects of creep, shrinkage, and temperature need not be considered except in unusual cases. The effects of shoring, or lack of shoring, on deflections and stresses shall be considered.

2504—Flexural design—Ultimate strength design (Part IV-B)

(a) *Design method*
In calculating the ultimate strength of a section, no distinction is made between shored and unshored members.

(b) *Limitations*

1. For beams designed on the basis of ultimate strength and built without shores, the effective depth of the composite section used in the calculation of the ultimate moment shall not exceed:

$$(1.15 + 0.24M_L/M_D)d_p$$

2. When the specified yield point of the tension reinforcement exceeds 40,000 psi, beams designed on the basis of ultimate strength should always be built with shores unless provisions are made to prevent excessive tensile cracking.

(c) *Construction loads*
The nonprestressed precast element shall be investigated separately to assure that the loads applied before the cast-in-place concrete has attained 75 per cent of its specified 28-day strength do not cause moment in excess of 60 per cent of the ultimate moment capacity of the precast section.

2505—Shear connection

(a) *Shear calculation*
The horizontal shear stress along the contact surface is given by:

$$v_h = \frac{VQ}{Ib'} \qquad (25\text{-}1)$$

(b) *Shear transfer*
Shear shall be transferred along the contact surface either by bond or by shear

keys. The capacity of bond at ultimate load may be taken as 1.9 times the values recommended below for service loads. Except as provided in 1, separation of the component elements in the direction normal to the surface shall be prevented by steel ties or other suitable mechanical anchorages.

1. When mechanical anchorages are not provided and the contact surface is rough and clean .... 40 psi
2. When minimum steel tie requirements of (c) are followed and the contact surface is smooth (troweled, floated, or cast against a form) .... 40 psi
3. When minimum steel tie requirements of (c) are followed and the contact surface is rough and clean .... 160 psi
4. When additional vertical ties are used the allowable bond stress on a rough surface may be increased at the rate of 75 psi for each additional area of steel ties equal to 1 per cent of the contact area.

(c) *Vertical ties*

When mechanical anchorage in the form of vertical ties is provided, spacing of such ties shall not exceed four times the thickness of the slab nor 24 in. A minimum cross-sectional area of ties of 0.15 per cent of the contact area shall be provided. It is preferable to provide all ties in the form of extended stirrups.

(d) *Web reinforcement*

Web reinforcement for the composite section shall be designed in the same manner as for an integral beam of the same shape. All stirrups so required shall be anchored into the cast-in-place slab, where their area may also be relied upon to provide some or all of the vertical tie steel required in (c).

CHAPTER 26—PRESTRESSED CONCRETE

2600—Notation

$a = A_s f_{su}/0.85 f'_c b$

A_b = bearing area of anchor plate of post-tensioning steel

A'_b = maximum area of the portion of the anchorage surface that is geometrically similar to and concentric with the area of the anchor plate of the post-tensioning steel

A_s = area of prestressed tendons

A_{sf} = area of reinforcement to develop compressive strength of overhanging flanges in flanged members

A_{sr} = area of tendon required to develop the web

A'_s = area of unprestressed reinforcement

A_v = area of web reinforcement placed perpendicular to the axis of the member

b = width of compression face of flexural member

b' = minimum width of web of a flanged member

d = distance from extreme compression fiber to centroid of the prestressing force

f'_c = compressive strength of concrete (see Section 301)

f'_{ci} = compressive strength of concrete at time of initial prestress

f_{cp} = permissible compressive concrete stress on bearing area under anchor plate of post-tensioning steel

f_d = stress due to dead load, at the extreme fiber of a section at which tension stresses are caused by applied loads

f_{pc} = compressive stress in the concrete, after all prestress losses have occurred, at the centroid of the cross section resisting the applied loads, or at the junction of the web and flange when the centroid lies in the flange. (In a composite member f_{pc} will be the resultant compressive stress at the centroid of the composite section, or at the junction of the web and flange when the centroid lies within the flange, due to both prestress and to the bending moments resisted by the precast member acting alone)

f_{pe} = compressive stress in concrete due to prestress only, after all losses, at the extreme fiber of a section at which tension stresses are caused by applied loads

f'_s = ultimate strength of prestressing steel

f_{se} = effective steel prestress after losses

f_{su} = calculated stress in prestressing steel at ultimate load

f_{sy} = nominal yield strength of prestressing steel

f_y = yield strength of unprestressed reinforcement (see Section 301)

F_{sp} = ratio of splitting tensile strength to the square root of compressive strength (see Section 505)

h = total depth of member

I = Moment of inertia of section resisting externally applied loads*

K = wobble friction coefficient per foot of prestressing steel

L = length of prestressing steel element from jacking end to any point x

M = bending moment due to externally applied loads*

M_{cr} = net flexural cracking moment

M_u = ultimate resisting moment

p = A_s/bd; ratio of prestressing steel

p' = A'_s/bd; ratio of unprestressed steel

q = pf_{su}/f'_c

s = longitudinal spacing of web reinforcement

T_o = steel force at jacking end

T_x = steel force at any point x

t = average thickness of the compression flange of a flanged member

V = shear due to externally applied loads*

V_c = shear carried by concrete

V_{ci} = shear at diagonal cracking due to all loads, when such cracking is the result of combined shear and moment

V_{cw} = shear force at diagonal cracking due to all loads, when such cracking is the result of excessive principal tension stresses in the web

V_d = shear due to dead load

* The term "externally applied loads" shall be taken to mean the external ultimate loads acting on the member, excepting those applied to the member by the prestressing tendons.

V_p = vertical component of the effective prestress force at the section considered

V_u = shear due to specified ultimate load

y = distance from the centroidal axis of the section resisting the applied loads to the extreme fiber in tension

α = total angular change of prestressing steel profile in radians from jacking end to any point x

ε = base of Naperian logarithms

μ = curvature friction coefficient

ϕ = capacity reduction factor (see Section 1504)

2601—Definitions

(a) The following terms are defined for use in this chapter:

Anchorage—The means by which the prestress force is permanently delivered to the concrete.

Bonded tendons—Tendons which are bonded to the concrete either directly or through grouting. Unbonded tendons are free to move relative to the surrounding concrete.

Effective prestress—The stress remaining in the tendons after all losses have occurred, excluding the effects of dead load and superimposed loads.

Friction:

Curvature friction—Friction resulting from bends or curves in the specified cable profile.

Wobble friction—Friction caused by the unintended deviation of the prestressing steel from its specified profile.

Jacking force—The temporary force exerted by the device which introduces the tension into the tendons.

Nominal yield strength—The yield strength specified by appropriate ASTM specification or as indicated by Section 405(f).

Post-tensioning—A method of prestressing in which the tendons are tensioned after the concrete has hardened.

Pretensioning—A method of prestressing in which the tendons are tensioned before the concrete is placed.

Tendon—A tensioned steel element used to impart prestress to the concrete.

Transfer—The operation of transferring the tendon force to the concrete.

2602—Scope

(a) Provisions in this chapter apply to flexural members prestressed with high-strength steel. Pavements, pipes, and circular tanks are not included.

(b) For prestressed concrete designs or constructions in conflict with, or not encompassed by the provisions of this chapter, see Section 104.

(c) All provisions of this code not specifically excluded and not in conflict with the provisions of this chapter are to be considered applicable to prestressed concrete.

(d) The following provisions shall not apply to prestressed concrete: Sections 906, 911, 913, Chapters 13 and 14, Section 1508, Chapter 18, Section 2001(a), Chapter 21, and Section 2504(b).

2603—General considerations

(a) Stresses and ultimate strength shall be investigated at service conditions and at all load stages that may be critical during the life of the structure from the time prestress is first applied.

(b) Stress concentrations due to the prestressing or other causes shall be taken into account in the design.

(c) The effects on the adjoining structure of elastic and plastic deformations, deflections, changes in length, and rotations caused by the prestressing shall be provided for. When the effect is additive to temperature and shrinkage effects, they shall be considered simultaneously.

(d) The possibility of buckling of a member between points of contact between concrete and prestressing steel and of buckling of thin webs and flanges shall be considered.

2604—Basic assumptions

(a) The following assumptions shall be made for purposes of design:

1. Strains vary linearly with depth through the entire load range.
2. At cracked sections, the ability of the concrete to resist tension is neglected.
3. In calculations of section properties prior to bonding of tendons, areas of the open ducts shall be deducted. The transformed area of bonded tendons may be included in pretensioned members and in post-tensioned members after grouting.
4. Modulus of elasticity of concrete shall be assumed as prescribed in Section 1102.
5. The modulus of elasticity of prestressing steel shall be determined by tests or supplied by the manufacturer.

2605—Allowable stresses in concrete

(a) Temporary stresses immediately after transfer, before losses due to creep and shrinkage, shall not exceed the following:

1. Compression . . . . $0.60f'_{ci}$
2. Tension stresses in members without auxiliary reinforcement (unprestressed or prestressed) in the tension zone . . . . $3\sqrt{f'_{ci}}$

   Where the calculated tension stress exceeds this value, reinforcement shall be provided to resist the total tension force in the concrete computed on the assumption of an uncracked section.

(b) Stresses at design loads, after allowance for all prestress losses, shall not exceed the following:

1. Compression . . . . $0.45f'_c$
2. Tension in the precompressed tension zone:

   Members, not exposed to freezing temperatures nor to a corrosive environment, which contain bonded prestressed or unprestressed reinforcement located so as to control cracking . . . . $6\sqrt{f'_c}$

   All other members 0

   These values may be exceeded when not detrimental to proper structural behavior as provided in Section 104.

(c) The bearing stress on the concrete created by the anchorage in post-tensioned concrete with adequate reinforcement in the end regions shall not exceed:

$$f_{cp} = 0.6f'_{ci}\sqrt[3]{A'_b/A_b} \tag{26-1}$$

but not greater than f'_{ci}.

2606—Allowable stresses in steel

(a) *Temporary stresses*

1. Due to temporary jacking force . . . . $0.80f'_s$
   but not greater than the maximum value recommended by the manufacturer of the steel or of the anchorages
2. Pretensioning tendons immediately after transfer, or post-tensioning tendons immediately after anchoring . . . . $0.70f'_s$

(b) *Effective prestress* . . . . $0.60f'_s$
or $0.80f_{sy}$
whichever is smaller

2607—Loss of prestress

(a) To determine the effective prestress, allowance for the following sources of loss of prestress shall be considered.

1. Slip at anchorage
2. Elastic shortening of concrete
3. Creep of concrete
4. Shrinkage of concrete
5. Relaxation of steel stress
6. Frictional loss due to intended or unintended curvature in the tendons

(b) Friction losses in post-tensioned steel shall be based on experimentally determined wobble and curvature coefficients,* and shall be verified during stressing operations. The values of coefficients assumed for design, and the acceptable ranges of jacking forces and steel elongations shall be shown on the plans. These friction losses shall be calculated:

$$T_o = T_x\varepsilon^{(KL+\mu\alpha)} \tag{26-2}$$

When $(KL + \mu\alpha)$ is not greater than 0.3, Eq. (26-3) may be used.

* Values of K (per lineal foot) and μ vary appreciably with duct material and method of construction. For metal sheathing the following table may be used as a guide.

Type of steel	*Usual range of observed values*		*Suggested design values*	
	K	μ	K	μ
Wire cables	0.0005–0.0030	0.15–0.35	0.0015	0.25
High strength bars	0.0001–0.0005	0.08–0.30	0.0003	0.20
Galvanized strand	0.0005–0.0020	0.15–0.30	0.0015	0.25

$$T_o = T_x(1 + KL + \mu\alpha) \qquad (26\text{-}3)$$

(c) When prestress in a member may be reduced through its connection with adjoining elements, such reduction shall be allowed for in the design.

2608—Ultimate flexural strength

(a) The required ultimate load on a member, determined in accordance with Part IV-B shall not exceed the ultimate flexural strength computed by:

1. Rectangular sections, or flanged sections in which the neutral axis lies within the flange:*

$$M_u = \phi[A_s f_{su} d(1 - 0.59q)] = \phi\left[A_s f_{su}\left(d - \frac{a}{2}\right)\right] \qquad (26\text{-}4)$$

2. Flanged sections in which the neutral axis falls outside the flange:†

$$M_u = \phi\left[A_{sr} f_{su} d\left(1 - \frac{0.59 A_{sr} f_{su}}{b' d f'_c}\right) + 0.85 f'_c (b - b')t(d - 0.5t)\right] \qquad (26\text{-}5)$$

where $A_{sr} = A_s - A_{sf}$
$A_{sf} = 0.85 f'_c (b - b')t/f_{su}$

3. Where information for the determination of f_{su} is not available, and provided that f_{se} is not less than $0.5 f'_s$, the following approximate values shall be used:

Bonded members

$$f_{su} = f'_s(1 - 0.5 p f'_s / f'_c) \qquad (26\text{-}6)$$

Unbonded members

$$f_{su} = f_{se} + 15{,}000 \text{ psi} \qquad (26\text{-}7)$$

4. Nonprestressed reinforcement, in combination with prestressed steel, may be considered to contribute to the tension force in a member at ultimate moment an amount equal to its area times its yield point, provided

$$\frac{p f_{su}}{f'_c} + \frac{p' f_y}{f'_c} \text{ does not exceed } 0.3$$

2609—Limitations on steel percentage

(a) Except as provided in (b), the ratio of prestressing steel used for calculations of M_u shall be such that

$$p f_{su}/f'_c \text{ is not more than } 0.30$$

For flanged sections, p shall be taken as the steel ratio of only that portion of the total tension steel area which is required to develop the compressive strength of the web alone.

*Usually where the flange thickness is *more* than $1.4\, d p f_{su}/f'_c$.
† Usually where the flange thickness is *less* than $1.4\, d p f_{su}/f'_c$.

(b) When a steel ratio in excess of that specified in (a) is used, the ultimate moment shall be taken as not greater than the following:

Rectangular sections, or flanged sections in which the neutral axis lies within the flange

$$M_u = \phi[0.25f'_c bd^2] \tag{26-8}$$

Flanged sections in which the neutral axis falls outside the flange

$$M_u = \phi[0.25f'_c b'd^2 + 0.85f'_c(b - b')t(d - 0.5t)] \tag{26-9}$$

(c) The total amount of prestressed and unprestressed reinforcement shall be adequate to develop an ultimate load in flexure at least 1.2 times the cracking load calculated on the basis of a modulus of rupture of $7.5\sqrt{f'_c}$.

2610—Shear

(a) Except as provided in (c), the area of shear reinforcement placed perpendicular to the axis of a member shall be not less than:

$$A_v = \frac{(V_u - \phi V_c)s}{\phi d f_y} \tag{26-10}$$

nor less than

$$A_v = \frac{A_s}{80} \cdot \frac{f'_s}{f_y} \cdot \frac{s}{d}\sqrt{\frac{d}{b'}} \tag{26-11}$$

The effective depth, d, used in Eq. (26-10) and (26-11) shall be as follows:

1. In members of constant over-all depth, d, equals the effective depth at the section of maximum moment, and the length of the stirrups at the section under consideration shall be at least equal to the length of the stirrups at the section of maximum moment.
2. In members of varying depth, d equals $h(d_m/h_m)$, where d_m and h_m are the effective depth and total depth respectively at the section of maximum moment, and h is the total depth at the section under consideration. The stirrups shall extend into the member a distance d from the compression face.

(b) The shear, V_c, at diagonal cracking shall be taken as the lesser of V_{ci} and V_{cw}, determined from Eq. (26-12) and (26-13).

1. For normal weight concrete

$$V_{ci} = 0.6b'd\sqrt{f'_c} + \frac{M_{cr}}{\dfrac{M}{V} - \dfrac{d}{2}} + V_d \tag{26-12}$$

but not less than $1.7b'd\sqrt{f'_c}$

where $M_{cr} = \dfrac{I}{y}(6\sqrt{f'_c} + f_{pe} - f_d)$

$$V_{cw} = b'd(3.5\sqrt{f'_c} + 0.3f_{pc}) + V_p \tag{26-13}$$

2. For lightweight aggregate concrete

$$V_{ci} = 0.1F_{sp}b'd\sqrt{f'_c} + \frac{M_{cr}}{\dfrac{M}{V} - \dfrac{d}{2}} + V_d \qquad (26\text{-}12A)$$

but not less than $0.25F_{sp}b'd\sqrt{f'_c}$

$$\text{where } M_{cr} = \frac{I}{y}(0.9F_{sp}\sqrt{f'_c} + f_{pe} - f_d)$$

$$V_{cw} = b'd\left[0.5F_{sp}\sqrt{f'_c} + f_{pc}\left(0.2 + \frac{F_{sp}}{67}\right)\right] + V_p \qquad (26\text{-}13A)$$

Alternatively V_{cw} may be taken as the live load plus dead load shear which corresponds to the occurrence of a principal tensile stress of $4\sqrt{f'_c}$ in normal weight concrete, or $0.6F_{sp}\sqrt{f'_c}$ in lightweight concrete, at the centroidal axis of the section resisting the live load. In flanged members, if the centroidal axis is not in the web, the principal tensile stress should be determined at the intersection of the flange and the web.

When applying Eq. (26-12) and (26-12A), the effective depth, *d*, shall be taken as the distance from the extreme compression fiber to the centroid of the prestressing tendons.

When applying Eq. (26-13) and (26-13A), the effective depth, *d*, shall be taken as the distance from the extreme compression fiber to the centroid of the prestressing tendons, or as 80 percent of the over-all depth of the member, whichever is the greater.

The value of (M/V) used in Eq. (26-12) and (26-12A) shall be that resulting from the distribution of loads causing maximum moment to occur at the section.

In a pretensioned prestressed beam in which the section distant $d/2$ from the face of the support is closer to the end face of the beam than the transfer length of the wire or strand used, the reduced prestress in the concrete at sections falling within the transfer length should be considered when calculating the diagonal cracking shear, V_{cw}. The prestress at the centroid of the section may be assumed to vary linearly from zero at the end face of the beam to a maximum at a distance from the end face equal to the transfer length, assumed to be 50 diameters for strand and 100 diameters for single wire.

(c) Web reinforcement between the face of the support and the section at a distance $d/2$ therefrom shall be the same as that required at that section.

Shear reinforcement shall be provided for a distance equal to the effective depth, *d*, of the member beyond the point theoretically required.

Web reinforcement shall be anchored at both ends in accordance with Section 919.

Shear reinforcement not less than determined from Equation (26-11) shall be provided at all sections and shall be spaced not farther apart than three-fourths the depth of the member, nor 24 in., whichever is the smaller, except when it is shown by tests that the required ultimate flexural and shear capacity can be developed when the web reinforcement is omitted.

A yield strength in excess of 60,000 psi shall not be considered for shear reinforcement.

2611—Bond

(a) Three or seven wire pretensioning strand shall be bonded to the concrete from the cross section under consideration for a distance in inches of not less than:

$$\left(f_{su} - \frac{2}{3} f_{se}\right) D$$

where D, the nominal strand diameter, is in inches and f_{su} and f_{se} are expressed in kips per square inch.

Investigation may be restricted to those cross sections nearest each end of the member that are required to develop their ultimate strength under the specified ultimate load.

2612—Repetitive loads

(a) The possibility of bond failure due to repeated loads shall be investigated in regions of high bond stress and where flexural cracking is expected at design loads.

(b) In unbonded construction subject to repetitive loads, special attention shall be given to the possibility of fatigue in the anchorages.

(c) The possibility of inclined diagonal tension cracks forming under repetitive loading at appreciably smaller stresses than under static loading shall be taken into account in the design.

2613—Composite construction

(a) General requirements for composite construction are given in Chapter 25.

2614—End regions

(a) End blocks shall be provided if necessary for end bearing or for distribution of concentrated prestressing forces safely from the anchorages to the cross section of the member.

(b) Reinforcement shall be provided in the anchorage zone to resist bursting and spalling forces induced by the concentrated loads of the prestressing steel. Points of abrupt change in section shall be adequately reinforced.

2615—Continuity

(a) For continuous girders and other statically indeterminate structures, moments, shears, and thrusts produced by external loads and prestressing shall be determined by elastic analysis. The effects of creep, shrinkage, axial deformation, restraint of attached structural elements, and foundation settlement shall be considered in the design.

(b) In the application of ultimate load factors where effects of dead and live loads are of opposite sign, the case of a dead load factor of unity shall be included in the investigation.

2616—Concrete cover

(a) The following minimum thicknesses of concrete cover shall be provided for prestressing steel, ducts and nonprestressed steel.

	Cover, in.
Concrete surfaces in contact with ground	2
Beams and girders:	
Prestressing steel and main reinforcing bars	1½
Stirrups and ties	1
Slabs and joists exposed to weather	1
Slabs and joists not exposed to weather	¾

(b) In extremely corrosive atmosphere or other severe exposures, the amount of protection shall be suitably increased.

2617—Placement of prestressing steel

(a) All pretensioning steel and ducts for post-tensioning shall be accurately placed and adequately secured in position.

(b) The minimum clear spacing between pretensioning steel at each end of the member shall be four times the diameter of individual wires or three times the diameter of strands, but at least 1⅓ times the maximum size of aggregate.

(c) Prestressing steel or ducts may be bundled together in the middle portion of the span, provided the requirements of (b) are satisfied.

(d) Ducts may be arranged closely together vertically when provision is made to prevent the steel, when tensioned, from breaking through the duct. Horizontal disposition of ducts shall allow proper placement of concrete.

(e) Where concentration of steel or ducts tends to create a weakened plane in the concrete cover, reinforcement shall be provided to control cracking.

(f) The inside diameter of ducts shall be at least ¼ in. larger than the diameter of the post-tensioning bar or large enough to produce an internal area at least twice the gross area of wires, strands, or cables.

2618—Concrete

(a) Suitable admixtures to obtain high early strength or to increase the workability of low-slump concrete may be used if known to have no injurious effects on the steel or the concrete. Calcium chloride or an admixture containing calcium chloride shall not be used. Sea water shall not be used.

(b) Concrete strength required at given ages shall be indicated on the plans. The strength at transfer shall be adequate for the requirements of the anchorages or of transfer through bond as well as meet camber or deflection requirements. For 7-wire strands, the minimum strength at transfer shall be 3,000 psi for ⅜-in. strands and smaller, and 3,500 psi for 7⁄16-in. and ½-in. strands.

2619—Grout

(a) Suitable admixtures, known to have no injurious effects on the steel or the concrete, may be used to increase workability and to reduce shrinkage. Calcium chloride shall not be used.

(b) Sand, if used, shall conform to "Specifications for Aggregate for Masonry Mortar" (ASTM C 144) except that gradation may be modified as necessary to obtain proper workability.

(c) Proportions of grouting materials shall be based on results of tests on fresh and hardened grout prior to beginning work. The water content shall be the minimum necessary for proper placement but in no case more than 5½ gal per sack. When permitted to stand until setting takes place, grout shall neither bleed nor segregate.

(d) Grout shall be mixed in a high-speed mechanical mixer and then passed through a strainer into pumping equipment which provides for recirculation.

(e) Just prior to grouting, the ducts shall be made free of water, dirt, and other foreign substances. The method of grouting shall be such as to ensure the complete filling of all voids between the prestressing steel and the duct and anchorage fittings.

(f) Temperature of members at the time of grouting must be above 50°F and at least this temperature shall be maintained for at least 48 hr.

2620—Steel tendons

(a) Prestressing steel shall be clean and free of excessive rust, scale, and pitting. A light oxide is permissable. Unbonded steel shall be permanently protected from corrosion.

(b) Burning and welding operations in the vicinity of prestressing steel shall be carefully performed, so that the prestressing steel shall not be subjected to excessive temperatures, welding sparks, or ground currents.

2621—Application and measurement of prestressing force

(a) Prestressing force shall be determined (1) by measuring tendon elongation and also (2) either by checking jack pressure on a recently calibrated gage or by the use of a recently calibrated dynamometer. The cause of any discrepancy which exceeds 5 per cent shall be ascertained and corrected. Elongation requirements shall be taken from average load-elongation curves for the steel used.

(b) If several wires or strands are stretched simultaneously, provision must be made to induce approximately equal stress in each.

(c) Transfer of force from the bulkheads of the pretensioning bed to the concrete shall be carefully accomplished, by proper choice of cutting points and cutting sequence. Release of pretensioning may be effected by gradual means or by burning of tendons. Long lengths of exposed strands shall be cut near the member to minimize shock to the concrete.

(d) The total loss of prestress due to unreplaced broken tendons shall not exceed 2 per cent of the total prestress.

(e) Where there is a considerable temperature differential between the concrete and the tendons, its effect shall be taken into account.

2622—Post-tensioning anchorages and couplers

(a) Anchorages, couplers, and splices for post-tensioned reinforcement shall develop the required ultimate capacity of the tendons without excessive slip. Couplers and splices shall be placed in areas approved by the Engineer and enclosed in housings long enough to permit the necessary movements. They shall not be used at points of sharp curvature.

(b) Anchorage and end fittings shall be permanently protected against corrosion.

2623—Formwork

(a) Forms for pretensioned members shall be constructed to permit movement of the member without damage during release of the prestressing force.

(b) Forms for post-tensioned members shall be constructed to minimize resistance to the shortening of the member. Deflection of members due to the prestressing force and deformation of falsework shall be considered in the design.

2624—Joints and bearings for precast members

(a) Design and detailing of the joints and bearings shall be based on the forces to be transmitted, and on the effects of dimensional changes due to shrinkage, elastic deformation, creep and temperature. Joints shall be detailed so as to allow sufficient tolerances for manufacture and erection of the members.

(b) Bearings shall be detailed to provide for stress concentrations, rotations, and the possible development of horizontal forces by friction or other restraints.

APPENDIX B

Excerpts from the report of ACI-ASCE Committee 512 are reprinted here with the permission of the American Concrete Institute. The complete report appeared in the August, 1964 issue of the *Journal of the American Concrete Institute* and is available from ACI as Title No. 61-51. It is recommended that engineers contemplating the design of precast structures obtain and study the complete report.

A discussion of the report, its application, and design examples are presented in Chap. 13.

Excerpts from Suggested Design of Joints and Connections in Precast Structural Concrete Reported by ACI-ASCE Committee 512

(Each excerpt is identified by the number of the section from which it is taken but the number does not necessarily mean that the entire section has been reprinted here.)

CHAPTER 1—INTRODUCTION

102—Scope

The intent of these recommendations is to help provide that all joints and connections perform their function at all stages of loading without overstress and with proper safety factors against failure due to overload.

These recommendations are intended to apply to joints and connections between precast members; between precast members and cast-in-place concrete members; and between precast members and structural steel members.

The connections considered in this report are those connecting columns to footings, multistory columns, beams to columns, beams to girders, and wall, floor or roof slabs to beams. Connections outside the scope of this report are those used to fasten appurtenances such as piping or other mechanical equipment to concrete.

Joints and connections may be made by welding steel reinforcement or structural steel inserts; by bolting; by use of pins; by transfer of tensile stress or compressive stresses by bond or anchorage; by use of clips and other devices which prevent separation of precast members from independently braced supporting members; by use of key-type devices; by use of bonding mediums which affect the adherence of one member to another; by use of friction between members induced by gravity forces; by prestressing; by combinations of the preceding methods; or by any other method which accomplishes the intent, making use of recognized means of design, fabrication, and erection, or methods determined by testing.

CHAPTER 2—DESIGN CONSIDERATIONS

201—Loading conditions

Loading conditions to be considered in the design of joints and connections are service loads including wind and earthquake forces, volume changes due to shrinkage, creep, and temperature change, erection loads, and loading encountered in stripping forms, shoring and removal of shores, storage, and transportation of members. Proper attention should be given to loads and the resulting stresses peculiar to the sequence of erection. Typical examples of construction in which the sequence and manner of erection affects the loading and stresses in the member are possible eccentric loading due to the erection of members on one side only of a member, installation of composite concrete toppings on shored or unshored slabs or beams, and continuity moment connections over supports. All significant combinations of loading should be considered, and the joints and connections should be designed for loadings consistent with these possible combinations of loading. For loadings other than those peculiar to precast concrete construction (decentering, handling, storage, and erection loads), loadings and load distributions as outlined in ACI 318-63 should be the minimum considered.

If it is not practical to provide for all possible temporary loading conditions which could occur during erection special erection procedures may be warranted. If so, complete erection instructions should be included in the plans and specifications which become part of the erection contract documents. Loading sequences, connection sequences and if necessary shoring or guying schedules should be clearly outlined. The disposition and strength of shoring should be stated and approved prior to construction.

202—Load factors

202.1—Design stress: Design stresses, except as noted hereafter, should not exceed those provided in ACI (318-63) "Building Code Requirements for Reinforced Concrete," ACI-ASCE "Tentative Recommendations for Prestressed Concrete," AISC "Specifications for the Design, Fabrication and Erection of Structural Steel for Buildings" or AWS "Standard Code for Arc and Gas Welding in Building Construction," whichever is applicable. No flexural tension stresses should be permitted in the concrete of prestressed joints and connections.

202.2—Ultimate load factors: The ultimate strength capacity of joints and connections should be at least 10 per cent in excess of that required of the members connected. This recommendation may be satisfied by proportioning the joint or connection to provide strength in the connection or joint 1.1 times the ultimate strength capacity required by ACI 318-63.

CHAPTER 3—DESIGN

301—Transfer of shear

301.1—General: The transfer of shear may be accomplished using reinforcing steel extended as dowels coupled with cast-in-place concrete placed between roughened concrete interfaces, mechanical devices such as embedded plates or shapes, brackets, prestressing force applied across the connecting surfaces, or any other ways which meet all accepted unit stress requirement for the materials involved and meet the ultimate strength requirements of Section 202.2 or which have been thoroughly tested and meet the requirements of Section 202.3. The entire shear should be considered as transferred through one type of device mentioned above, even though a

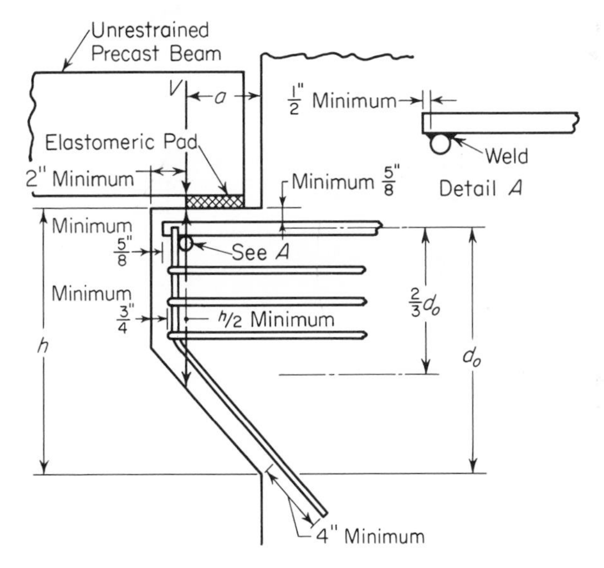

FIG. 301.3a. Corbels subject to vertical load only.

combination of devices may be available at the joints or supports being considered. The device should be designed to resist the maximum shear in the section at the connecting surfaces.

301.3—Brackets: A bracket may be a corbel, i.e., a protrusion cast on to the side of a column or wall to serve as a beam seat, the top of a column, or a ledge on a column.

Positive means should be taken to prevent the reaction from bearing on the outermost edge of the bracket. However, the vertical reaction should be assumed to act through the outer edge of the bearing pad as shown in Fig. 301.3a and 301.3b.

From the standpoint of the structural behavior of the precast members a bracket serves as a shear transfer device, but the bracket itself must be designed for flexure, shear, bearing, and the splitting forces accompanying bearing.

301.3.1—Flexure: The flexural design of a bracket with a/d_o ratio greater than 1.0 should follow the procedures of flexural design for ordinary reinforced concrete beams (ACI 318). The ultimate load of a corbel with a ratio of $a/d_o < 1$ should not exceed:

$$V = \phi[6.5bd_o\sqrt{f'_c}(1 - 0.5^{d_o/a})(1000p)^{1/3}]$$

where V = total vertical load at ultimate
b = width of bracket
d_o = distance from extreme compression fiber to centroid of tension reinforcement at the column face

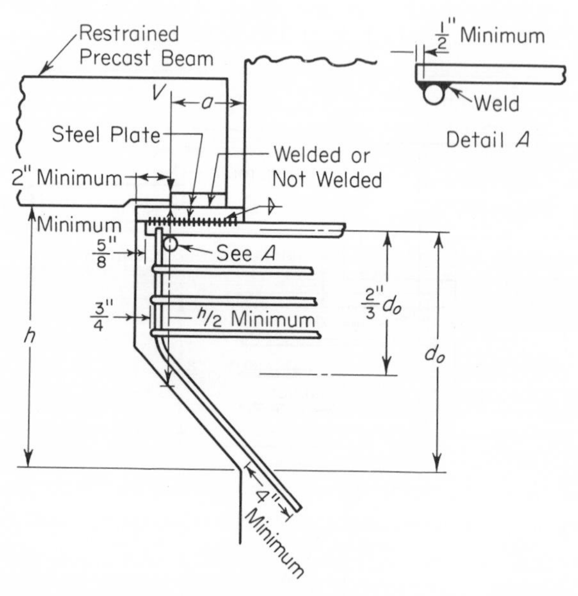

FIG. 301.3b. Corbels subject to vertical load and restrained creep and shrinkage (steel plates welded or not welded).

f'_c = compressive strength of concrete
a = lever arm of vertical reaction
p = A_s/bd_o, reinforcement ratio at the column face
ϕ = capacity reduction factor (recommended value = 0.85 for shear transfer device)

The reinforcement index q for all corbels should have a maximum of 0.15 and a minimum of 0.04 where $q = pf_y/f'_c$. The depth of the bracket outer face should be at least from 0.4 to 0.5, the total bracket depth.

There is frequently insufficient room to develop the tensile reinforcement through bond; therefore adequate anchorage provisions should be made.

Additional reinforcement to resist horizontal forces which may develop from volume changes due to shrinkage, creep, and temperature change should be considered in the design. The horizontal friction force may be estimated or taken at 0.5 times the vertical force.

301.3.2—Diagonal tension: Diagonal tension reinforcement for a/d_o ratios (see notation under Section 301.3.1) greater than 1.0 should be computed in accordance with the provisions of ACI 318-63 with a minimum stirrup requirement equal to $0.005bd_o$. For a/d_o ratios less than 1.0, stirrups in amount of $0.005bd_o$ should be used. This reinforcement should be placed perpendicular to the normal component of the reaction. The reinforcement should be distributed in pairs as close to the side faces of the bracket as regulations for cover permit. Furthermore, they should be

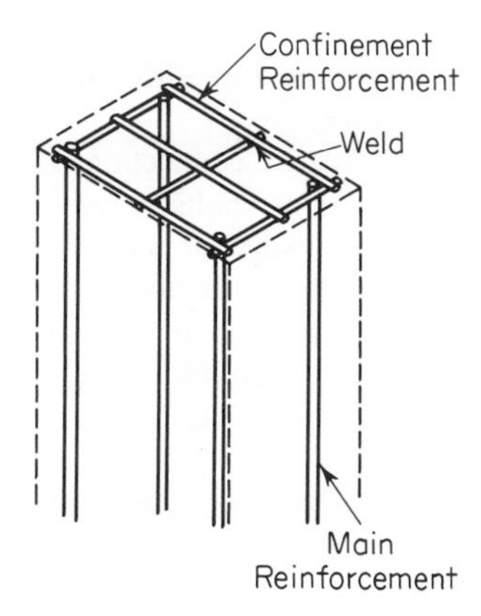

FIG. 301.3.4. Reinforcement to resist splitting.

distributed equally over the depth of the bracket with at least three pairs employed. These bars should either be bent in a "U" shape, bent to a closed tie, or welded to cross pieces at the outer face of the bracket as close as regulations for cover will permit. Anchorage at the other end (into supporting member) should be adequately provided for.

301.3.3—Bearing: It is desirable to insure that bearing be located away from the edge of a bracket. Bearing stresses at ultimate loads on corbels should not exceed $f'_c/2$ and on column heads:

$$69\sqrt{f'_c}\left(\frac{s}{w}\right)^{1/3} \tag{2}$$

where s = distance from outer edge of column head to the center of bearing plate
w = width of bearing plate

To avoid local spalling the edge of the bracket should either be armored with steel anchored substantially, provided with liberal chamfer, or other means provided to relieve pressure against the edge.

Direct concrete to concrete bearing is not recommended, between units unless the bearing is accomplished by cast-in-place concrete or grout.

301.3.4—Splitting: Reinforcement for confinement (see Fig. 301.3.4) should be provided to resist the tensile stresses which accompany bearing if the bearing stresses exceed $0.1f'_c$. This reinforcement should be placed parallel to the bearing surface in both directions and should be considered for both the supporting and supported member. The reinforcement lateral to the direction of the longitudinal axis of the supported member may be eliminated in the bracket if the width of the bracket is greater than four times that of the supported member. A maximum concrete cover of 1 in. is recommended. The steel quantity required in each direction for both the supporting member and the supported member may be computed as follows:

$$A_s = \frac{V}{100{,}000} \tag{3}$$

where V = total design reactions in pounds supported by the bracket

This reinforcement should be supplied in addition to that required for flexure or tensile stresses from any other source.

301.3.5—Other tensile stresses: Positive connections at the bearing of both ends of simply supported members by welding, bolts or any other means which prevent movements arising from creep, shrinkage or temperature change are not recommended. One end should be free to accommodate such movement. However, if such connections are made at both ends, reinforcement should be provided to resist the tensile forces which may develop in both the supporting and supported member from horizontal forces, and the bearing plates or other such devices should be designed to resist these forces.

301.5—Embedded structural steel shapes or plates: Structural steel shapes or plates used for the purpose of transferring shear (see Fig. 301.5), should be designed for the following considerations:

1. Bearing stress in embedded portion
2. Shear in the steel between faces of connected concrete members
3. Flexural stress in the steel between faces of connected concrete members
4. Confinement of concrete embedding shapes or plates

302—Transfer of moment

302.1—General: The transfer of moment through connections between precast members or a precast member and structural steel member or a precast member and cast-in-place member may be accomplished by: reinforcing steel extended as dowels; composite construction, embedded plates or structural steel shapes; or prestressing force applied to the joint and properly developed by the connecting members. The entire moment should be considered as transferred through one of the types of device mentioned above, even though a combination of devices may be available at the connection.

Reinforcing bars which are lapped or welded or steel plates or shapes which are welded to reinforcing bars, or to other steel plates or shapes, should be detailed so that there is a minimum eccentricity of force which is transferred through the connection. If small eccentricities cannot be avoided and more than one reinforcing bar or other element is to be connected, the disposition of the laps or welds should be symmetrical with respect to the center of gravity of the transferred force if such is practical.

The design of the connection and the adjacent members should consider all stresses due to eccentric loading conditions, creep, shrinkage, and temperature changes. Provisions of Section 301 should be followed to insure proper capacity for shear transfer at the connection.

The members at the connection shall be reinforced properly to withstand unavoidable stress concentrations. Care should be taken that any portions of existing con-

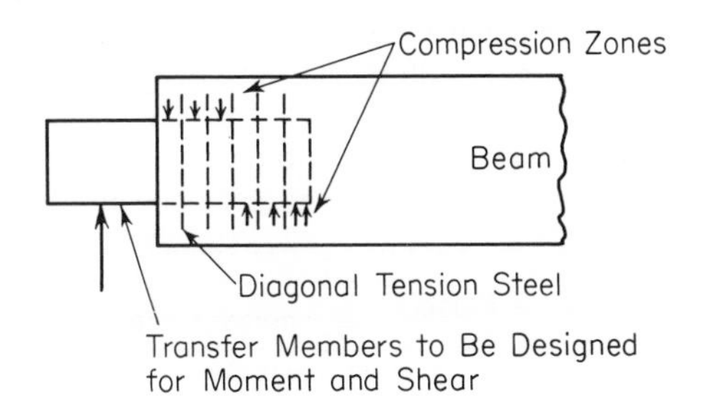

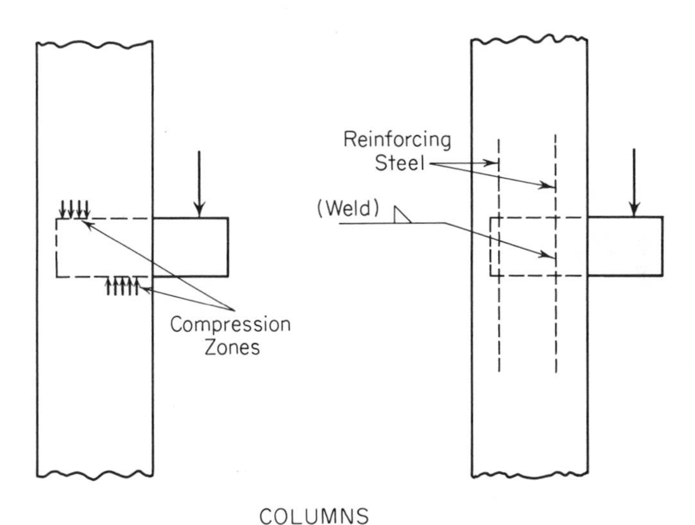

FIG. 301.5. Shear transfer devices for beams and columns.

crete that are to be in contact with cast-in-place connection concrete are clean before the concrete at the connection is placed.

302.3—Reinforcing bars

302.3.1—Extension of reinforcing bars: The transfer of tension through the protruding reinforcing bars may be accomplished by sufficient lap, by welding which may or may not employ auxiliary hardware, or by other mechanical devices. These direct tension connections should be designed and made to develop the tensile force in accordance with the requirements of Section 1506, ACI 318-63.

302.3.2—Additional reinforcing bars: If the precast members are designed with a cast-in-place structural composite concrete topping, continuity may be obtained by placing reinforcing bars in the cast-in-place concrete at the connection (see Fig. 302.3.2.). Composite action should be assured by designing in accordance with the provisions of Chapter 25, ACI 318-63.

302.5—Prestressing force: The cast-in-place concrete used at the connection should be of a design strength at least equal to that of the connected members. Filling the spaces at the ends of scarfed joints with mortar improves the behavior of joint subjected to axial loads. When the cast-in-place concrete has reached the required strength, the prestressing force may be applied to the connection. The design should provide that no tensile stresses be permitted at any point in cast-in place concrete

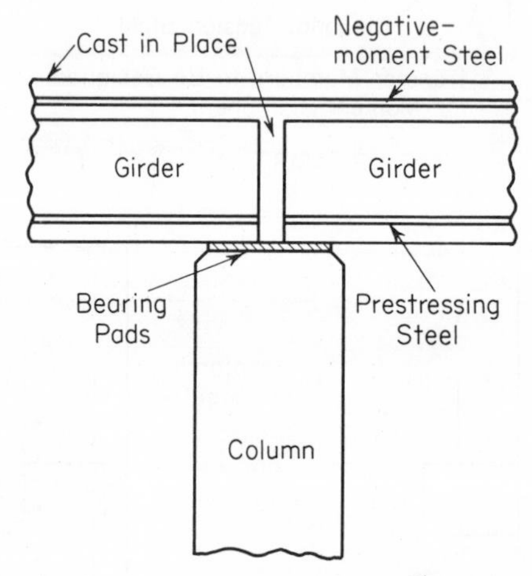

FIG. 302.3.2. Cast-in-place concrete continuity connection for prestressed members.

across the connection for any stage of loading, if there is to be any subjection to weather or corrosive environment, otherwise tensile stresses up to $3\sqrt{f'_c}$ may be permitted. The bolts of scarfed joint connections should be prestressed at least to the level required by working loads. The effect of the joint on the flexural rigidity of the precast members can be determined by assuming that the depth in the joint region is reduced by one-third.

Ultimate strength design of connections should be by the applicable provisions of ACI 318-63. Connections including post-tensioned systems subject to both flexural and prestress compression in the same region should be designed by load factors increased 20 per cent. In connections involving pretensioning members with reinforcement index "q" below 0.25 no increase of load factor need be made. For "q" values above 0.25 the load factor should be increased 20 per cent.

302.7—Stresses due to shrinkage, creep, and temperature change: In a moment connection, the shortening of a precast member after erection due to shrinkage, creep, and temperature change may cause high stresses in the members and the connection. The stresses produced by this shortening should be provided for in design.

A suggested method of reducing the potential shortening is to specify a minimum age of a precast member at the time of erection.

In a prestressed member, it may be necessary to extend reinforcing bars from the compression zone of the member into the cast-in-place concrete of the connection, to prevent separation of the member from the connection. Such bars should be properly anchored in both the member and the connection to develop the full strength of the bars.

303—Transfer of torsion

303.1—General: The design of connections should always include an investigation of the possibility of torsional stresses due to unequal loading, wind, and seismic forces, differential settlement or temporary erection loading conditions. In general members should be detailed so that torsion is held to a minimum at the connection.

304—Transfer of axial tension

304.1—General: Axial tension forces carried through a connection should not produce stress in any part of the connection which combined with bending stresses or torsional stresses exceed permissible stresses for the materials involved. Tensile stresses in concrete should never be relied upon to transfer axial tension. In designing connections which are to transfer tensile forces, special attention should be given to possible eccentricities.

305—Transfer of axial compression

305.1—General: Compression should be transferred across a connection without eccentricity whenever possible. The stresses resulting from axial compression combined with any other stresses on the section should be kept within the allowable stresses for the materials involved.

APPENDIX C

Elstner Method—Simplified Procedure for Shear Analysis of Prestressed Concrete Building Members

The shear analysis of prestressed concrete building members in strict accordance with Sec. 2610 of ACI 318-63 involves a great deal of computation. Most of the work is associated with solving Eq. (26-12 ACI) for V_{ci}. For the benefit of the many designers who still do not have electronic computers Richard Elstner, a partner in Wiss, Janney, Elstner and Associates of Northbrook, Illinois, has developed a very much simplified procedure for computing V_{ci}. Elstner began with fundamental equations for principal tensile stress and for combined shear and flexure and, using assumptions that were conservative, developed his recommendations.

After studying Elstner's derivation Robert West, a partner in West, Preston and Sollenberger of Lawrenceville, New Jersey, developed the simplified derivation presented here. West began with the equations in Sec. 2610 of ACI 318-63 and, making only assumptions that were on the conservative side, came out with exactly the same values as those developed by Elstner.

The following derivation applies to a simple span beam with a uniform load. For other conditions the basic equations in Sec. 2610 should be used. Notations used are those from ACI 318-63, Sec. 2600 plus

w_{S+L} = superimposed loads per unit length = Total design load minus weight of beam

w_G = weight of beam per unit length

M_{S+L} = moment due to w_{S+L}

From ACI 318-63,

$$V_{ci} = 0.6b'd\sqrt{f'_c} + \frac{M_{cr}}{\dfrac{M}{V} - \dfrac{d}{2}} + V_d \qquad \text{(26-12 ACI)}$$

in which

$$M_{cr} = \frac{I}{y}(6\sqrt{f'_c} + f_{pe} - f_d) \qquad \text{(C-1)}$$

This expression for M_{cr} is exactly equal to the bending moment caused by superimposed loads if the member is designed to the full allowable stresses of Sec. 2605. If the member has less tensile stress than the $6\sqrt{f'_c}$ permitted, it simply means that the shear steel computed by this analysis will be on the conservative side. Therefore

$$M_{cr} = M_{S+L} = \frac{(w_{S+L})(L^2)12}{8} \tag{C-2}$$

in which M_{cr} is in inch-pounds, w_{S+L} is in pounds per foot, and L is in feet.
Substituting in Eq. (26-12 ACI),

$$V_{ci} = 0.6b'd\sqrt{f'_c} + \frac{(w_{S+L})(L^2)12}{8\left(\dfrac{M}{V} - \dfrac{d}{2}\right)} + V_d \tag{C-3}$$

Since $d/2$ is small compared with M/V (especially at locations where V_{ci} is critical) the term $d/2$ can be dropped from Eq. (C-3). In any case eliminating $d/2$ is conservative because the effect is to reduce the computed value of V_{ci}. Then

$$V_{ci} = 0.6b'd\sqrt{f'_c} + \frac{(w_{S+L})(L^2)12}{8\dfrac{M}{V}} + V_d \tag{C-4}$$

For any simple span, uniformly loaded beam the values for shear and moment at a distance X from the end of the beam are

$$V_x = w\left(\frac{L}{2} - X\right)$$

$$M_x = \frac{wX}{2}\left(L - X\right)$$

from which

$$\frac{M_x}{V_x} = \frac{wX(L - X)/2}{w(L - 2X)/2} = \frac{X(L - X)}{L - 2X} \tag{C-5}$$

Since the ratio M/V does not contain any term except the span and the distance along the span to the section being investigated, it may be expressed in feet as follows by substituting in Eq. (C-5):

At $X = L/8$, $\quad \dfrac{M}{V} = \dfrac{7L}{48} = 0.1458L$

At $X = L/4$, $\quad \dfrac{M}{V} = \dfrac{3L}{8} = 0.3750L$

At $X = L/3$, $\quad \dfrac{M}{V} = \dfrac{2L}{3} = 0.6667L$

At $X = L/2$, $\quad \dfrac{M}{V} = \text{infinity}$

Substituting these values for M/V in Eq. (C-4) and expressing V_d in terms of w_G and L we get:

At $X = L/8$,

$$V_{ci} = 0.6b'd\sqrt{f'_c} + \frac{(w_{S+L})(L^2)12}{8(0.1458L)12} + 0.375w_GL$$

from which

$$V_{ci} = 0.6b'd\sqrt{f'_c} + 0.857(w_{S+L})L + 0.375w_GL$$

At $X = L/4$,

$$V_{ci} = 0.6b'd\sqrt{f'_c} + 0.333(w_{S+L})L + 0.250w_G L$$

At $X = L/3$,

$$V_{ci} = 0.6b'd\sqrt{f'_c} + 0.1875(w_{S+L})L + 0.167w_G L$$

At $X = L/2$,

$$V_{ci} = 0.6b'd\sqrt{f'_c}$$

APPENDIX D

Bending Moment Chart for AASHO Loadings

Figure D-1* was developed to facilitate computation of maximum live-load bending moment at any point on a simple-beam span of a bridge. It is based on H20-S16-44 loading of the American Association of State Highway Officials.

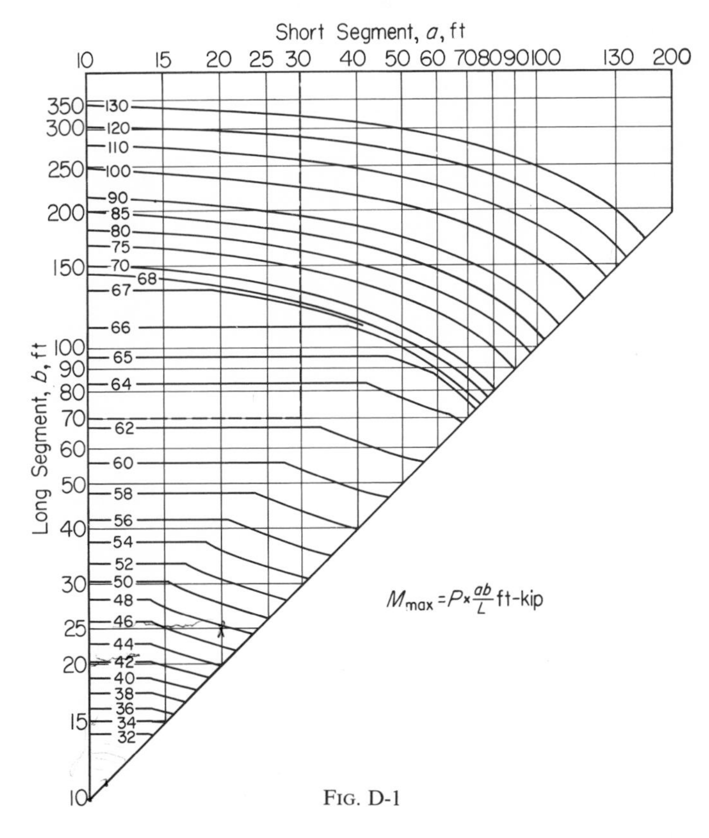

Fig. D-1

* Reprinted by permission from Gerald K. Gillan, Professor of Civil Engineering, Pennsylvania State University. From *Eng. News-Record,* Oct. 7, 1954. Copyright 1954 by McGraw-Hill Publishing Company, Inc.

While maximum moments at the center of various spans are given in an appendix to the AASHO Standard Specification for Highway Bridges, maximum moments at other locations, though important in design, are not presented. This chart is intended to fill the gap.

By means of the chart an equivalent concentrated load is obtained, from which the required bending moment can easily be computed. An equivalent concentrated load was used because it leads to small interpolation intervals and is quite accurate, even when used within the usual limitations of a small diagram.

To find the bending moment for a single-lane load at a point that is at a distance of a ft from one end of a span and b ft from the other end, enter the chart horizontally at b and vertically at a. The intersection determines the equivalent load P in kips, which may be read by interpolating between the heavy-line curves. The required moment is found to be the product Pab/L, where L is equivalent to the span.

For example, find the maximum moment in a 100-ft span at a point 30 ft from one end. If you enter the chart at the left side with 70 ft and at the top with 30 ft, the intersection of the two ordinates falls between the curves marked 62 and 64 and may be interpolated to be 62.3. The required moment then is $62.3 \times 30 \times 70/100 =$ 1,310 ft-kips.

APPENDIX E

The Ratio of M/V Is the Same for Design and Ultimate Loads

The ratio M/V is used in Eq. (26-12 ACI) and (26-12A ACI) of ACI 318-63, Sec. 2610(b). These values are defined as follows in Sec. 2600:

M = bending moment due to externally applied loads
V = shear due to externally applied loads

Section 2600 says: "The term *externally applied loads* shall be taken to mean the external ultimate loads acting on the member, excepting those applied to the member by the prestressing tendons." The authors interpret this as meaning all loads on the member except the dead weight of the member itself.

The purpose of this appendix is to demonstrate that the ratio of M/V is the same for both the design-loading condition and the ultimate-loading condition. For this demonstration we will use the following notation:

w_S = load per unit length due to additional dead load such as wearing surface, roofing, etc.
w_L = live load per unit length
V_D = shear due to externally applied design loads which means shear due to $w_S + w_L$
M_D = moment due to externally applied design loads which means moment due to $w_S + w_L$
M = moment due to externally applied ultimate loads which means moment due to $1.5w_S + 1.8w_L$
V = shear due to externally applied ultimate loads which means shear due to $1.5w_S + 1.8w_L$

We wish to demonstrate that

$$\frac{M_D}{V_D} = \frac{M}{V} \tag{E-1}$$

Since M_D and M are being considered at the same point in the same beam, the same formula would be used for computing their numerical values and all the symbols in the formula except w would have the same numerical value. Therefore, no matter what the formula for bending moment is, we can remove w from it and replace all of the other symbols and numbers with a constant C. The bending moment at the section under investigation then becomes

$$M = wC \tag{E-2}$$

If a simple beam is considered as an example, the basic formula is

$$M = \frac{wX}{2}(L - X)$$

Remove w and let

$$C = \frac{X}{2}(L - X)$$

then

$$M = wC$$

In the same manner take w out of the formula for V and replace the remaining symbols with K. Then

$$V = wK \tag{E-3}$$

Using the foregoing notations we can write the following formulas:

$$M_D = w_S C + w_L C$$
$$V_D = w_S K + w_L K$$
$$M = 1.5w_S C + 1.8w_L C$$
$$V = 1.5w_S K + 1.8w_L K$$

Then

$$\frac{M_D}{V_D} = \frac{w_S C + w_L C}{w_S K + w_L K} = \frac{C(w_S + w_L)}{K(w_S + w_L)} = \frac{C}{K}$$

and

$$\frac{M}{V} = \frac{1.5w_S C + 1.8w_L C}{1.5w_S K + 1.8w_L K} = \frac{C(1.5w_S + 1.8w_L)}{K(1.5w_S + 1.8w_L)} = \frac{C}{K}$$

Hence

$$\frac{M_D}{V_D} = \frac{M}{V}$$

Index